ADVANCED
ECOLOGICAL THEORY

Advanced Ecological Theory

Principles and Applications

EDITED BY

Jacqueline McGlade
Director, Centre for Coastal and Marine Sciences,
Natural Environment Research Council,
Plymouth Marine Laboratory,
Prospect Place, The Hoe,
Plymouth PL1 3DH

b
Blackwell
Science

Editorial Offices:
Osney Mead, Oxford OX2 0EL
25 John Street, London
WC1N 2BL
23 Ainslie Place, Edinburgh
EH3 6AJ
350 Main Street, Malden
MA 02148 5018, USA
54 University Street, Carlton
Victoria 3053, Australia
10, rue Casimir Delavigne
75006 Paris, France

Other Editorial Offices:
Blackwell Wissenschafts-Verlag
GmbH
Kurfürstendamm 57
10707 Berlin, Germany

Blackwell Science KK
MG Kodenmacho Building
7–10 Kodenmacho Nihombashi
Chuo-ku, Tokyo 104, Japan

First published 1999

Set by Excel Typesetters Co.,
Hong Kong
Printed and bound in
Great Britain by
MPG Books Ltd, Bodmin,
Cornwall

A catalogue record for this title
is available from the British
Library

ISBN 0-86542-734-8

Library of Congress
Cataloging-in-publication Data

McGlade, J. M. (Jacqueline
Myriam), 1955–
Advanced ecological theory:
principles and applications/
Jacqueline McGlade.
p. cm.
Includes bibliographical
references and index.
ISBN 0-86542-734-8
1. Ecology. I. Title.
QH541 M3855 1999
577—dc21

98-34024
CIP

DISTRIBUTORS
Marston Book Services Ltd
PO Box 269
Abingdon, Oxon OX14 4YN
(*Orders*: Tel: 01235 465500
Fax: 01235 465555)

USA
Blackwell Science, Inc.
Commerce Place
350 Main Street
Malden, MA 02148 5018
(*Orders*: Tel: 800 759 6102
781 388 8250
Fax: 781 388 8255)

Canada
Login Brothers Book Company
324 Saulteaux Crescent
Winnipeg, Manitoba R3J 3T2
(*Orders*: Tel: 204 837-2987)

Australia
Blackwell Science Pty Ltd
54 University Street
Carlton, Victoria 3053
(*Orders*: Tel: 3 9347 0300
Fax: 3 9347 5001)

For further information on
Blackwell Science, visit our
website:
www.blackwell-science.com

Contents

Contributors, vii

Foreword, ix

Preface, xi

Acknowledgements, xvii

1 *J.M. McGlade*: Individual-based models in ecology, 1

2 *E. Renshaw*: Stochastic effects in population models, 23

3 *M. Keeling*: Spatial models of interacting populations, 64

4 *D.A. Rand*: Correlation equations and pair approximations for spatial ecologies, 100

5 *R. Law*: Theoretical aspects of community assembly, 143

6 *S.L. Pimm*: The dynamics of the flows of matter and energy, 172

7 *W.M. Getz*: Population and evolutionary dynamics of consumer-resource systems, 194

8 *P.H. Harvey and A. Purvis*: Understanding the ecological and evolutionary reasons for life history variation: mammals as a case study, 232

9 *M.L. Rosenzweig*: Species diversity, 249

10 *E.J. Milner-Gulland*: Ecological economics, 282

11 *J.M. McGlade*: Ecosystem analysis and the governance of natural resources, 309

Index, 343

Contributors

W.M. Getz
Division of Insect Biology, Department of Environmental Science, Policy and Management, University of California at Berkeley, California 94720-3112, USA

P.H. Harvey
Department of Zoology, University of Oxford, South Parks Road, Oxford OX1 3PS, UK

M. Keeling
Department of Zoology, University of Cambridge, Downing Street, Cambridge CB2 3EJ, UK

R. Law
Department of Biology, University of York, Heslington, York YO10 5YW, UK

J. McGlade
Centre for Coastal and Marine Sciences, Natural Environment Research Council, Plymouth Marine Laboratary, Prospect Place, The Hoe, Plymouth PL1 3DH, UK

E.J. Milner-Gulland
Department of Biological Sciences, University of Warwick, Coventry CV4 7AL, UK

S.L. Pimm
Department of Ecology and Evolutionary Biology, University of Tennessee, Knoxville, TN 37996, USA

A. Purvis
Department of Biology, Imperial College, Silwood Park, Ascot SL5 7PY, UK

D.A. Rand
Mathematics Institute, University of Warwick, Coventry CV4 7AL, UK

E. Renshaw
Department of Statistics and Modelling Science, University of Strathclyde, Livingstone Tower, 26 Richmond Street, Glasgow G1 1XH, UK

M.L. Rosenzweig
Department of Ecology and Evolutionary Biology, University of Arizona, Tucson, Arizona 85721, USA

Foreword

This book both is, and is not, an update of the earlier edited volumes on *Theoretical Ecology: Principles and Applications* which I put together for Blackwell Science in 1976 and again–with a somewhat different cast of authors and topics—in a second edition in 1981.

The present volume is similar in its basic organization, beginning with a focus on the behavioural ecology of individual organisms, and building up through populations to communities and ecosystems, ending with two chapters which examine how broader political and economic considerations roil together with ecological constraints in problems of resource management.

The present volume, however, differs mainly in that a lot has happened in the more than two decades since 1976. Then, advances in theoretical ecology, which often had origins going back to Lotka—Volterra and others in the 1920s, were beginning to find specific and useful engagement with ecological data from field and laboratory.

Even so, considerable scepticism still prevailed in some circles as to the role of mathematical models in ecology: some still had difficulty seeing that theoretical ecology–much of it ineluctably mathematical–was as much part of a mature ecological science as theoretical physics is of physics. At the risk of exaggeration, it could be said that the 1976 volume helped define the end of the 'romantic age' of theoretical ecology. Today this subject is clearly seen as an integral part of the discipline of ecology, and a fair bit of what in 1976 was seen as specialized material is now embedded firmly in undergraduate texts. Against this background, the present collection of chapters sets its mathematical standards somewhat higher than the earlier ones of 1976 and 1981. But the book remains squarely in the idiom of its predecessors by its focus on grounding the theory on field and laboratory observations, and using the models to address empirical questions.

In short, over the past two decades and more we have come a long way in understanding the dynamical principles which govern the way individuals, populations, communities, and ecosystems respond to disturbance, natural or human created. But we still have a lot to learn. This book does much to help consolidate past advances, and at the same time it highlights unanswered questions. Few, if any, areas of science are more important or more relevant to our future in an ever more crowded and disturbed world.

Robert M. May
Zoology Department
Oxford University

Preface

Anticipating change in the planetary system and especially the biosphere has become a modern preoccupation. The need to know, and to know instantaneously, what is happening to different ecosystems across the globe has brought ecologists out from the corners of research departments into the limelight. Mass-media and the culture of open access to information has exposed environmental issues to far higher levels of scrutiny than even the most progressively minded activist could ever have dreamed of. Saving endangered species and protecting the environment has now become a national pastime in many countries and of significant global institutional concern. But simple 'knee-jerk' reactions to change will not be enough to mitigate against any long-term ecological damage. There has to be a concerted effort to understand how ecosystems function under conditions that have yet to be experienced—we can only run the planetary experiment once—and for this a solid theoretical foundation is needed.

As our interest in the future of the planet intensifies, the number of people actively engaged in studying ecosystems is also increasing. In order to make the most use of their experience and knowledge, hooks into a theoretical and analytical framework are needed. Mathematics is an important part of this framework but it can be off-putting, even to those who have had an analytical training. One role of theoretical ecologists is thus to help bridge the gap between observation (either real or virtual), mathematical description and the analytical foundations of ecology by providing models and methods for practitioners coming in from other fields to use. The other role is to develop the field of theoretical ecology itself. What is encouraging is the growing number of international, national and local debates and research activities that are aimed at questioning and hence building up the fundamental tenets of ecology.

With so many people trying to integrate their own efforts into what is an extremely diverse field, it is important that we have literature which can provide signposts to the relevant areas for field workers, experimentalists and natural scientists. Much of what is written about theoretical ecology in journals and texts is very technical and highly specific. Thus the aim of this volume is to bring together areas of literature and new research in order to address particular issues which are of concern to a wide audience. The authors of this volume are all internationally recognized researchers in their own fields, but are also individuals who actively promote the interdisciplinary and varied nature of ecology. Thus their contributions provide not only new insights into particular areas of ecology, but also reflect how these might influence thinking in other areas. In choosing topics for this volume, I have not attempted to cover the breadth of ecology—that is far better done in some of the new texts in ecology (e.g. Begon *et al.* 1996)—but rather to address some of the areas proving to be of greatest technical or conceptual difficulty for ecologists. Thus it is that the contributors include mathematicians and statisticians as well as biologists, physicists and resource economists. Some areas, such as epidemiology and dynamical systems, are not covered explicitly but are referred to in many chapters; the case for epidemiology is especially relevant as it provides many examples where data and models are rapidly accumulating.

The structure to the book is largely one of scale. It starts out with individuals, populations, communities and ecosystems and then ends up with the institutional and societal requirements to manage the world's natural resources. However, throughout each chapter is an unswerving theme that without well-tried mathematical and physical evidence, theoretical ecology cannot proceed. Much of the mathematics is presented in a nonspecialist way, so that even nonanalytical ecologists can see the reason behind a particular avenue of investigation. But of course the details of proofs, etc. are accessible through the literature that is cited alongside the text.

Chapter 1 begins with an overview of individual-based models (IBMs), which have gained wide credence amongst many, but which have potential shortcomings if used without reference to scale. IBMS are looked at critically because they represent an approach which is theoretically easy to grasp but sometimes technically difficult to execute and for which there is still relatively limited mathematical understanding. Chapter 2 addresses the need to incorporate stochastic effects into biological and ecological models. This chapter uses a set of basic models (birth–death, single-species processes, simple epidemics, spatial colonies and reaction-diffusion) to demonstrate the discrepancies between predictions and observations using a deterministic approach and those including stochastic effects versus models. As an associated stochastic process can be written down for any deterministic system, and as ecosystems are inherently stochastic, the author concludes that theoretical studies should always include such effects.

Chapter 3 expands on advances in modelling individuals and populations and their spatial interactions and asks fundamental questions about how best to quantify the dynamical behaviours of individuals and populations and how to relate these ideas to the body of mathematical knowledge concerning nonlinear dynamical systems and interacting particle systems.

Chapter 4 examines the use of pair approximations and correlation equations in a wide range of ecological applications. The author shows how certain phenomena, such as the way in which parasites can drive the dynamics of a system and how spatial correlations enable altruism and cooperation to evolve, do not occur in the equivalent mean-field systems. Instead they rely on the discrete nature of organisms and the significant role of stochastic effects on correlations and fluctuations. The analytical insights referred to in Chapters 3 and 4, derived from studying ecological and biological systems, are also providing contributions in their own right to mainstream mathematics and physics.

In Chapter 5 we look at how ecosystems function, from the perspective of fluxes in species and the effect this has on community assembly. Using dynamical systems as a fundamental underpinning, we see how important stochastic effects and historical accident can be in determining the path of community assembly that the field ecologists observe. From the analyses presented we are given a theoretical understanding of why certain species are doomed to disappear as communities of interacting species change over time. The novel concept of a permanent endcycle also has some interest for mathematicians in its relationship to an equilibrium point.

Chapter 6 takes us further into the details of species interactions by examining the trophic dynamics of consumers and their resource base using the theoretical underpinning of evolutionary stable strategies (ESSs). Two paradigms are examined: continuous systems, exemplified by Lotka–Volterra models, and a discrete-time logistic food web. Spatial aspects are also included as are considerations of storage in food webs, which up until now have received almost no attention. Critically this chapter shows the potential importance of adopting an ESS approach to gain insights not only into the impacts of harvesting and the immunity of populations to macroparasites, but also to understand phenomena such as the 'ghost of predation past', i.e. where predators which disappear play a critical role in promoting the evolution of diversity that is maintained once the predator is gone, also alluded to in Chapter 5.

The structure of food webs, both the specific patterns and overall complexity, provide the basis for Chapter 7. Questions about the resilience and resistance of ecosystems to transient shocks, the impacts of the loss of ecosystem components and the role of complexity are addressed through analyses of the dynamical stability of different trophic configurations. We find that some systems may never recover from some previous shock, even though it may be short and sharp, but that adding a trophic level can increase a system's resilience. On the other hand removing what might be a keystone predator can cause surprising results, such as a decrease in its prey. Species richness may indeed increase resistance, but the question asked, and answered to some extent in the text, is how many species are really needed to preserve a particular ecosystem. This theme is picked up in Chapter 9, in an extensive analysis of the measures and approaches used to measure species diversity. Once again, the underlying assumption is one of a dynamical system and how species diversity varies according to three rates: the rate of species production, the rate of extinction and the time that the rates have been in effect. The chapter creates strong links between measures of diversity, species–area curves, log-normal distributions, island biogeography and other parts of theoretical ecology, such as population, metapopulation and patch dynamics, to answer the question of having once collected samples what procedures can be used to determine their true underlying diversity?

Chapter 8 provides the reader with an analysis of how ideas about life history variation evolved over this century and the consequences of this on our understanding of the correlation between life history patterns and selection in mammals. The authors provide a clear case study of the critical nature of the importance of interactions between theory and observation, and present a detailed commentary on the present ability of scientists to build explicitly formulated, testable models which can predict a wide range of phenomena such as body size distributions.

The two final chapters look at ecosystems from the perspective of resource exploitation and how best to manage and govern the process both locally and globally. In 1976 Colin Clark presented an overview of bioeconomic modelling, showing the very clear parallel to mathematical modelling that was characteristic of theoretical ecology and biology. In Chapter 10 we see that nearly 17 years later the field of ecological economics is only just emerging from being seen as a branch of mainstream economics. The fundamental problem is of course that most economists still view the system as essentially open and so capable of infinite growth; ecologists on the other hand understand the concepts of carrying capacity, leading to a closed economy. A number of steps are needed to make the links between ecological and economic theory stronger and these are clearly laid out in Chapter 10. Chapter 11 addresses the fundamental problem of trying to govern a system of resources, including their exploitation, conservation and development, in the face of intrinsic uncertainties. The uncertainty arises from nature itself as well as from our ability to measure or perceive what is present in a system. The use of fuzzy logic to describe and capture the uncertainties in information and knowledge gathered about ecosystems is presented as a more appropriate method for ecologists and resource managers than Bayesian probability. The inclusion of uncertainty into the decision-making process is taken up in a discussion about expert systems and their use in providing alternatives to simple risk assessment models.

The contributions in this volume demonstrate that theories as well as patterns in nature are needed to ask the right questions. And the success of theoretical ecology will be measured by its ability to provide the right answers.

Jacqueline M. McGlade

Acknowledgements

I would first like to thank all the authors for their written contributions; without their endeavours in the field of theoretical ecology such a book would not have been possible. The book has benefited considerably from all those who have reviewed sections and earlier drafts of chapters, but special thanks must go to Bryan Grenfell and Charles Godfray for reviewing the entire manuscript and Dave Woods for his help on mathematical details. I would also like to extend special thanks to John Marshall, Susan Taylor, Andrew Yool and Glenys Hicks for their help in processing the manuscripts and to Ian Sherman at Blackwell Science for his consistent encouragement and support. The Department of Biological Sciences, Warwick University and the Centre for Coastal and Marine, Natural Environment Research Council Sciences provided financial and administrative support.

This book does not cover every aspect of theoretical ecology. The choice of what was included and the responsibility for areas not included is completely mine.

1: Individual-based models in ecology

J.M. McGlade

Introduction

Individual-based models (IBMs) are, as their name suggests, mathematical models in which individual organisms, or groups of individuals with the same characteristics, are explicitly studied. Over recent years there has been an increasing focus on their use in ecology, as documented by various authors (e.g. Huston *et al.* 1988; Łomnicki 1988; Hogeweg & Hesper 1990; McGlade 1993; and in the book edited by DeAngelis and Gross 1992); however the rise in their use has been accompanied by a shared belief that they are a 'magic bullet' for all of ecology.

Contrary to the other chapters in this volume, I will thus explore the use of IBMs as a tool, their advantages and disadvantages, and review some of the key areas where they have been applied, rather than try to answer any one specific ecological question. However, many of the ideas and arguments presented here are taken up in more detail in Chapters 3 and 4 to address key problems. The emphasis in this chapter is rather to provide guidelines as to where IBMs might be successfully used and why, and where not.

From a historical perspective, Judson (1994) argues that IBMs are a natural outcome of advances in ecological modelling, especially May's work on chaos (1974). She also cites other driving forces behind the development of IBMs such as:

1 the advent of more powerful computing systems;

2 the debate between advocates of simple abstract models and supporters of more detailed simulations;

3 the knowledge that most individual interactions occur locally rather than globally; and

4 that individual differences are inadequately dealt with in other models.

Whilst some have criticized Judson's view about the importance of chaos (e.g. Bullock 1994) it is true to say that one of the key motivating forces behind the development of IBMs has come from the need to understand the dynamics of spatially extended systems and the nature of complex behaviours in ecosystems.

There are many advantages to IBMs including the fact that they are intuitively accessible to biologists, the parameters required are typically those measured (e.g. individual movements, diets, numbers), the models are generally robust and fewer assumptions have been made about the underlying processes in comparison with population models. The disadvantages are that they tend to be computer-intensive and highly dependent on computational methodology, are nonanalytical and hence more difficult to assess, have macroscopic behaviours that are often complex and require additional analysis and place limitations on the number of individuals included.

Ecologists and epidemiologists often comment that many of the results arising from IBMs merely confirm what is already known from observations. For example, that coexistence is more pervasive than suggested by mean-field models (e.g. as observed by Levin & Paine 1974), that altruists can more frequently invade a system, that diversity of a system can be more easily sustained and that the cycles of epidemics are less extreme. Many modellers would of course see this as an achievement! However, when the models become spatially explicit, a number of new phenomena arise which do not occur in their standard mean-field counterparts and which are potentially of great interest. Examples include the large effect that parasites can have on host genetic diversity (see Chapter 3), or the fact that systems evolve towards critical states (e.g. Rand *et al.* 1995; Ghandi *et al.* 1997). What is not necessarily clear is the extent to which these observations can provide real, novel insights or are artefacts created by the treatment of space and individuals.

To explore this point further, I first examine the background to IBMs, highlighting the key problems that the models have attempted to address and discuss these in relation to new approaches that are emerging. Where these are described in detail in other chapters of the book I give only a brief overview, e.g. spatial dynamics (Chapter 3) and pairwise approximations (Chapter 4). I then go on to examine different applications of IBMs to establish where new insights have been obtained or observations made clearer and try to assess whether it does indeed make a difference to use IBMs rather than other analytical methods (Uchmański & Grimm 1996).

Background to the development of individual-based models

The level at which individuals are incorporated into a model is more than a matter of simple mathematical expedience. Often individuals are best considered in models when they are at low densities or the numbers are small throughout the system, especially if extinction represents a real concern. As systems can be modelled using different degrees of detail, it is important to clarify not only the aims of building a model, but also the extent to which individuals need to be represented.

The simplest approach is to assume that all the individuals in a system are homogeneous, evenly mixed in space and that they interact equally—what in physics is termed a *mean-field*. Such assumptions are the basis of most population models, and are generally represented through a system of ordinary differential or difference equations.

Following on from Metz & Diekmann (1986) we can define the state of an individual (its *i-state*) as the information needed to specify an individual's response to different environments; typical variables include age, size and physiology. The population state, or *p-state*, comprises a system of interacting individuals. We can summarize the population as an *i-state distribution* (*sensu* Caswell & John 1992), where there is an assumption of mixing so that all individuals have the same dynamics and can be treated collectively as in the mean-field approach. In an *i-state configuration*, the mixing assumption fails, i.e. local interactions are strong and each individual must be tracked separately. The *i-state configuration* models are equivalent to IBMs (*sensu* Huston *et al.* 1988; McGlade 1993) and *i-state distribution* models equivalent to state variable models; the important distinction is that the former generally rely on simulation and the latter on analytical techniques.

If movement of individuals and equal interactions are introduced, any mathematical description is best seen as a *reaction-diffusion system*. Much has been written about the degree to which local interactions and individual diffusion can be accounted for in these models, especially when the terms describing them are nonlinear (see Murray 1990). Part of the reason for the success of reaction-diffusion equations stems from their ability to easily reproduce patterns that are prevalent in nature, such as the wave-like phenomena of travelling chemical concentrations and insect invasions, or the pattern of markings on animals. However, these techniques do not lend themselves to the treatment of local stochasticity or individuals as discrete units. In order to achieve this, the continuum of space scales in the reaction-diffusion models needs to be replaced, either by patches or by a subdivision of space containing discrete individuals (see Durrett & Levin 1994).

One of the first examples of a patch model with no further specification of spatial arrangements or accessibility was the metapopulation model of Levins (1969, 1970). He assumed that the habitat of a species was subdivided into patches that could only occur as two states: occupied or empty. Occupied patches became empty when the local population of species went extinct; empty patches became occupied when they were colonized from occupied patches. The conceptual clarity of the Levins model was such that by the beginning of the 1990s, metapopulation dynamics had been established in conservation biology, landscape ecology, pest control, etc. (see overview by Gilpin & Hanski 1991; Hanski & Gilpin 1991).

Chesson (1981) took Levins' work forward and examined the interplay between spatial localization and stochasticity, including both nonlinear interactions and discrete individuals into the model. He found that local stochastic effects not only introduced systematic effects into the average population density, but that these did not disappear at large population sizes. He thus rightly concluded that if local interactions were an important element in a system, mean-field models would fail to adequately describe the dynamics.

Durrett & Levin (1994) subsequently undertook a comparative analysis of four approaches (mean-field, patch models, reaction-diffusion and interacting particle systems) to examine the difference between the predictions made arising from interactions in spatially distributed systems. Where individuals had to compete for the same resource or where extinctions were likely to occur, the spatial models disagreed with the nonspatial models. In the case where extinctions could occur, their results supported the importance of spatial refugia for fugitive species to maintain their viability in a system.

When the assumption about equal accessibility is relaxed, it is possible to model the interactions amongst individuals via a range of models including interacting particle systems (IPS) (Durrett & Levin 1994), coupled map lattices (CML), (probabilistic) cellular automata (PCA), other IBMs and artificial ecologies (AE) (see McGlade 1993; Rand *et al.* 1995). The main characteristics of these models are as follows. CML have a map (as opposed to a node) to evolve the states at each lattice site. The sites are coupled using finite-difference approximations of the Laplacian operator (the sum of the second partial derivatives of a function). In IPS, PCA and AE, individuals are constrained to sites on a lattice and their movements and birth/death governed by rules and stochastic rates. The difference between a PCA and an AE is that in a given neighbourhood in a PCA, events only change the central site, whereas in an AE other sites in the neighbourhood can be changed. By allowing the time evolution to be defined in terms of stochastic rates, IPS extend cellular automata by making time continuous. Lagrangian IPS extend IPS by allowing particles to assume any location in space. I will look in more detail at some of these models in the section below, where examples are given.

Between the mean-field and lattice models lies another, newer class of general models known as pairwise approximation (see Chapter 4 for a detailed description of the mathematics behind this approach). The technique is derived from statistical physics where it has been used to study turbulence, chemical reactions, stochastic processes and the time evolution of complex structures (Dickman 1986; Ben-Naim & Krapivsky 1994; Filipe & Rodgers 1995). The models involve the formulation of a system of deterministic equations that captures the essential dynamics of the full stochastic system. The state variables are the number of pairs of direct neighbours, so that although space is not mapped directly it is captured through the neighbourhood of every individual.

Not only is this method analytically tractable, it is also of direct relevance to studying biological systems because of the dominance of interactions between individuals (which can be captured very effectively by low-order correlations) and because the stochastic background of biological systems effectively destroys high-order correlations. As a consequence, these models are often superior to mean-field approximations both in terms of predicting transient as well as long-run stationary behaviour. There are now a wide range of applications using pairwise approximations covering vegetative propagation and seed dispersal in plant populations (Harada & Iwasa 1994), forest gap dynamics (Kubo *et al.* 1996), host–pathogen interactions (Satō *et al.* 1994) and epidemics (Levin & Durrett 1997; Keeling *et al.* 1997a), as well as a number of more general papers about interacting populations (see Chapter 4 and Harada *et al.* 1995).

The literature on all types of individual-based approaches has expanded rapidly over the past few years, as can be seen below and in Chapters 3 and 4, so before going on to discuss some specific examples, it is important to look again at the criticisms that have been levelled at IBMs. The main criticism is that many of these explicit individual-based, spatial population models are notoriously difficult to analyse and hence many studies are based on simulations alone. Consequently, while being excellent for developing intuitions about a system, the models often have a number of deficiencies. There is also a lack of any real mathematical understanding of many of the models, so that in most cases it is impossible to say whether the behaviour observed is reasonable without some direct input from empirical studies. However, as model structures are often inconsistent with biological reality, it is very difficult to make simple comparisons. One of the main conclusions must be therefore that it is important for mathematicians and ecologists to work closely together when constructing these types of IBMs for different applications.

Examples of individual-based models

To understand how an IBM can be developed, readers should refer to DeAngelis and Gross (1992). However, there are now a number of examples of IBMs that can be accessed on the World Wide Web. What is important in each case is that the model structure need not be excessively complicated, but care should be taken to address such issues as having a system of sufficient size to guarantee that the dynamics are not under- or over-sampled, or how the events are updated (see Chapter 3). The applications and examples described below are diverse enough to give a sense of how different IBMs have been built and tested. Readers should refer to the modelling details in each case. My aim in this section is to show how the interplay of empirical studies with theory has been used to address particular problems (e.g. succession, competition, etc.).

Plant monocultures

A number of IBM models have been developed to study the growth of plants and monocultures (e.g. Crawley & May 1987; Colosanti & Grime 1993). In this first example, a CML is used to demonstrate neighbourhood effects and the importance of different forms of competition between plants (Hendry *et al.* 1996). In a CML, the variables can vary continuously on a discrete grid through space and time. The formalism follows Kaneko (1990) in which a lattice of points is considered. Each cell has a continuous variable describing plant mass, which is updated each time-step as a function of its own value and the cells in a specified neighbourhood. A von Neumann neighbourhood of five cells is used, so that each cell has a fixed area equal to one-fifth of the maximum area attainable by a plant (Fig. 1.1). The plant mass is given as a proportion of the maximum possible plant size attainable by an uninhibited plant. The lattice is toroidal, i.e. the boundaries are periodic. The model for plant growth is an extension of Aikman & Watkinson (1980), where the change in mass Δm_i from one time-step to another is given as:

$$\Delta m_i = g(a_i - l_i) - b m_i^2 \Delta t \tag{1.1}$$

with $a_i = cm_i^{2/3}$, a_i given by the 3/2 self-thinning rule, l_i the growing area lost by competitors, g the intrinsic growth rate of the plant, b and c constants and Δt the time-step. The lost growing area depends on the competition: the

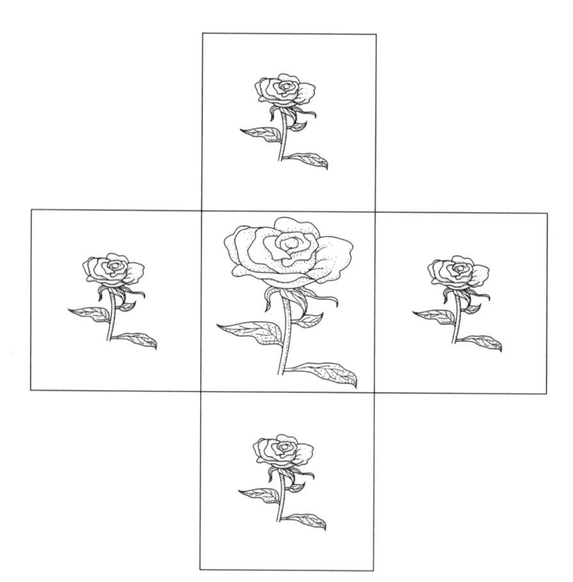

Fig. 1.1 Configuration of the lattice neighbourhood.

maximum size per cell can be derived by setting l_i and the change in mass to zero, and dividing over the five sites, to give $\alpha = (c/5)\sqrt{gc/b}$. If a plant overlaps into its four neighbouring sites equally, then the total overlap Ω_{ij} between two neighbours *i* and *j* is:

$$\Omega_{i,j} = \max\left(\frac{a_i - \alpha}{4}, 0\right) + \max\left(\frac{a_j - \alpha}{4}, 0\right) \tag{1.2}$$

and for the four-cell neighbourhood *nhd* is:

$$\Omega_i = \sum_{i \in nhd} \max\left(\frac{a_i - \alpha}{4}, 0\right) + \max\left(\frac{a_j - \alpha}{4}, 0\right) \tag{1.3}$$

Four types of competition are examined: *absolute asymmetry*, i.e. the larger plant of two plants takes resources from the entire overlap area: equal-sized plants share the area equally; *absolute symmetry*, i.e. plants divide the area equally; *relative symmetry*, i.e. the overlap is weighted by the relative masses linearly for the *relative symmetry* and quadratically for *relative asymmetry*.

The CML is run on grids of 20 × 20, 50 × 50 and 100 × 100 cells, with densities ranging from 0.1 to 1.0. Statistical analysis of the numerical simulations include derivation of the mean mass μ over the whole lattice/plot, the coefficient of variation and the Gini coefficient. The results show clearly that:

1 mean mass decreases as density increases;

2 growth is sigmoidal, so that the early phase is exponential. Variation in growth rates only occurs when competition sets in, i.e. after the exponential stage;

3 the mean mass of plants is the same for all competition models except asymmetric competition. In this case the even spacing over the lattice is suppressed; higher variability at higher densities thus implies greater asymmetry in neighbourhood interactions.

The key results are shown in Fig. 1.2, where the variability in models increases from symmetric to asymmetric competition, and in Fig. 1.3 in which the variation in plants can be seen. The results show that size hierarchies can be used as evidence that asymmetric competition is the dominant process determining size variation in plant populations as opposed to simple neighbourhood effects.

Forest models

One of the most successful models of forest dynamics is the IBM SORTIE, developed by Pacala and colleagues (Pacala *et al.* 1993, 1996). SORTIE characterizes the relationships among site variation in sapling performance, interactions with tree species in the community and overall species composition. It consists of four species-specific submodels that quantitatively describe the behaviour of each individual tree-growth as a function of light

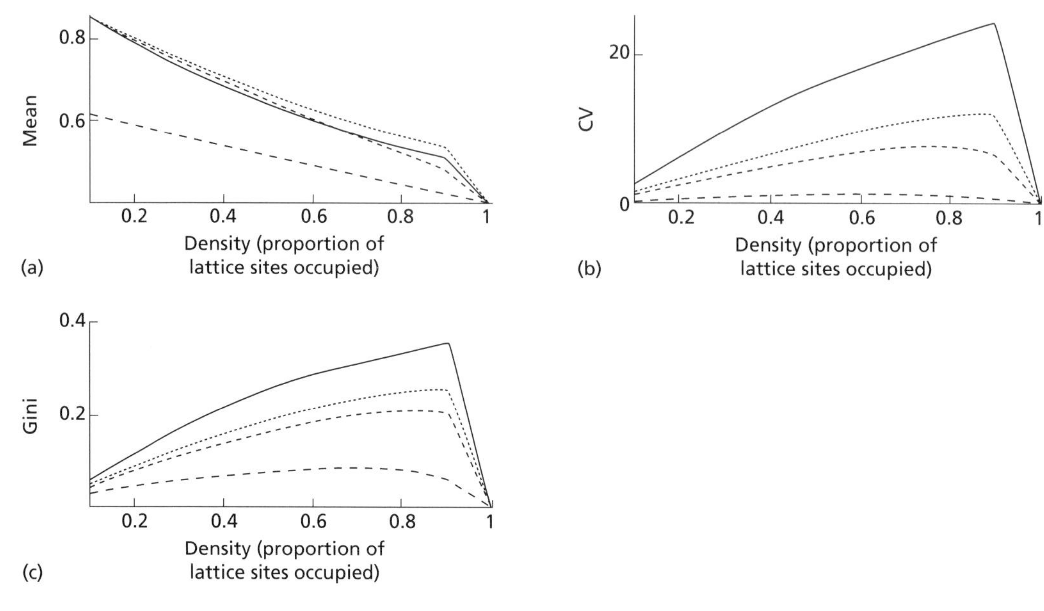

Fig. 1.2 Results from the coupled of map lattice model after 200 days as a function of density: (a) mean mass; (b) coefficient of variation of mass; (c) Gini coefficient of mass (redrawn, with permission from Hendry 1995; Hendry *et al.* 1996).

availability (Pacala *et al.* 1994), fecundity and dispersal of seedlings as functions of adult tree size (Ribbens *et al.* 1994), death as a function of recent growth (Kobe *et al.* 1995) and modifications of light as a function of tree-crown interception (Canham *et al.* 1994). All the models were calibrated directly from field data at Great Mountain Forest in north-western Connecticut, USA.

SORTIE simulations have been run for a wide variety of forests (e.g. Kobe 1996) with an increasing degree of success in terms of being able to provide a wide range of predictions. For example, Kobe (1996) was able to use SORTIE to apply among-site differences in only juvenile survivorship and growth to predict dominant species in the adult canopy, showing the importance of sapling stages in community dynamics (see 'Forest mosaic' below).

The basic model underpinning SORTIE is as follows: a seedling produced by a parent tree on a particular site establishes itself; the spatial distribution densities of seedlings are a function of distance from and size of parent trees (Ribbens *et al.* 1994). The seedling develops and survives with a certain probability, as described by the submodels indicated above. As the individual tree grows it influences the local light environment and begins to reproduce and produce its own seedlings, as in the plant monoculture and mosaic models described in this section. The lattice then repeats itself. What is important then is that from these rather straightforward empirical inputs

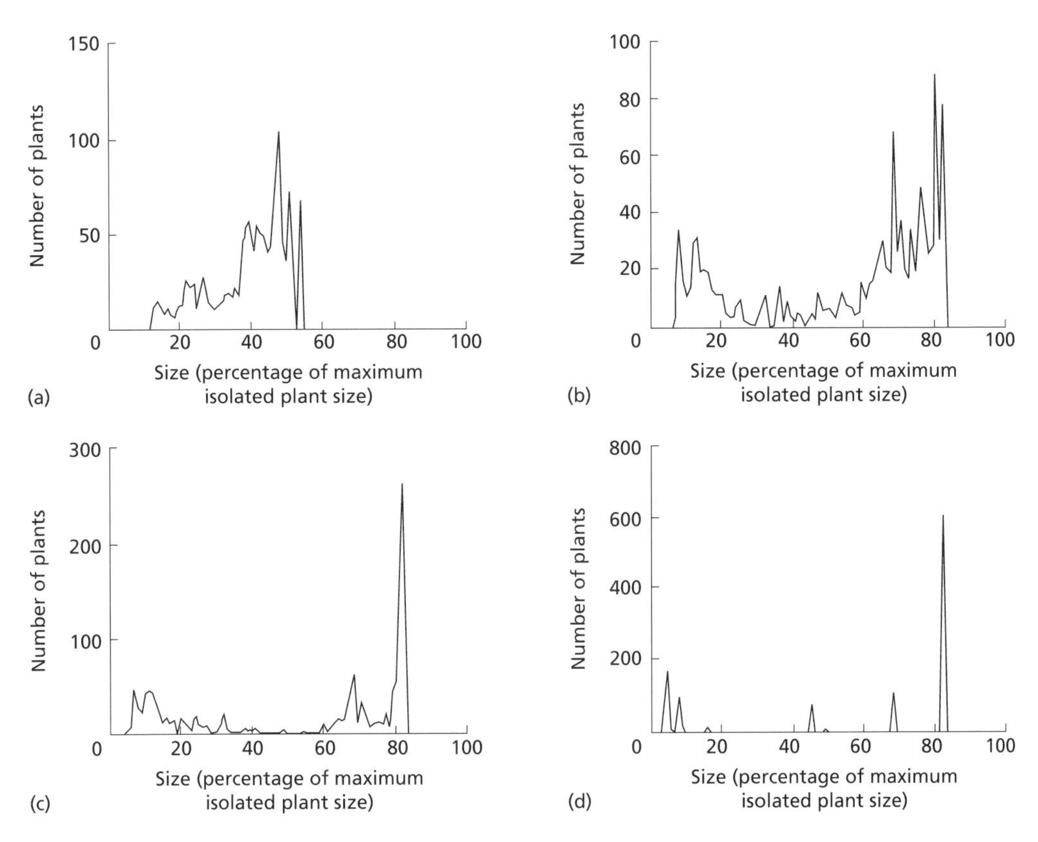

Fig. 1.3 Distribution of plant sizes at time 500 and density 0.5 for the coupled map lattice, as a percentage of the maximum size of an isolated plant under different conditions of competition: (a) absolute symmetry; (b) relative symmetry; (c) relative asymmetry; (d) absolute asymmetry (redrawn, with permission, from Hendry 1995; Hendry *et al.* 1996).

SORTIE correctly predicts shifts in dominance at different sites, thus making very good use of a 'bottom–up' approach of empirically determined relationships.

Forest mosaic cycle

The next example of an individual-based approach refers to the modelling of mosaic cycles. The concept of the mosaic cycle was revived by Remmert (1991) in his work on the middle-European beech forest. Contrary to the idea in classical ecology that an ecosystem reaches a climax state or fixed equilibrium after a certain period, mosaic cycle theory asserts that an ecosystem is in a constant state of flux. The mosaic derives from patches or 'stones' that cycle through a set of states.

The idea of a mosaic cycle has been used in studies of temperate forests, tropical and subalpine forests and marine kelp forests (see contributions in

Remmert 1991). In this example, the dominant long-lived species is beech: gaps created by fallen trees are invaded by an early successional monoculture of birch which survives for approximately 50 years (Fig. 1.4). A mixed forest then appears consisting of oak, cherry, ash and maple and survives for up to 150 years. Beech gradually succeeds this forest, initially as young thicket and then gradually as a mature stand lasting up to three centuries. The cycle can be interrupted by losses of trees and understorey through disease and poor climatic conditions, and stages can be missed by invasion of later successional forms from neighbouring sites. Autosuccession (i.e. when a species replaces itself) does not occur in the case of beech because the soil nutrients have generally become too depleted for another beech tree to survive. Another important process is radiation death. Beech trees are particularly sensitive to solar radiation, and their smooth black bark splits open if exposed to intense sunlight. Thus in the northern hemisphere gaps appearing to the south of a beech tree will generally lead to its death.

In this analysis, the roles of memory and gap formation in the forest mosaic cycle are examined (Hendry & McGlade 1995); the ecology is based on Wissel's (1992) model. There are 55 states, each corresponding to a 10-year period, giving an overall cycle of 550 years. A nine-cell Moore neighbourhood is used. At each iteration, the lattice is updated synchronously and the states progress by 1, reverting to state 1 when they reach 55. Early colonization is modelled by letting the gap state (1–2) move immediately to birch (3) if the neighbouring states are birch (3–8). If however one of the neighbouring states is mixed (9–22) then the gap will move to a mixed forest (9). There is an increased probability of die-back (in this case in the beech forest only) in states 51–54, given by P_0. Beech trees can also die through radiation death. To achieve this the lattice is given an orientation, and beech

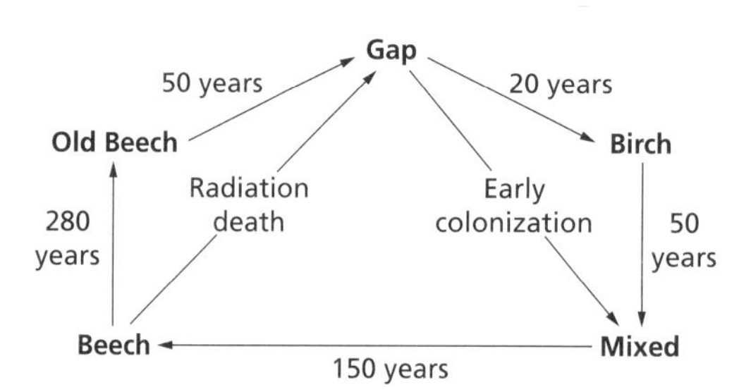

Fig. 1.4 A representation of the mid-European beech forest mosaic cycle. Gaps caused by fallen beech are invaded by two early-successional communities. The first pioneers are birch, which survive for about 50 years. This monoculture is followed by a mixed forest, with species such as oak, cherry, ash and maple, which live up to 150 years. This intermediate community is eventually succeeded by beech, initially as young thicket, followed by thinning and growth of survivors into mature trees over three centuries or longer (redrawn from Hendry & McGlade 1995).

trees with gaps in cells to the south-west, south and south-east of their cells die with probability P_{sw}, P_s, P_{se}.

Wissel showed that the cyclic cellular automata, together with the local neighbourhood effects, were sufficient to produce a mosaic. The pattern was also insensitive to the parameter values used. He also claimed that no patches were formed in the absence of local radiation effects; however in the ecology described above a second mechanism—memory—is also shown to be vital in the creation of spatial patterns. Memory in this context is the way the past history of the system affects its present structure and behaviour. From the earlier example of plant monocultures, it was shown that for sessile organisms, current growth and size is strongly related to the effects induced by their neighbours. The history of current spatial patterns also goes back well beyond the lifetime of the plants currently established because the nutrient status of the soil, water levels, presence of fauna, etc. all have an effect. After a beech forest has been growing for a number of centuries, the water table is substantially lower and certain essential nutrients are absent. This exhaustion of the soil and water is taken into account in the IBM above by the absence of autosuccession in the long-lived beech stands.

If we mirror the longevity of different species in the system by the age of individuals in the beech stands, then the importance of memory in the system and its impact on the structure of the mosaic can be studied through analyses of the beech trees. To do this, four models are constructed relating to *no memory/no radiation*; *no memory/radiation*; *memory/no radiation*; *memory/radiation*. For *no memory* the gap, birch and mixed states are converted into single states with transition probabilities equal to the reciprocal of the expected lifetimes:

$$P_{gap} = 1/2,\ P_{birch} = 1/5,\ P_{mixed} = 1/15 \text{ and } P_{beech} = (33 - 10P_0^2 - 5P_0^3 + P_0^4)^{-1} \qquad (1.4)$$

to take account of early death in states 51–54. No radiation is simply implemented by setting $P_w = P_s = P_e = 0$. Contrary to what Wissel found, i.e. that removal of radiation death would prevent patches forming, the presence of the memory mechanism (*memory/radiation* and *memory/no radiation*) gives rise to a mosaic pattern whilst the other two become randomly mixed.

A mean-field approximation to the mosaic cycle can be derived in the form of a Markov process on the set of configurations. Comparisons between analytical solutions of the mean-field models and numerical results of the spatial ecology above show that the root mean square errors increase when local radiation effects are included as expected. However, what is more interesting is the amplification of the local spatial effects when memory is included. Thus memory is an important component in any definition of characteristic length scales.

Within real beech forests, a number of characteristic spatial scales exist. Wissel concluded from his model and from field observations that the fundamental cell size was 30 m^2. Thus it is appropriate to model these systems at the level of individual trees. Analysis of the cellular automata described above, using a technique of fluctuation analysis (Keeling *et al.* 1997b; see also Chapter 3) shows that the memory mechanism causes the spatial coupling to extend over a larger distance than when it is absent, and so a larger lattice is needed to see the dynamics as statistically stationary (*sensu* Keeling *et al.* 1997b).

An extension to the standard method of evaluating maximum positive Lyapunov exponents (a measure of how a system evolves dynamically from the steady state) for the full ecology (see details in Hendry & McGlade 1995 & Chapter 5), indicates that the system is chaotic. Removal of *memory* removes the chaos; however, removal of *radiation* does not, suggesting again that memory has a fundamental effect on the dynamics of the ecology. (One interesting point is that despite the fact that the average Lyapunov exponent is simpler to evaluate and provides a measure of global predictability, it fails to detect the chaotic nature of the system.) Analysis of the lattices resulting from the model using a singular value decomposition combined with embedding techniques (see Keeling *et al.* 1997b) shows that the *memory* model has a much lower dimensionality than the *no memory* model. Finally, the stability of the mosaic cycle can be examined by removing subregions of the lattice; results show that all but the smallest disturbances cause large oscillations in species distributions and increase the degree of clumping. This is because the range of spatial coupling is temporarily increased. However, after sufficient time, the system re-establishes itself even after large disturbances, suggesting that this structure is very resilient and persistent.

Territorial behaviour in red grouse

Many organisms compete for space or for resources that are linked to territories, and territorial behaviour is an expression of this process. The final example of an IBM is one developed to examine territorial behaviour and the population consequences of kinship in red grouse (*Lagopus lagopus scoticus*) (Hendry *et al.* 1997). The model was constructed using data from observations of grouse in Scotland where the species has been studied for many years because it is an important game-bird of heather moorlands. The data come from experiments and from long-term records of numbers of grouse shot. These provide reliable information to show that the population dynamics of grouse are driven by the cocks or male birds and that cyclic fluctuations occur.

Two conflicting explanations of grouse cycles have emerged:

1 that they are caused by the spread of a disease (resulting from a caecal nematode); and

2 that they are due to overcrowding and the effect of kin-selection.

An IBM was constructed to test the latter, using two behavioural phenotypes: intolerant cocks which were equally aggressive to all neighbours, and kin-tolerant cocks which were less aggressive to their close kin. The 100 × 100 lattice was constructed as a two-dimensional torus to avoid any boundary effects; each square was owned by just one male bird, with contiguous sets of squares comprising an individual's territory. On the assumption that a typical territory at an intermediate grouse density was approximately 3 ha, then each cell in the lattice represented about 0.05 ha of moorland. The overall lattice was therefore much larger than any field study site. The annual cycle comprised the production of chicks in spring, summer rearing, philopatric acquisition of territories in autumn and overwintering. The acquisition of territories depended on the kin-tolerance and fighting rank strength of individual males. In this context birds were related if they were fathers, brothers (full-brothers and half-brothers), uncles and grandfathers. During the autumn related birds were reassembled on the fathers' territories and pairwise contests run between neighbouring individuals.

The results showed that pure populations of intolerant birds reached low but stable densities whereas pure populations of kin-tolerant cocks attained higher densities but with larger fluctuations. Populations of kin-tolerant cocks, able to disperse, showed even larger fluctuations. Mixed populations coexisted at intermediate frequencies but these also had marked fluctuations. Kin-tolerant behaviour was concluded to destabilize the model, giving it the potential to bifurcate, although within the parameter regime used, cycles were not observed. However, when kin-tolerance or the number of dispersal attempts was altered, cycles were in fact produced with a period of 6–15 years. The main conclusions were that differential aggressive behaviour between kin and nonkin during territorial disputes could destabilize the population, as could dispersal, and could produce density cycles with periods and amplitudes comparable with those observed in studies of wild grouse.

Other applications

A wide range of IBMs now exists in the literature (see especially the contributions in DeAngelis & Gross 1992). It is therefore worthwhile giving a brief overview of some of these to highlight the breadth of topics and problems that ecologists are tackling with these approaches. For convenience these have been grouped into four areas: behavioural studies; plant community and forest dynamics; food webs and trophic structure; and population studies and environmental impacts.

Behavioural studies

In some models there is the assumption that a patch or geographic area contains a number of individuals that have the same *i-state*, but that there are differences between patches or sites; to the extent that individual profiles differ, these should thus be considered as a type of IBM. So for example, work by Bernstein *et al.* (1991) on the role of individual decisions in relation to the energetic costs of moving between patches and the influence of the environment on these movements is relevant to the discussion. In this particular case it was shown that in semicontinuous environments, the more coarse-grained the distribution of prey items the poorer the adjustment of predators to achieve an ideal free distribution. The role of individual decision-making was found to be crucial in determining these rates. Wolff (1994) also modelled individual decision-making in relation to behaviour in a wading bird community (i.e. foraging, bioenergetics, movements, interactions with conspecifics and reproduction), but then tracked assemblages of individuals. The model was subsequently modified to look at mercury contamination of the wood stork community and showed that the birds were indeed at risk due to sublethal effects (Hallam *et al.* 1996). Neuert *et al.* (1995) also developed an IBM of a bird community—the green wood hoopoe (*Phoeniculus purpureus*)—to clarify how the behaviour and ecological factors might determine group size. They discovered that without scouting flights the population was likely to disintegrate into smaller isolated subpopulations that would be unable to survive.

Many of the interesting studies using IBMs come from attempts to model population-level processes by projecting up from very simple individual behaviours (e.g. as in ant colonies, flocks of birds and fish schools) (Gordon 1997). As shown above for the red grouse, modelling the territorial behaviour of individuals not only yields insights into the spatial patterning of kin groups, but also the long-term dynamics of fluctuations and cycles observed at the population level. The influence of nest site selection on population dynamics has been explored by Le Page & Cury (1996) in their IBM model known as SeaLab. The model tracks artificial fish in a heterogeneous environment, using hexagonal patches. The effect of behaviour (obstinate or opportunistic) in relation to the lattice composition and configuration directly affects the success of reproduction, with opportunistic individuals becoming 'stuck' in local spatial optima. This approach is very similar to that developed by Allen & McGlade (1986, 1987) in an earlier paper on individual strategies in fishing fleets, where successful individuals would remain to discover and exploit areas of high densities of fish.

The role of individual variation has also been studied in an IBM in spider predatory foraging strategies on prey captures and spider fecundity (Provencher & Reichert 1995). The model used 18 generalist intraspecific spider phenotypes created by crossing different levels of aggressiveness with strategies of prey species selection. Prey selection strategy had a highly significant effect on predator success, whilst surprisingly aggressiveness had no direct effect on its own. Habitat fragmentation caused the decline of prey numbers, and thus prey captures and spider fecundity; increased prey richness increased spider fecundity. The phenotype that achieved the highest fecundity and captured the most prey exhibited high aggressiveness while specializing on the prey that was increasing in prominence. The phenotype that did least well was not aggressive and specialized on a randomly chosen prey.

A final example comes from a paper on the self-motivated hare, analysed through a physiologically based IBM (Focardi 1997). The model is based on an artificial intelligence, object-oriented framework within which the author looks at a set of behavioural rules that move the animals around, according to a simulation of the decision-making system itself. The model is used to look at the effects of landscape and the presence of foxes on the hare's movements and status. The simulated hare exhibits a stable ethogram under variable environmental conditions and performs both adaptive behavioural strategies as well as state-conditioned activities in order to maintain a constant and positive fitness level.

Plant community and forest dynamics

Earlier papers by Bak and co-workers (Bak *et al.* 1988; Bak & Chen 1990) sparked off a large debate about the use of lattice models to look at the large-scale effects of forest fires. This interesting theoretical paper, directed as much to exploring the idea of self-organized criticality and turbulence as to forest fires, caused much concern amongst ecologists who began to see that the very process of controlling forest fires could in fact be helping to create structures whereby they could spread more easily. Ratz (1995) constructed a cellular automata to investigate these ideas further, and to try to gain insights into specific forms of age-dependent flammability and the effect of the fine structure of single fires. The model results shared many features with natural fires that had been analysed from empirical studies and showed that large system-wide catastrophic fires occurred without explicit need for a stochastic (e.g. weather) input.

The next example refers to another form of death in forests—that of die-back. Between 1954 and 1977, large areas of the Hawaiian rainforest died or suffered large-scale defoliation of the canopy-forming dominant tree species. Because no obvious cause could be ascribed to the changes, the theory that the die-back might be part of the natural process of succession was proposed. Jeltsch & Wissel (1994) developed a cellular automata for die-back for this example and four other similar monospecific forests around the world, taking into account the fact that abiotic factors could trigger premature senescence which could then be accelerated by biotic agents. The results showed that a combination of such processes could indeed cause a synchronized die-back.

At the community level, IBMs have been used to test ideas about the processes underlying the distribution of species in forest and plant communities. As indicated above SORTIE is one such example. Williams (1996) has also developed a forest development model, called Arcadia, which is an individual-based, fine-scaled, three-dimensional forest stand model. He has used the model to examine the effects of tree size and distribution on gap phase dynamics, the role of spatial location on the nature of competition and the effect of morphology on spatial occupancy. He showed that the model could successfully replicate the stand structure and productivity relationships in the old-growth forest in the north-eastern USA, as derived from tree rings. Experiments with the model also revealed that the more productive conifers were out-competed by hardwoods due to the flexible crown morphology and greater shade tolerance of the hardwoods.

Humphries *et al.* (1996) also used an IBM to study plant communities, in this case to look at plant distributions in alpine communities. The distributions were analysed with respect to recruitment, growth and mortality of individual plants and in relation to environmental factors such as average daily maximum temperature during the growing season, soil water availability, snow depth and disturbances. The patterns of species composition and above-ground biomass which were generated were in general agreement with observations.

Food webs

Theoretical studies of food webs have generally not incorporated space as a contingency affecting the coexistence of species. Keitt (1997) examined the consequences of introducing spatial heterogeneity into uniform and nonuniform lattice models and found that local interactions resulted in more species-rich webs. The introduction of spatial heterogeneity further added to this and altered web connectance, interaction strengths and increased the coexistence among species. An inverse relationship between web connectance and species richness suggested that the product of these could govern the stability of real, finite webs.

Population dynamics and environmental impacts

In many instances the number of individuals that can be modelled in an IBM is limited by the computation time required. To help overcome this, Scheffer *et al.* (1995) devised the concept of a superindividual, which they used to model recruitment on an individual basis in a large population of striped bass and starvation in a size-distributed cohort. Their approach effectively allows the model to move from an individual-by-individual basis to a cohort representation without changing model formulation. They achieve this through the addition of an extra variable feature to create the superindividual. The observed behaviour of this 'individual' in response to a wide variety of processes can be checked to see whether it is real or an artefact of lumping individuals together.

The effect of environmental pollutants on organisms has also been explored using an IBM by Koh *et al.* (1997). Using a set of inputs describing the physiological status of a model *Daphnia* population, the authors were able to test the stressor effects of a nonpolar organic lipohilic chemical. They concluded that population health is reflected in population age and size structure, and that the population effects can be more damaging than expected from known levels of effects on the individual. They were also able to show that the combined effects of environmental and chemical stressors can reduce the parameter ranges of population survival. Similar results were found in another study, undertaken by Jaworska *et al.* (1997) to show the effects of another contaminant, polychlorinated biphenyl (PCB), on young-of-the-year largemouth bass.

IBMs have also been developed for environmental and resource management purposes. For example, Schmutz *et al.* (1997) have looked at the relative effects of survival and reproduction on the population dynamics of emperor geese using an IBM, in order to improve the knowledge required by managers to help in the recovery of this resource. Jager *et al.* (1997) have recently developed an IBM to predict instream flow effects on smolt production for fall chinook salmon in regulated rivers. The principal aim was to create a model that could serve as a management tool to evaluate the effects on salmon of instream releases from upstream reservoirs. The model included effects of the riverine habitat at each lifestage. Prediction of development, growth and survival showed a good agreement with empirical studies and clearly demonstrated that flow-related redd mortality and temperature-related juvenile mortality were the major limitations to smolt production. Results from another IBM of salmonids showed that density-dependence was very important for instream survival; this mainly operated during the fry and juvenile stages. Species also showed marked differences in their sensitivities to the crowding of fry (Clark & Rose 1997).

Comments

Do IBMs really constitute something new for ecology? Clearly if the number of studies involving ecologists in the process of modelling is any indication, then the conclusion would have to be yes. But are we really observing a new type of dynamics born out of modelling ecosystems from the perspective of individuals, or are we simply becoming more adept at detecting phenomena that we know exist in nature but have been unable to extract from the more traditional models?

Uchmański & Grimm (1996) believe that IBMs are a new form of ecological modelling, and that they have the potential to induce a paradigm shift in ecology. The four criteria they consider as defining IBMs are: the high degree to which the complexity of individual life-cycles is represented; explicit modelling of resource dynamics; representation of populations by natural or real-numbers; and the inclusion of individual variation. In defending their conclusions they reiterate the link between IBMs with nonequilibrium dynamics, because they contend that narrowly defined IBMs are unable to produce equilibrium results (Bolker *et al.* 1997; Uchmański & Grimm 1997). However, in reviewing the examples of IBMs in the literature it is not clear to what extent this is really the case, or simply a reflection of the extent to which they have been tested, both internally and against empirical data.

Several features do differentiate IBMs from the models of classical ecology; these include the need to formulate ideas as phenomenological rules reflecting processes rather than as simple external forcing functions, an overall trend towards stabilizing effects on the various components of the system and the emergence of counterintuitive behaviours. In particular the evolution of rules, rather than their outcomes must be seen as one of the key driving forces in biological systems. How else would we see the huge range of behaviours in individuals, generally termed instinct? It is therefore clear that IBMs can contribute extensively to a better understanding of the uniqueness of rule-sets and the effects that they have on large-scale spatial and temporal dynamics. However, without the widespread analytical tractability found in some of the more traditional models, it is important that we continue to look for ways in which IBMs can be assessed effectively and not just simply used because they produce results which look similar to observations. Bart (1995) has described a set of guidelines for evaluating IBMs which includes a description of the model and estimates of the reliability of its predictions. However, these are really only an interim step until more robust measures can be formulated. The chapters which follow contain many new ideas that will help in this process, but the most significant challenge that still remains for ecologists is to:

1 elucidate the rule-sets that potentially underpin the behaviour of individuals and which can then be tested in IBMs; and

2 develop data sets at appropriate scales for parameterizing and testing the wide range of ecological models that now exist.

References

Aikman, D.O. & Watkinson, A.R. (1980) A model for growth and self-thinning in even-aged monocultures of plants. *Annals of Botany*, **45**, 419–427.

Allen, P.M. & McGlade, J.M. (1986) Dynamics of discovery and exploitation: the case for the Scotian Shelf groundfish fisheries. *Canadian Journal of Fisheries and Aquatic Sciences*, **43**, 1187–1200.

Allen, P.M. & McGlade, J.M. (1987) Modelling complex human systems: a fisheries example. *European Journal of Operational Research*, **30**, 147–167.

Bak, P. & Chen, K. (1990) A forest-fire model and some thoughts on turbulence. *Physics Letters*, **A147**, 297–300.

Bak, P., Tang, C. & Wiesenfeld, K. (1988) Self-organized criticality. *Physics Review*, **A38**, 364.

Bart, J. (1995) Acceptance criteria for using individual-based models to make management decisions. *Ecological Applications*, **5**, 411–420.

Ben-Naim, E. & Krapivsky, P.L. (1994) Cluster approximation for the contact process. *Journal of Physics*, **A27**, L481–487.

Bernstein, C., Kacelnik, A. & Krebs, J.R. (1991) Individual decisions and the distribution of predators in a patchy environment. *Journal of Animal Ecology*, **60**, 205–225.

Bolker, B.M., Deutschman, D.H., Hartvigsen, G. & Smith, D.L. (1997) Individual-based modelling: what is the difference? *Trends in Evolution and Ecology*, **12**, 111.

Bullock, J. (1994) Correspondence: Individual-based models. *Trends in Evolution and Ecology*, **9**, 299.

Canham, C.D., Finzi, A.C., Pacala, S.W. & Burbank, D.H. (1994) Causes and consequences of resource heterogeneity in forests: interspecific variation in light transmission by canopy trees. *Canadian Journal of Forest Research*, **24**, 337–176.

Caswell, H. & John, A.M. (1992) From the individual to the population in demographic models. In: *Individual-based Models and Approaches in Ecology. Populations, Communities and Ecosystems* (eds D.L. DeAngelis & L. Gross). Chapman & Hall, London.

Chesson, P.L. (1981) Models for spatially distributed populations: the effect of within-patch variability. *Theoretical Population Biology*, **19**, 288–325.

Clark, M.E. & Rose, K.A. (1997) Individual-based model of stream-resident rainbow trout and brook charr: model description, corroboration, and effects of sympatry and spawning season duration. *Ecological Modelling*, **94**, 157–175.

Colasanti, R.L. & Grime, J.P. (1993) Resource dynamics and vegetation processes: a deterministic model using two-dimensional cellular automata. *Functional Ecology*, **7**, 169–176.

Crawley, M.J. & May, R.M. (1987) Population dynamics and plant community structure: competition between annuals and perennials. *Journal of Theoretical Biology*, **125**, 475–489.

DeAngelis, D.L. & Gross, L.J. (eds) (1992) *Individual-based Models and Approaches in Ecology Populations, Communities and Ecosystems*. Chapman & Hall, London.

Dickman, R. (1986) Kinetic phase transition in a surface-reaction model: mean-field theory. *Physics Review*, **A34**, 4246–4250.

Durrett, R. & Levin, S.A. (1994) Stochastic spatial models: a user's guide to ecological applications. *Philosophical Transactions of the Royal Society of London*, **B343**, 329–350.

Filipe, J.A.N. & Rodgers, G.J. (1995) Theoretical and numerical studies of chemisorption on a line with precursor layer diffusion. *Physics Review*, **E52**, 6044–6054.

Focardi, M.R. (1997) A physiologically-based model of a self-motivated hare in relation to its ecology. *Ecological Modelling*, **95**, 191–209.

Ghandi, A., Levin, S.A. & Orszag, S. (1998) 'Critical slowing down' in time-to-extinction: an example of critical phenomena in ecology. *Journal of Theoretical Biology*, **192**, 363–376.

Gilpin, M. & Hanski, I. (1991) *Metapopulation Dynamics, Empirical and Theoretical Investigations*. Academic Press, London.

Gordon, D.M. (1997) The population consequences of territorial behaviour. *Trends in Evolution and Ecology*, **12**, 63–65.

Hallam, T.G., Trawick, T.L. & Wolff, W.F. (1996) Modelling effects of chemicals on a population: application to a wading bird nesting colony. *Ecological Modelling*, **92**, 155–178.

Hanski, I. & Gilpin, M. (1991) Metapopulation dynamics: brief history and conceptual domain. *Biological Journal of the Linnaean Society*, **42**, 3–16.

Harada, Y. & Iwasa, Y. (1994) Lattice population dynamics for plants with dispersing seeds and vegetative propagation. *Research in Population Ecology*, **36**, 237–249.

Harada, Y., Ezoe, H., Iwasa, Y., Matsuda, H. & Satō, K. (1995) Population persistence and spatially limited social interaction. *Theoretical Population Biology*, **48**, 65–91.

Hendry, R.J. (1995) *Spatial modelling in plant ecology*. PhD dissertation, University of Warwick.

Hendry, R.J. & McGlade, J.M. (1995) The role of memory in ecological systems. *Proceedings of the Royal Society of London*, **B259**, 1153–1159.

Hendry, R., McGlade, J.M. & Weiner, J. (1996) A coupled map lattice model of the growth of plant monocultures. *Ecological Modelling*, **84**, 81–90.

Hendry, R., Bacon, P.J., Moss, R., Palmer, S.C.F. & McGlade, J. (1997) A two-dimensional individual-based model of territorial behaviour: possible population consequences of kinship in Red Grouse. *Ecological Modelling*, **105**, 23–39.

Hogeweg, P. & Hesper, B. (1990) Individual-oriented modelling in ecology. *Mathematics and Computer Modelling*, **13**, 83–90.

Humphries, H.C., Coffin, D.P. & Lauernroth, W.K. (1996) An individual-based model of alpine plant distributions. *Ecological Modelling*, **84**, 99–126.

Huston, M., DeAngelis, D.L. & Post, W. (1988) New computer models unify ecological theory. *BioScience*, **38**, 682–691.

Jager, H.I., Cardwell, H.E., Sale, M.J., Bevelhimer, M.S., Countant, C.C. & Van Winkle, W. (1997) Modelling the linkages between flow management and salmon recruitment in rivers. *Ecological Modelling*, **103**, 171–191.

Jaworska, J.S., Rose, K.A. & Brenkert, A.L. (1997) Individual-based modelling of PCBs effects on young-of-the-year largemouth bass in southeastern USA reservoirs. *Ecological Modelling*, **99**, 113–135.

Jeltsch, F. & Wissel, C. (1994) Modelling dieback phenomena in natural forests. *Ecological Modelling*, **75/76**, 111–121.

Judson, O. (1994) The rise of the individual-based model in ecology. *Trends in Evolution and Ecology*, **9**, 9–14.

Kaneko, K. (1990) Simulating physics with coupled map lattice. In: *Formation, Dynamics and Statistics of Patterns* (eds K. Kawasaku, M. Suzuki & A. Onuki). World Scientific, Singapore.

Keeling, M.J., Rand, D.A. & Morris, A.J. (1997a) Correlation models for childhood epidemics. *Proceedings of the Royal Society of London*, **B264**, 1149–1156.

Keeling, M.J., Mezić, I., Hendry, R.J., McGlade, J.M. & Rand, D.A. (1997b) Characteristic length scales of spatial models in ecology via fluctuation analysis. *Philosophical Transactions of the Royal Society of London*, **B352**, 589–1601.

Keitt, T.H. (1997) Stability and complexity on a lattice: co-existence of species in an individual-based food web model. *Ecological Modelling*, **102**, 243–258.

Kobe, R.K. (1996) Intraspecific variation in sapling mortality and growth predicts geographic variation in forest composition. *Ecological Monographs*, **66**, 181–201.

Kobe, R.K., Pacala, S.W., Silander, J.A. Jr & Canham, C.D. (1995) Juvenile tree survivorship as a component of shade tolerance. *Ecological Applications*, **5**, 517–532.

Koh, H.L., Hallam, T.G. & Lee, H.L. (1997) Combined effects of environmental and chemical stressors on a model *Daphnia* population. *Ecological Modelling*, **103**, 19–32.

Kubo, T., Iwasa, Y. & Furomuto, N. (1996) Forest spatial dynamics with gap expansion: total gap area and gap size distribution. *Journal of Theoretical Biology*, **180**, 229–246.

LePage, C. & Cury, P. (1996) How spatial heterogeneity influences population dynamics: simulations in SeaLab. *Adaptive Behaviour*, **4**, 255–281.

Levin, S.A. & Durrett, R. (1997) From individuals to epidemics. *Philosophical Transactions of the Royal Society of London*, **B351**, 1615–1621.

Levin, S.A. & Paine, R.T. (1974) Disturbance, patch formation, and community structure. *Proceedings of the National Academy of Sciences USA*, **71**, 2744–2747.

Levins, R. (1969) Some demographic and genetic consequences of environmental heterogeneity for biological control. *Bulletin of the Entomological Society of America*, **15**, 237–240.

Levins, R. (1970) Extinction. *Lectures on Mathematics in Life Sciences*, **2**, 77–107.

Łomnicki, A. (1988) *Population ecology of individuals*. Princeton University Press, Princeton, NJ.

McGlade, J.M. (1993) Alternative ecologies. *New Scientist*, (Suppl.), **137**, 14–16.

May, R.M. (1974) Biological population with non-overlapping generations: stable points, stable cycles and chaos. *Science*, **186**, 645–647.

Metz, J.A.J. & Diekman, O. (1986) *The dynamics of physiologically structured populations*. Springer-Verlag, New York.

Murray, J. (1990) *Mathematical Biology*. Springer-Verlag, Berlin.

Neuert, C., Du Plessis, M.A., Grimm, V. & Wissel, C. (1995) Welche ökologischen Faktoren bestimmen die Gruppengrösse bei *Phoeniculus purpureus* (Gemeiner Baumhopf) in Südafrika? Ein individuen basiertes Modell. *Verhandlungen der Gesellschaft für Ökologie*, **24**, 145–149.

Pacala, S.W., Canham, C.D. & Silander, J.A. Jr (1993) Forest models defined by field measurements: I. The design of a northeastern forest simulator. *Canadian Journal of Forest Research*, **23**, 1980–1988.

Pacala, S.W., Canham, C.D., Silander, J.A. Jr & Kobe, R.K. (1994) Sapling growth as a function of resources in a northern temperate forest. *Canadian Journal of Forest Research*, **24**, 2172–2183.

Pacala, S.W., Canham, C.D., Saponara, J., Silander, J.A. Jr & Kobe, R.K. (1996) Forest models defined by field measurements: estimation, error analysis and dynamics. *Ecological Monographs*, **66**, 1–43.

Provencher, L. & Reichert, S.E. (1995) Theoretical comparisons of individual success between phenotypically pure and mixed generalist predator populations. *Ecological Modelling*, **82**, 175–191.

Rand, D.A., Keeling, M. & Wilson, H.B. (1995) Invasion, stability and evolution to criticality in spatially extended, artificial host-pathogen ecologies. *Proceedings of the Royal Society of London*, **B259**, 55–63.

Ratz, A. (1995) Long-term spatial patterns created by fire: a model oriented towards boreal forests. *International Journal of Wildland Fire*, **5**, 25–34.

Remmert, H. (ed.) (1991) *The mosaic-cycle concept of ecosystems*. Springer-Verlag, New York.

Ribbens, E., Silander, J.A. Jr. & Pacala, S.W. (1994) Recruitment in northern hardwood forests: a method for calibrating models that describe patterns of tree seedling dispersal. *Ecology*, **75**, 1794–1806.

Satō, K., Matsua, H. & Sasaki, A. (1994) Pathogen invasion and host extinction in lattice structured populations. *Journal of Mathematical Biology*, **32**, 251–268.

Schmutz, J.A., Rockwell, R.F. & Peterson, M.R. (1997) Relative effects of survival and reproduction on the population dynamics of emperor geese. *Journal of Wildlife Management*, **61**, 191–201.

Scheffer, M., Baveco, J.M., DeAngelis, D.L., Rose, K.A. & van Nes, E.H. (1995) Super-individuals a simple solution for modelling large populations on an individual basis. *Ecological Modelling*, **80**, 161–170.

Uchmański, J. & Grimm, V. (1996) Individual-based modelling: what makes the difference? *Trends in Evolution and Ecology*, **11**, 437–441.

Uchmański, J. & Grimm, V. (1997) Reply from J. Uchmański and V. Grimm. *Trends in Evolution and Ecology*, **12**, 112.
Williams, M. (1996) A three-dimensional model of forest development and competition. *Ecological Modelling*, **89**, 73–98.
Wissel, C. (1992) Modelling the mosaic cycle of a middle European beech forest. *Ecological Modelling*, **63**, 22–45.
Wolff, W. (1994) An individual-oriented model of a wading bird nesting colony. *Ecological Modelling*, **72**, 75–114.

2: Stochastic effects in population models

E. Renshaw

Introduction

The remarkable variety of dynamic behaviour exhibited by many species of plants, insects and animals has stimulated great interest in the development of both biological experiments and mathematical models. After a relatively slow start in the 1920s and 1930s, the pace of research quickened dramatically throughout the remainder of the 20th century, though mathematicians and biologists have tended to take widely diverging paths. Theoreticians often profess an interest in biology purely to give them access to an intriguing set of stochastic (i.e. probability) equations, with the occasional biological reference being thrown in merely to perpetuate the mirage of potential applicability. Whilst, supposedly grass-roots 'mathematical biologists' are often tempted to develop vaguely plausible deterministic models which reflect mathematical hope rather than biological reality.

Many researchers still use one approach to the total exclusion of the other. The reasons are two-fold. First, pioneering biological studies were greatly influenced by deterministic mathematics and reluctance to accept the importance of stochastic ideas is still deeply ingrained. Second, mathematicians are often taught in a practical vacuum, with the result that instead of using mathematics to interpret and understand biological phenomena they become transfixed by the models themselves. Renshaw (1991) presents a unifying approach between these two extremes, showing that both deterministic and stochastic models have important roles to play and should therefore be considered together. Popular deterministic ideas of even simple systems involving logistic, predator–prey and competition relationships can change markedly when viewed from a stochastic viewpoint. In biology we are often asked to infer the nature of population development from a single data set, yet different realizations of the same process can vary enormously. Even theoretical stochastic solutions are only of limited help here, and simple computer simulation procedures need to be constructed which provide much needed insight into the underlying biological generating mechanisms. Renshaw (1991) also advocates full recognition that the environment has a spatial dimension, since individual population members rarely mix homogeneously over the territory available to them but develop instead within separate subregions. Subsequent interaction between these subregions, whether in the form of migration of individuals, or cross-infection of disease, can vary from being purely local to involving the entire area under study. Fortunately, fairly simple models can be developed which highlight the effects that geographic restrictions and species/infection mobility may have on population development. These models can provide vital knowledge about the dispersal and control of many natural populations, not just of animals, insects and plants, but also of diseases such as malaria, rabies and AIDS (see Chapters 3 and 4).

Although the development of exact theoretical stochastic solutions to all but the simplest of models generates at best difficult and at worst impossible mathematical demands, construction of computer-simulated realizations of even highly complex modelling scenarios is intrinsically easy and of potentially great benefit to biologists. The purpose of this chapter is therefore:

1 to show how deterministic and stochastic approaches relate to each other;

2 to show how they may be combined into a mathematically tractable 'halfway house'; and

3 to illustrate how easily understood, and simply implemented, simulation procedures can be developed for a variety of model structures developing in both time and space.

The necessity of incorporating such a holistic approach is becoming ever more vital, since with user-friendly software for solving systems of deterministic equations now easily available, new problems can be supposedly 'solved' in little more time than it takes to input the equations into a personal computer. Production of nice graphical output adds to the illusion.

It should always be assumed that stochastic effects play an important role in any given process unless proved otherwise. If it transpires that over the range of parameter interest there is little practical distinction to be drawn between deterministic and stochastic realizations, then the user has at least gained confidence in the robustness of the deterministic solution. Conversely, if substantial differences are seen, or if effects are produced that had not been previously anticipated, then appropriate biological investigations can be undertaken. Either way, the investigator has nothing to lose and potentially a great deal to gain from such an approach.

The basic birth–death process

As stressed in Renshaw (1991), recent interest in population dynamics has become polarized to an undesirable extent: the great tragedy is that far too few researchers realize that *both* deterministic and stochastic models have important roles to play in the analysis of any particular system. Slavish obedience to one particular approach can lead to disaster. One advantage of continuous deterministic models is the relative ease with which they can be handled numerically. This may make it feasible to allow the inclusion of more realistic features into the model, and to analyse and interpret the sensitivity of results to changes in the parameter values. A major disadvantage is that they cannot provide any information on the likely variability of predicted population sizes.

Provided that population numbers never become too small, a deterministic model *may* enable sufficient biological understanding to be gained about the system; if at any time population numbers do become small then a stochastic analysis is vital. So pursuing both approaches simultaneously ensures that we do not become trapped either by deterministic fantasy or unnecessary mathematical detail. Mollison (1986) strengthens this philosophy by arguing that if the underlying process is stochastic then the only way to derive a deterministic model that is approximately correct may be to *start* from an explicitly stochastic standpoint.

To illustrate the underlying ideas let us consider a simple scenario of population growth, in which individuals develop without interacting with each other, they have unlimited food resource, and the birth rate λ and death rate μ are the same for all individuals regardless of their age.

Deterministic approach

Taking the classic deterministic approach first, let $N(t)$ denote the population size at time t. Then in the subsequent *small* time interval of length h the increase in population size due to a single individual is $\lambda \times h$ (i.e. rate $\times$ time), so the total increase is $\lambda \times h \times N(t)$. Similarly, the decrease is $\mu \times h \times N(t)$. Thus

$$N(t+h) - N(t) = \lambda h N(t) - \mu h N(t) \tag{2.1}$$

Dividing both sides by h and letting h approach zero then yields the differential equation

$$\mathrm{d}N(t)/\mathrm{d}t = (\lambda - \mu)\, N(t) \tag{2.2}$$

which integrates to give one of the oldest results in population dynamics, namely that

$$N(t) = N(0)\exp\{(\lambda - \mu)t\} \tag{2.3}$$

The expression in Equation 2.3 tells us that if births predominate over deaths then the population size will explode exponentially fast, whilst if deaths predominate then exponential decay leads to inevitable extinction. For large $N(0)$ and realistically small t this is indeed an excellent description, but let us reflect for a moment on what happens if $N(0)$ is small. Suppose that $N(0) = 1$ and $\lambda = 2\mu$ so that births are twice as likely to occur as deaths. Then the expression in Equation 2.3 predicts exponential growth with $N(t) = \exp(\mu t)$. But the first event to occur may be a death, with probability $\mu/(\mu + \lambda) = 1/3$, which results in immediate extinction. Thus the probability of eventual extinction must exceed 1/3, in direct contradiction to the deterministic prediction of guaranteed exponential growth. The situation becomes even more absurd when $\lambda = \mu$. Then $N(t)$ remains absolutely constant at $N(t) = 1$ even though the actual process involves substantial population size change due to births and deaths.

An apparently hard lesson to learn is the extent to which different realizations of the *same process* can vary. For example, Fig. 1.1 shows 20 simulations of a simple birth–death model with $\lambda = 1.0$, $\mu = 0.8$ and $N(0) = 3$. Though 11 simulations exhibit exponential growth, in line with deterministic theory, one grows linearly, whilst eight die out completely. Such simulations are clearly valuable in highlighting the degree of variability that we might expect to observe in practice. If a large number of simulations all lie reasonably close to the deterministic curve then we can be satisfied that a deterministic approach will provide an adequate description of population development. If they do not, then a stochastic description is required. Moreover, such model-based simulations can highlight hitherto unforeseen features of a process and thereby suggest further profitable lines of biological investigation. It is in this continued interplay between modelling and experimentation that population studies are at their most powerful.

Stochastic analysis

For simple types of population model, such as this linear birth–death process, theoretical development is possible. In our short time interval $(t, t + h)$ the probabilities (Pr) that a given individual gives birth or dies are λh and μh, respectively, so the probabilities of a birth or death within the population are λNh and μNh, respectively. Thus on denoting

$$p_N(t) = \Pr\{\text{population is of size } N \text{ at time } t\}$$

and taking h sufficiently small to ensure that the probability of more than one event occurring in $(t, t + h)$ is negligible, we have

$$\begin{aligned} p_N(t+h) = {} & p_N(t) \times \Pr\{\text{no birth or death in } (t,\ t+h)\} \\ & + p_{N-1}(t) \times \Pr\{\text{one birth in } (t,\ t+h)\} \\ & + p_{N+1}(t) \times \Pr\{\text{one death in } (t,\ t+h)\} \end{aligned} \tag{2.4}$$

i.e.

$$\begin{aligned} p_N(t+h) = {} & p_N(t) \times [1 - N(\lambda + \mu)h] + p_{N-1}(t) \times [(N-1)\lambda h] \\ & + p_{N+1}(t) \times [(N+1)\mu h] \end{aligned} \tag{2.5}$$

Rearranging Equation 2.5, dividing through by h, and letting h approach zero, then yields the set of equations

$$dp_N(t)/dt = \lambda(N-1)p_{N-1}(t) - (\lambda+\mu)Np_N(t) + \mu(N+1)p_{N+1}(t) \tag{2.6}$$

over $N = 0,1,2,\ldots$ and $t \geq 0$. Although these may be solved using standard differential equation techniques (see Cox & Miller 1965), the 'solution' is usually expressed in terms of the generating function $G(z;t) = p_0(t) + p_1(t)z + \ldots + p_N(t)z^N + \ldots$. The probabilities $p_N(t)$ are then determined by expanding $G(z;t)$ as a power-series in z and extracting the coefficient of z^N. Yet even in this very simple case the resulting expression (see Bailey 1964) is really too messy to be of much practical use unless $N(0) = 1$.

However, as individuals are assumed to develop independently of one another, we may consider the development of a population of initial size $N(0) = n_0$ as being equivalent to the development of n_0 *separate* populations each of initial size 1. Thus in order to understand the stochastic behaviour of the process it is sufficient just to consider $n_0 = 1$. In this case the population size $N(t)$ follows a geometric distribution with mean

$$m(t) = n_0 e^{(\lambda-\mu)t} \tag{2.7}$$

which is the same as the deterministic value (Equation 2.3) and variance

$$V(t) = n_0[(\lambda+\mu)/(\lambda-\mu)]e^{(\lambda-\mu)t}(e^{(\lambda-\mu)t} - 1) \tag{2.8}$$

(Feller 1939). Note that unlike $m(t)$, $V(t)$ depends not only on the difference between the birth and death rates but also on their absolute magnitudes. This is what we should expect, since predictions about the future size of a population will be less precise if births and deaths occur in rapid succession than if they occur only occasionally.

Simulation

Whilst mathematically one may regard the derivation of the probabilities $\{p_N(t)\}$, together with the moments which may be obtained from them, as being a full solution, this is not the case from a biological perspective. For probabilities do not relate to specific single realizations of a process, but represent instead average behaviour over all possible realizations. Consider, for example, a single toss of a fair coin scoring 1 for a head and 0 for a tail. Then the expected score is $1 \times \Pr(\text{head}) + 0 \times \Pr(\text{tail}) = 1 \times 1/2 + 0 \times 1/2 = 1/2$, which is totally unrepresentative of what actually happens. So for our simple birth–death process what do $m(t)$ and $V(t)$ actually tell us about the development of individual population paths? If $\lambda > \mu$, does the variance in Equation 2.8 relate almost wholly to the random times taken for the process to 'get going', with individual realizations broadly mimicking $m(t)$ thereafter; or does $V(t)$ relate to individual realizations perpetually 'wandering' about $m(t)$? The only real way to find out is to examine simulated realizations.

No matter how complicated the process under consideration appears to be, it can always be completely described as a series of events $E_1, E_2, \ldots$ which occur at times $t_1, t_2, \ldots$. So given that we have just observed the nth event-time (E_n, t_n), all that immediately concerns us is determining:

1 the next event E_{n+1}; and

2 the inter-event time $(t_{n+1} - t_n)$.

In essence all we need to do is to employ the following procedure (see Renshaw 1991 for specific details). Let $B(N)$ and $D(N)$ denote the population birth and death rates, respectively; so for the simple birth–death process $B(N) = \lambda N$ and $D(N) = \mu N$. Then

1 enter the functions $B(N)$ and $D(N)$;

2 generate two random uniform (0,1) random numbers Y_1 and Y_2;

3 if $0 \leq Y_1 \leq B(N)/[B(N) + D(N)]$ take the next event to be a birth, otherwise a death;

4 evaluate the inter-event time $s = -[\log_e(Y_2)]/[B(N) + D(N)]$, and update t to $t + s$;

5 change N to $N + 1$ (birth) or $N - 1$ (death);

6 return to step **2**.

Any 'reasonable' pseudo-random number generator can be used for Y_1 and Y_2, but care in selection does need to be taken. For example, although the FORTRAN NAG-routine G05CAF is well recommended, the early RANDU generator (see Ripley 1987) is a horror-story since a generated Y_i-value is highly dependent on the two previous values. Readers interested in the generation and testing of pseudo-random numbers should consult the excellent introductory text of Morgan (1984).

QBASIC provides an excellent medium for the initial interrogation (and viewing) of a stochastic process for potential types of behavioural structure; Fig. 2.1 shows a screen image of 20 simulations for a simple birth–death process. Eight of the realizations result in extinction, with the most persistent one dying out at $t = 13.3$. The others exhibit exponential growth (in line with deterministic theory), but they are far less 'smooth' than one might anticipate. We note four different behaviour patterns:

1 early extinction;

2 persistence at low population levels followed by extinction;

3 initial 'linear' growth (to be followed later by exponential growth);

4 immediate exponential growth.

If these four sequence-types were observed in the laboratory, how many applied researchers would seriously consider the possibility that the same process was responsible for each one?

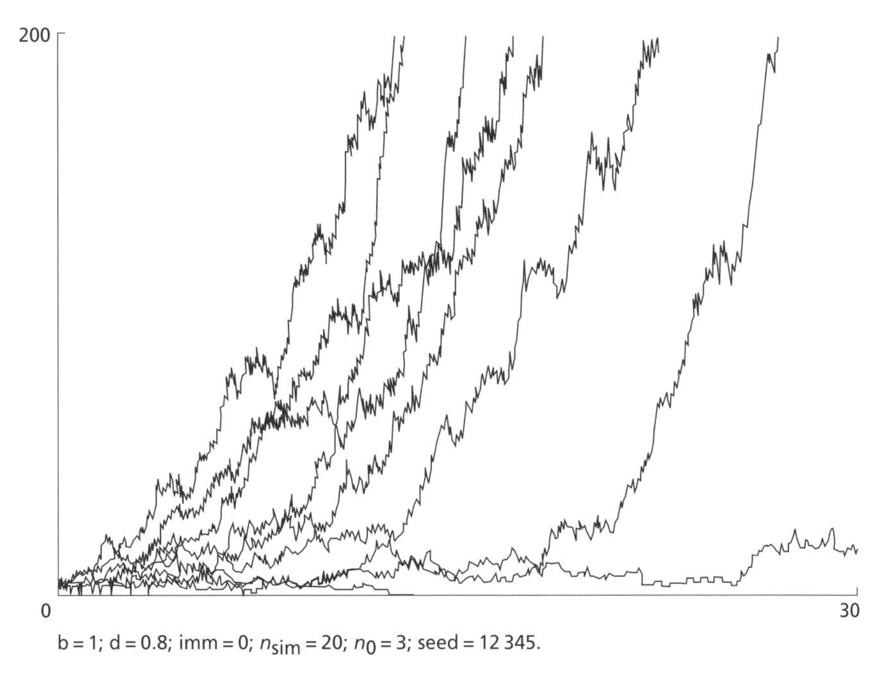

Fig. 2.1 Screen dump of 20 stochastic realizations of the simple birth–death process with λ = 1.0, μ = 0.8 and n_0 = 3 over $0 \leq t \leq 30$.

For this simple model we can write down the probability of extinction by time t as

$$p_0(t) = \{[\mu - \mu e^{-(\lambda-\mu)t}]/[\lambda - \mu e^{-(\lambda-\mu)t}]\}^{n_0} \tag{2.9}$$

(see Renshaw 1991). Thus if $\lambda < \mu$ (i.e. negative net growth rate) then $p_0(t)$ approaches 1 as t increases so ultimate extinction is certain. Conversely, if $\lambda > \mu$ then $p_0(t) \rightarrow (\mu/\lambda)^{n_0}$. So for our example $p_0(\infty) = (0.8)^3 = 0.51$, in rough agreement with our crude simulated estimate of 0.4. Moreover, since the exp $\{-(\lambda - \mu)t\}$ term in Equation 2.9 decays to 0.01 once t reaches 23, virtually all realizations that lead to extinction will have done so by this time. To determine how large an initial population size has to be before ultimate extinction is highly unlikely, we could put $(\mu/\lambda)^{n_0} = 0.001$ (say). Whence $0.8^{n_0} = 0.001$, which gives $n_0 = 31$.

Immigration

In the preceding model, the population becomes either extinct or infinite. Introducing immigration into the system allows us to avoid these two extremes, and also acknowledges the fact that few populations develop in true isolation. Suppose that immigrants arrive randomly with mean rate α. Then the deterministic Equation 2.2 becomes

$$dN(t)/dt = (\lambda - \mu) N(t) + \alpha, \tag{2.10}$$

and for $\lambda < \mu$ the solution approaches the constant value $\alpha/(\mu - \lambda)$ at the exponential rate $\exp\{(\lambda - \mu)t\}$. Extinction (though not an empty population) is clearly impossible, since a new immigrant will always arrive to start the population off again. Thus when $\lambda < \mu$ a stochastic approach will give rise to an equilibrium distribution of population size, denoted by $\pi_N = \Pr$ (population is ultimately of size N). Since in equilibrium successive states must be in 'balance', in the sense that the probability of a death in a population of size N must match the probability of a birth in a population of size $N - 1$, we can write

$$(\mu N)\pi_N = [\alpha + \lambda(N-1)]\pi_{N-1} \tag{2.11}$$

Solving for π_N then yields the negative binomial distribution

$$\pi_N = \binom{N-1+(\alpha/\lambda)}{N}\left(\frac{\lambda}{\mu}\right)^N\left(1-\frac{\lambda}{\mu}\right)^{\alpha/\lambda} \quad (N = 0, 1, 2, \ldots) \tag{2.12}$$

(see Renshaw 1991). This has mean $\alpha/(N - \lambda)$ (in line with the deterministic solution) and variance $\alpha\mu/(\mu - \lambda)^2$. If $\lambda = 0$, then the expression in Equation 2.12 reduces to the Poisson distribution with parameter (α/μ).

Note that we can also generate an equilibrium distribution through the simple device of placing a new individual in the population immediately it becomes empty, instead of allowing random immigration at rate α. This means that $\pi_0 = 0$, whence the balance equations now solve to give

$$\pi_N = \pi_1 (\lambda/\mu)^{N-1}/N \tag{2.13}$$

over $N = 1, 2, \ldots$. Using $\pi_1 + \pi_2 + \ldots = 1$ leads to $\pi_1 = -(\lambda/\mu)/\ln[1 - (\lambda/\mu)]$

General single-species processes

The underlying assumptions for the simple birth–death process can only hold true if there is no interference amongst individual population members. However, in a restricted environment the growth of any expanding population must eventually be limited by a shortage of resources, which means that *individual* birth and death rates *must* depend on population size N (see a discussion of this in Chapter 1). The general deterministic approach extends immediately from Equation 2.2, with λN and μN being replaced by the functions $B(N)$ and $D(N)$, respectively, giving

$$dN(t)/dt = B[N(t)] - D[N(t)] \tag{2.14}$$

Though the feasibility of integrating Equation 2.14 depends on the form of B and D, it's always worthwhile trying to generate a theoretical solution before resorting to numerical procedures. For only through a theoretical analysis can one be confident of discovering the full qualitative structure of a process.

Equilibrium probabilities

Similarly, the equation for the probabilities $\{p_N(t)\}$ is the immediate extension of the birth–death Equation 2.6, namely

$$\mathrm{d}p_N(t)/\mathrm{d}t = D(N+1)p_{N+1}(t) - [B(N)+D(N)]\,p_N(t) + B(N-1)p_{N-1}(t) \qquad (2.15)$$

for $N = 0, 1, 2, \ldots$ and $B(-1) = 0$. However, $B(N)$ and $D(N)$ will often be nonlinear, thereby rendering analytic solution of Equation 2.15 either very difficult or impossible. Fortunately, limiting resource usually places an effective brake on indefinite population growth, thereby allowing an equilibrium situation to develop. In this situation the initial development of a population is often of far less importance than its long-term properties, so the mathematical difficulty surrounding the solution of Equation 2.15 is really irrelevant. Generalizing the 'balance equation' Equation 2.11 gives

$$D(N)\pi_N = B(N-1)\pi_{N-1}, \qquad (2.16)$$

whereby repeated application of this relation over $N = 1, 2, \ldots$, combined with $\pi_0 + \pi_1 + \ldots = 1$, yields the general equilibrium solution (Renshaw 1991)

$$\pi_N = q_N \Big/ \sum_{i=0}^{\infty} q_i \text{ where } q_0 = \frac{1}{B(0)},\ q_1 = \frac{1}{D(1)} \text{ and}$$
$$q_N = \frac{B(1)B(2)\cdots B(N-1)}{D(2)D(3)\cdots D(N)} \quad (N \geq 2) \qquad (2.17)$$

To illustrate this approach consider a birth–death process with a severe density-dependent death rate, namely $B(N) = \lambda N$ and $D(N) = \mu N(N-1)$. Since $D(1) = 0$ the population can never become extinct. The equilibrium solution Equation 2.17 immediately yields

$$\pi_N = \frac{(\lambda/\mu)^N}{N!}(\mathrm{e}^{\lambda/\mu} - 1)^{-1} \qquad (2.18)$$

which is a censored Poisson distribution over $N = 1, 2, \ldots$. Note that although this is a totally different scenario to the immigration–death process, both lead to Poisson forms.

In many biological situations, a sufficiently violent downward fluctuation in population size is bound to occur which will drive the population extinct. However, if a long time has to elapse before the probability of this happening becomes nonnegligible, then although an equilibrium distribution $\{\pi_N\}$ will not formally exist there will be a *quasi-equilibrium* distribution $\{\pi_N^{(Q)}\}$ defined by the proportion of nonextinct realizations at some large time t that contain exactly $N > 0$ individuals. Thus $\{\pi_N^{(Q)}\}$ is effectively a true equilibrium distribution over ecologically relevant periods of time. The mathematical detail of this is given in Renshaw (1991), but the key result is that the solution for $\{\pi_N^{(Q)}\}$ is identical to the expression given in Equation 2.17 except that q_0 is replaced by zero.

Also provided are expressions for the probability of ultimate extinction (usually 1 unless the system is subject to some kind of external input), and the mean time to extinction T_E, though these are slightly cumbersome to work with. Large values of T_E relate to populations which are ecologically stable, and small values to those which are unstable, and since the order of magnitude of T_E represents the degree of stability we define the *stability index* $\xi = \log_e(T_E)$. Though exact results for the probability of extinction $p_0(t)$ are much harder to determine, a useful simplification exists for processes which, prior to extinction, have settled down to a quasi-equilibrium state, namely that

$$p_0(t) \simeq 1 - \exp(-\tau) \qquad (2.19)$$

Here τ denotes the normalized time $\tau = t/T_E$. Inverting the expression in Equation 2.19 to obtain $\tau \simeq -\log_e[1 - p_0(t)]$ shows that $p_0(t)$ reaches 0.01, 0.5 and 0.99 when τ equals 0.01, 0.69 and 4.61, respectively. Thus extinction is highly unlikely to occur before $\tau = 0.01$ and is very likely to have occurred by $\tau = 4.6$. Note that the skewness of the exponential distribution of Equation 2.19 results in the median time to extinction ($\tau = 0.69$) being considerably less than the mean time to extinction ($\tau = 1$).

Logistic process

Although the number of possible choices for the birth and death rates $B(N)$ and $D(N)$ is clearly unlimited, pragmatism demands that selection should be made on sensible biological grounds. That is, the simplest possible functions should be chosen that are consistent with the known biological characteristics of the system. One such scenario is when the individual birth and death rates respectively decrease and increase linearly with increasing N; crowding might reduce mating capability, whilst scarcity of food might increase the chance of death. So let us put

$$B(N) = N(a_1 - b_1 N) \quad \text{and} \quad D(N) = N(a_2 + b_2 N) \qquad (2.20)$$

for some positive constants a_1, a_2, b_1 and b_2. On writing $r = a_1 - a_2$ and $s = b_1 + b_2$, the associated deterministic Equation 2.14 reduces to

$$\mathrm{d}N/\mathrm{d}t = N(r - sN) \qquad (2.21)$$

which is the well-known Verhulst–Pearl logistic equation with the solution

$$N(t) = K/[1 + \{(K - N(0))/N(0)\}\exp(-rt)] \tag{2.22}$$

Here $K = r/s$ denotes the 'carrying capacity' of the system. Note that although this classic result was produced by Verhulst as early as 1938 to describe the growth of human populations, it was virtually ignored until Pearl and Reed resurrected it in 1920 (see Pearl 1930).

One of the surprising features of this solution is that population growth often closely follows it even when the underlying biological growth mechanism patently does not satisfy the Verhulst–Pearl assumptions. Sang (1950), for example, points out that in Pearl's (1927) laboratory experiments on the growth of *Drosophila melanogaster* the yeast that was the source of food was itself a growing population, yet the logistic still fits the data reasonably well. It is such robustness that has led to the logistic curve Equation 2.22 gaining the credence of a universal biological *law*.

Although the data of Carlson (1913) on yeast growth in laboratory cultures are described almost perfectly by deterministic logistic growth, in general, population development will not only exhibit some degree of variability about this in the initial sigmoid growth phase but it will also often show large fluctuations once the carrying capacity K has been attained. Renshaw (1991) shows plots of two classic field data sets which exhibit this behaviour pattern, namely the number of sheep recorded in South Australia from 1838 to 1936 (Davidson 1938a) and Tasmania from 1818 to 1936 (Davidson 1938b). Estimates are obtained for the parameters r, s and K, and whilst both sheep populations show good agreement between the data and fitted logistic curves over the initial sigmoid period, if we are to describe the subsequent large oscillations in population size, then a stochastic analysis has to be undertaken.

Given that we have already highlighted the considerable mathematical difficulties likely to be encountered in constructing the full time-dependent probabilities $\{p_N(t)\}$, there is clearly little biological point in pursuing such a strongly mathematically oriented approach. Far better is to concentrate on the oscillatory behaviour about the asymptote, since substituting the rates given in Equation 2.20 into Equation 2.17 immediately yields the quasi-equilibrium probabilities

$$\pi_N^{(Q)} = q_N \Big/ \sum_{i=1}^{\infty} q_i \text{ where } q_1 = 1/(a_2 + b_2) \text{ and } q_N = \frac{[a_1 - b_1]\cdots[a_1 - (N-1)b_1]}{[a_2 + b_2]\cdots[a_2 + Nb_2]N} \tag{2.23}$$

Local stochastic approximation

Although the result in Equation 2.23 is easy to compute numerically, it is certainly not user-friendly when it comes to evaluating algebraic expressions for means, variances, skewness, etc. I shall therefore now demonstrate an extremely useful technique which represents a compromise between the fully deterministic and stochastic representations. In essence we write

$$\text{stochastic} = \text{deterministic} + \text{noise} \tag{2.24}$$

where the 'noise' component reflects local variation about the equilibrium value. Note that although the following analysis relates specifically to the logistic scenario, it can be applied to *any* deterministic representation $dN/dt = f(N)$, with appropriate growth function $f(N)$, that possesses a quasi-equilibrium state.

First parallel Equation 2.1 by writing the Verhulst–Pearl Equation 2.21 in the form

$$N(t+h) - N(t) \simeq N(t)[r - sN(t)]h \tag{2.25}$$

where h is a small time increment. To turn this into the stochastic representation Equation 2.24 we write the random population change in time h as

$$X(t+h) - X(t) = X(t)[r - sX(t)]h + Z(t)h \tag{2.26}$$

where the stochastic population size is given by the variable $X(t)$. Here the noise component $Z(t)h$ represents the chance fluctuation in time h, and has mean $E[Z(t)] = 0$, where E denotes expectation. Now if the population fluctuates around the mean value m, then both $E[X(t)]$ and $E[X(t+h)]$ equal m. So taking expectations of both sides of Equation 2.26 yields

$$0 = rmh - sE[X^2(t)]h, \quad \text{i.e. } rm - s\{m^2 + \text{Var}[X(t)]\} = 0 \tag{2.27}$$

Hence, as a first approximation to m is given by the carrying capacity $K = r/s$, an improved approximation is given by

$$\hat{m} = (r/s) - (s/r)\text{Var}[X(t)] \tag{2.28}$$

Thus unlike the case of the simple birth–death process, for which the mean and expected value are identical, here they are different. This disparity is the rule rather than the exception, and forms a nasty trap for unwary determinists. Note that for fixed r and s, $\hat{m}$ decreases with increasing $\text{Var}[X(t)]$, and so when large fluctuations are present $\hat{m}$ may be substantially different from K.

An extension of this argument (detailed in Renshaw 1991) shows that

$$\text{Var}[X(t)] \simeq r\gamma/2s^2 \quad \text{where} \quad \gamma \simeq (s/r)(a_1 + a_2) - (b_1 - b_2) \tag{2.29}$$

whence substituting back into the expression in Equation 2.28 leads to $\hat{m} = (2r - \gamma)/(2s)$. Thus given that $r = a_1 - a_2$ and $s = b_1 + b_2$ can be estimated from the initial sigmoid development, this variance approximation can be used to obtain a rough estimate for γ. If one can argue that $b_2 = 0$, i.e. logistic death can be replaced by simple death, then we have three equations which can be solved for a_1, a_2 and b_1. This, however, minimizes the maximum population size a_1/b_1, and if larger upward surges are judged possible then $b_2 > 0$ should be chosen to accommodate them.

Simulation

To simulate the different types of stochastic behaviour possible under this logistic structure, we simply replace $B(N) = \lambda N$ and $D(N) = \mu N$ in the basic birth–death simulation program by Equation 2.20. Figure 2.2 shows four such stochastic realizations; each has $K = 100$, yet exhibits behaviour ranging from almost deterministic growth to wild oscillations. Figure 2.2(a): pure logistic birth $(D(N) \equiv 0)$ gives rise to sigmoid growth towards the steady state K, in line with deterministic prediction. Figure 2.2(b): simple death $(D(N) = 0.5N)$ produces similar initial development, but with mild oscillations about K thereafter. Note that since $B(N) = N(a_1 - b_1N)$ is a birth rate, and hence cannot be negative, N is restricted to the range $1, \ldots, a_1/b_1$; so here $N \leq 150$. Figure 2.2(c): for simple birth $(B(N) = N)$ $b_1 = 0$, so N is totally unrestricted and the oscillations have larger amplitude. Figure 2.2(d): to display extreme oscillations we take $b_1 < 0$, so that $B(N)$ and $D(N)$ both increase as N^2. Thus whilst Fig. 2.2(c) and (d) are deterministically identical, stochastically the underlying biological processes are quite different since in Fig. 2.2(d) the individual birth rate *rises* with the number of possible mating pairs.

Although the first-order variance approximation given in Equation 2.29 appears to be rather crude, it predicts behaviour surprisingly well. In the first three cases the rough 95% confidence interval $K \pm 2\sqrt{\{\mathrm{Var}[X(t)]\}}$ yields the ranges (100 100), (85 115) and (72 128), which are in good agreement with the output. Whilst in the last case we have (6194), which suggests the possibility of extinction in the medium term. Not only does this seem likely from Fig. 2.2(d), but here extinction also nearly occurs in the initial sigmoid phase with $X(t) = 1$ at times $t = 15.9$ and 20.0.

Note that the carrying capacity K plays two very different roles in deterministic and stochastic analyses. In the former it represents the maximum sustainable population size; whilst in the latter it reflects a long-term 'average-value' around which the population can exhibit large fluctuations. Once $X(t)$ reaches the vicinity of K, a *long* single realization can (in principle) provide information on the quasi-equilibrium probability structure since $\pi_N^{(Q)}$ may be estimated by the proportion of time spent in state N, given that extinction has not occurred during this time. Moreover, an easy way of assessing the probability of extinction over an ecologically relevant period of time is to simulate the process say 100 times and to record the times at which extinction occurs together with the number of times it does not occur.

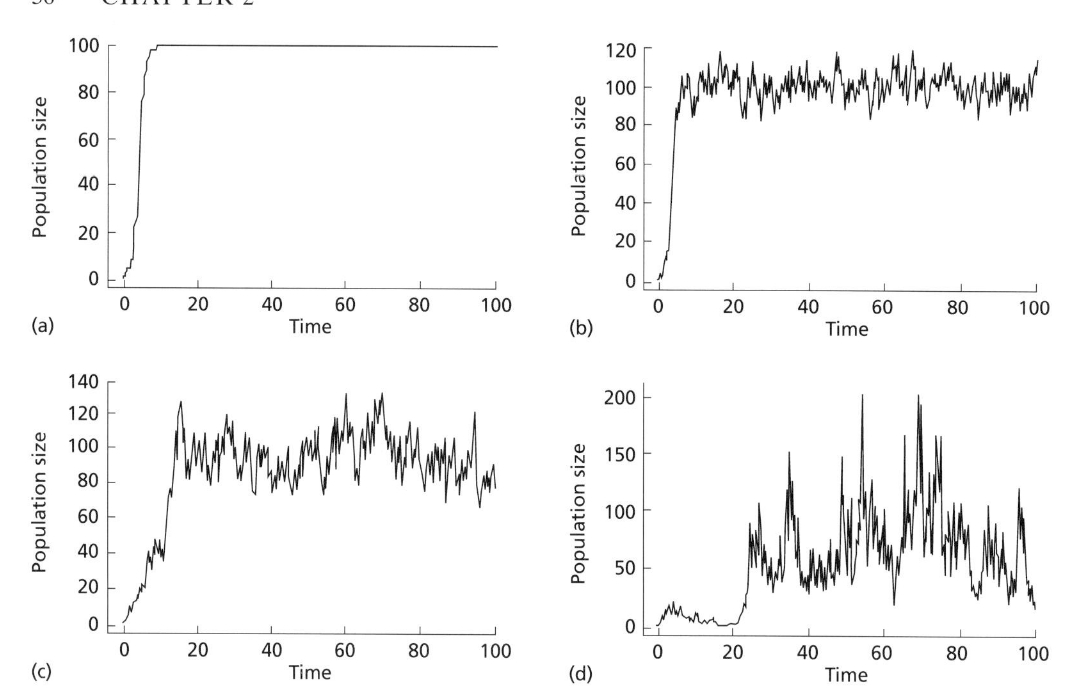

Fig. 2.2 Four stochastic logistic realizations with parameter values: (a) $a_1 = 1$, $b_1 = 0.01$, $a_2 = b_2 = 0$; (b) $a_1 = 1.5$, $b_1 = 0.01$, $a_2 = 0.5$, $b_2 = 0$; (c) $a_1 = 1$, $b_1 = 0$, $a_2 = 0.5$, $b_2 = 0.005$; and (d) $a_1 = 1$, $b_1 = -0.1$, $a_2 = 0.5$, $b_2 = 0.105$.

This simulated distribution may then be used to check the accuracy of theoretical approximations for the probability of extinction by time t ($p_0(t)$), and the mean time to extinction (T_E). Care must be taken though to check that the underlying biological parameters do remain constant during this long time period; environmental or economic change can affect field experiments, whilst genetic change can affect laboratory experiments.

Multispecies interaction

In nature organisms do not generally exist in isolated populations but they live in tandem with many other species. Whilst a large number of these species will be unaffected by the presence or absence of one another, in some cases two or more species will interact. For example, if there is direct competition for a common resource, such as food or space, then the single-species logistic Equation 2.21 extends immediately to the two-species competition process

$$dN_1/dt = N_1(r_1 - s_{11}N_1 - s_{12}N_2) \quad \text{and} \quad dN_2/dt = N_2(r_2 - s_{21}N_1 - s_{22}N_2) \tag{2.30}$$

If the second population lives on the waste products of the first then we have the scavenging process

$$dN_1/dt = N_1(r_1 - s_{11}N_1) \quad \text{and} \quad dN_2/dt = N_2(-r_2 + s_{21}N_1 - s_{22}N_2) \qquad (2.31)$$

whilst if the second population farms the first we have the symbiotic relationship

$$dN_1/dt = N_1(-r_1 + s_{12}N_2) \quad \text{and} \quad dN_2/dt = N_2(-r_2 + s_{21}N_1) \qquad (2.32)$$

The total number of possible configurations is clearly large, though few will have any real biological significance. One that does have great ecological importance is predation, and in its most basic form

$$dN_1/dt = N_1(r_1 - s_{12}N_2) \quad \text{and} \quad dN_2/dt = N_2(-r_2 + s_{21}N_1) \qquad (2.33)$$

Readers are recommended to consult Renshaw (1991) for a detailed deterministic and stochastic study of such systems. Here we just note that the deterministic form Equation 2.33 gives rise to closed population cycles (Lotka 1925; Volterra 1926) which contrast directly with very early stochastic extinction, whilst including the logistic component $(-s_{11}N_1)$ in the prey equation produces damped deterministic cycles (Volterra 1931), but sustained (albeit weak) stochastic cycles. To produce true limit cycle behaviour in both deterministic and stochastic modes necessitates the introduction of *nonlinear* predation rates. For example, the deterministic Holling–Tanner process (Holling 1965; Tanner 1975)

$$\begin{aligned} &dN_1/dt = N_1[r_1 - cN_1 - \{wN_2/(D + N_1)\}] \quad \text{and} \\ &dN_2/dt = N_2[r_2 - (b_2N_2/N_1)] \end{aligned} \qquad (2.34)$$

shows a strong approach to a limit cycle (Fig. 2.3a). However, when N_1 and N_2 are both small, all cycles lie very close to one another, and so stochastic perturbations in this region may cause substantial change in amplitude (Fig. 2.3b). This is where a local stochastic analysis is highly relevant. We first place $dN_1/dt = dN_2/dt = 0$ and solve Equations 2.34 for the equilibrium values (N_1^*, N_2^*), and then determine approximations for the ratios $k_1 = \sqrt{\{\mathrm{Var}[N_1]\}}/N_1^*$ and $k_2 = \sqrt{\{\mathrm{Var}[N_2]\}}/N_2^*$. If k_1 or $k_2 \simeq 3$ then early extinction is very likely, whilst if k_1 and k_2 are both less than 1 then long-term persistent cycles are likely to occur. The large amplitude variability shown in Fig. 2.3(b) is directly related to the ease with which stochastic trajectories can switch across the associated deterministic paths.

The general epidemic

The basic Lotka–Volterra Equation 2.33, together with its subsequent

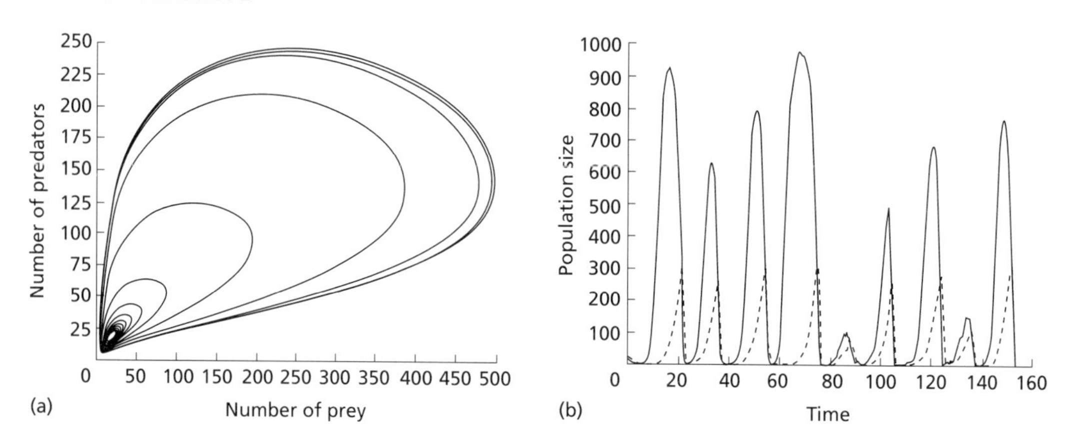

Fig. 2.3 (a) Approach to the limit cycle for the deterministic Holling–Tanner process

$dN_1/dt = N_1[1 - 0.001N_1 - (1.8N_2)/(15 + N_1)]$

$dN_2/dt = N_2[0.5 - 0.5N_2/N_1]$

with $N_1(0) = N_2(0) = 15$ (reproduced from Renshaw 1991, by permission of Cambridge University Press). (b) Simulation corresponding to the deterministic Holling–Tanner process (a) for prey (——) and predators (---) (reproduced from Renshaw, 1991, by permission of Cambridge University Press).

enhancements, forms one of the fundamental building blocks in the development of biological reaction systems. Moreover, only a slight variation on this theme gives rise to a whole new field of study, namely the modelling of infectious diseases. Suppose we let the number of prey ($N_1(t)$) and predators ($N_2(t)$) correspond to the number of individuals susceptible to ($x(t)$) and infected by ($y(t)$) a given disease. Then on placing $r_1 = 0, s_{12} = s_{21} = \beta$ and $r_2 = \gamma$, Equations 2.33 take the epidemic form

$$dx(t)/dt = -\beta x(t)y(t) \quad \text{and} \quad dy(t)/dt = \beta x(t)y(t) - \gamma y(t) \qquad (2.35)$$

Thus we assume that the population mixes freely, so that the total infection rate rises in direct proportion to the number of possible contact-pairs between susceptibles and infectives, and that individual infectives are removed or die at rate γ. Note that in spite of the close similarity between these two deterministic representations, there is a wealth of difference in their interpretation. For in Equations 2.35 the 'death' of a susceptible automatically gives rise to the birth of an infective, so $(x,y) \to (x - 1,y + 1)$; whilst in Equations 2.33 prey do not become predators but merely act as food for them, so $(x,y) \to (x - 1,y)$ and $(x,y) \to (x,y + 1)$ are separate events. In principle, an infinite number of processes can give rise to the same deterministic form, so the ability of deterministic representations to provide precise descriptions of dynamic behaviour is clearly questionable. However, in spite of their naivety they do provide useful insight into qualitative features of epidemic development (see Bailey 1975).

Since the means of the stochastic system satisfy

$$\begin{aligned} \mathrm{d}E[X(t)]/\mathrm{d}t &= -\beta E[X(t)]E[Y(t)] - \beta \mathrm{Cov}\{X(t), Y(t)\} \\ \mathrm{d}E[Y(t)]/\mathrm{d}t &= (\beta E[X(t)] - \gamma)E[Y(t)] + \beta \mathrm{Cov}\{X(t), Y(t)\} \end{aligned} \tag{2.36}$$

(Isham 1993), it is immediately clear that Equation 2.35 does not yield the mean curves for the stochastic system. If $X(0)$ is close to n and $Y(0)$ is small, then for small t $Y(t)$ will increase as $X(t)$ decreases so their covariance is negative. Thus the stochastic mean number infected will increase more slowly than the corresponding deterministic curve at the start of the epidemic. The existence of such differences between deterministic and mean behaviour is often ignored by modellers, with potentially serious consequences if the covariance term plays a dominant role in population development.

If individuals are not removed from circulation by recovery, isolation or death, then $\gamma = 0$. Ultimately all individuals susceptible to the disease must therefore become infected. So if $n = x(t) + y(t)$ denotes total population size, we may replace Equation 2.35 by the single equation

$$\mathrm{d}y/\mathrm{d}t = \beta y(n - y) \tag{2.37}$$

i.e. a pure logistic birth process. The fact that infective death does not occur enables the construction of useful expressions for the mean and variance of the duration time of the epidemic (see Renshaw 1991). Though it is possible to construct the probability $\Pr\{Y(t) = y\}$ (see Bailey 1975), the form of solution is sufficiently opaque to be of limited biological use; even the expression for the mean value $E[X(t)]$ is cumbersome. Yet this is the simplest of epidemic processes. So although many important mathematical results have been developed relating to general epidemic theory, their relevance is generally confined to the practising mathematician. Simulation provides the pragmatic key to biological understanding.

Suppose now that $\gamma > 0$, and for convenience define $\rho = \gamma/\beta$ as the relative removal rate. At the start of the epidemic let $x(0) = x_0$ and $y(0) = y_0$. Then we see from Equation 2.35 that at time $t = 0$

$$\mathrm{d}y/\mathrm{d}t = \beta y_0(x_0 - \rho) \tag{2.38}$$

so an epidemic can only build up (i.e. have $\mathrm{d}y/\mathrm{d}t > 0$) if $x_0 > \rho$. Thus $x_0 = \rho$ defines a deterministic threshold density of susceptibles below which an epidemic cannot develop, since existing infectives are removed at a faster rate than new infectives can be produced. If $x_0 = \rho + \nu$ where $\nu > 0$, then the celebrated *threshold* theorem of Kermack and McKendrick (1927) shows that the number of susceptibles is eventually reduced to $\rho - \nu$, i.e. to a value as far below the threshold as it was initially above. Although their argument consistently underestimates the contact-pair infection rate β, and hence the total size of the epidemic, their pioneering paper provides considerable insight into the mechanics governing the control of epidemic disease.

However, is the concept of a precise deterministic threshold, with epidemic behaviour changing character immediately the number of susceptibles exceeds a specific value, really appropriate? Far more likely is that the *probability* that an outbreak occurs will change. Let $X(t)$ and $Y(t)$ denote the stochastic number of susceptibles and infectives, respectively, at time t. Then in the small time interval $(t, t+h)$ we have the transition probabilities

$$\begin{aligned} &\Pr\{(X,Y)\to(X-1,Y+1)\}=\beta XYh \text{ (infection)} \\ &\Pr\{(X,Y)\to(X,Y+1)\}=\gamma Yh \text{ (removal)} \end{aligned} \tag{2.39}$$

A standard approach for constructing the probabilities $p_{ij}(t) = \Pr\{X(t) = i, Y(t) = j\}$ leads to

$$p_{ij}(t+h)=\beta h(i+1)(j-1)p_{i+1,j-1}(t)+\gamma h(j+1)p_{i,j+1}(t)+[1-(\beta ij+\gamma j)h]p_{ij}(t) \tag{2.40}$$

whence letting $h \to 0$ produces a set of differential equations for the $\{p_{ij}(t)\}$. Unfortunately these equations are extremely laborious to work with, and their solution provides little real insight into the stochastic structure of the process. However, numerical solutions can be obtained fairly easily by the crude means of evaluating Equation 2.40 over successive time points $t = 0, h, 2h, \ldots$ for appropriately small h. Surface plots of the probabilities $\{p_{ij}(t)\}$ over $i, j = 0, \ldots, n$ at specific times t, together with time-series plots over $t = 0, \ldots, T$ (say) for specific probabilities $p_{ij}(t)$, can then be made by inputting these numerical solutions into a graphics package such as Splus. Alternatively, one can employ a software package (such as SOLVER) which will both solve the differential equations accurately and permit direct graphical representation of the solution.

Suppose we wish to assess the error involved in replacing the expected number of susceptibles and infectives by the deterministic solutions $x(t)$ and $y(t)$ over a range of values for β and γ. Provided the total population size n is not too large we can derive numerical solutions for the $\{p_{ij}(t)\}$ at selected times t, use these to compute $E[X(t)] = \Sigma^n_{i,j=0}\, ip_{ij}(t)$ and $E[Y(t)] = \Sigma^n_{i,j=0}\, jp_{ij}(t)$, and compare them with the equivalent values for $x(t)$ and $y(t)$. Alternatively, we can develop a theoretical bivariate Normal approximation to the joint distribution of $X(t)$ and $Y(t)$ (Isham 1991). This method is of general applicability, and goes back to Whittle (1957). In essence it consists of taking the equations for the means, variances and covariances of the variables and replacing any higher-order moments in these equations by their multivariate Normal representations in terms of first- and second-order moments. For the general stochastic epidemic this produces a set of five simultaneous differential equations which are simple to solve numerically.

Another approach is to 'bound' the process by say a simple birth–death process for which theoretical results are readily available (see above). For example, suppose we wish to assess Kermack and McKendrick's deterministic threshold result. With one initial infective the transition rates βXY, which successively equal $\beta(n-1), 2\beta(n-2), \ldots$, lie close to $n\beta Y$ for small Y. Thus in the opening stages of the epidemic the number of infectives $Y(t)$ behaves like a simple birth–death process with parameters $\lambda = n\beta$ and $\mu = \gamma$. Now we know from Equation 2.9 that for this process the probability of ultimate extinction is given by 1 if $\lambda \leq \mu$, i.e. if $n \leq \rho$, and by $\mu/\lambda = \rho/n$ if $\lambda > \mu$, i.e. if $n > \rho$. Moreover, if $\lambda > \mu$ and the population has already grown to size a, say, then the probability of extinction from then on, namely $(\mu/\lambda)^a$, is substantially reduced. So the future of the epidemic is determined in its opening stages, when its similarity with the simple birth–death process is strongest. We therefore have the basic *stochastic threshold theorem*:

1 if $n \leq \rho$ then a major outbreak cannot occur,

2 if $n > \rho$ then a minor or major outbreak occurs with probability ρ/n and $1 - \rho/n$, respectively.

Note that if a infectives are initially present, instead of just one, then these probabilities change to $(\rho/n)^a$ and $1 - (\rho/n)^a$. Whittle (1955) strengthens this result by determining the probability that an epidemic of not more than a given *intensity* takes place, where intensity denotes the total number of susceptibles who eventually contract the disease. This birth–death approximation is clearly a powerful and general device for determining initial population development.

Simulation provides a good way of checking such results. For example, for large n, numerical computation of the expected values of $X(t)$ and $Y(t)$ may not be feasible, yet it may be perfectly practical to simulate the process say $n_{\text{sim}} = 10000$ times. Keeping a running tally of the sum of $X_i(t), Y_i(t), X_i^2(t)$ and $Y_i^2(t)$ at times $t = 0,1,2,\ldots$ over $i = 1,\ldots,n_{\text{sim}}$ then enables us to construct the sample means $\bar{X}(t)$ and $\bar{Y}(t)$, together with the sample variances $s_X^2(t)$ and $s_Y^2(t)$. Forming the confidence limits $\bar{X}(t) \pm 2\sqrt{\{s_X^2(t)/n_{\text{sim}}\}}$ and $\bar{Y}(t) \pm 2\sqrt{\{s_Y^2(t)/n_{\text{sim}}\}}$ allows us to assess whether n_{sim} is large enough to provide the required precision.

Constructing simulated realizations is particularly easy since all that is required is a minor modification to the single-species algorithm. Two types of event are now possible, so on letting $\{W\}$ denote a sequence of uniform (0,1) pseudo-random variables, we have:

1 if $W \leq \beta XY/(\gamma Y + \beta XY)$, i.e. $W \leq \beta X/(\gamma + \beta X)$, then $X \to X - 1, Y \to Y + 1$ (infection), otherwise $Y \to Y - 1$ (removal);

2 the inter-event time $= -[\log_e(W)]/\{Y(\gamma + \beta X)\}$.

Renshaw (1991) illustrates this procedure by simulating a general epidemic with $\beta = 0.01$ and $\gamma = 0.5$, starting from $X(0) = 99$ and $Y(0) = 1$. Since $n = 100$ is twice the threshold value $\rho = \gamma/\beta = 50$, the deterministic threshold result suggests that major outbreaks will be severe. The stochastic threshold result tells us that these have only an even chance of occurring, since the extinction probability $\rho/n = 0.5$. The simulations show that minor outbreaks (which occur the other 50% of the time) soon fizzle out, in marked contrast to the major outbreaks in which about three-quarters of the susceptible population become infected.

Recurrent epidemics

These stochastic ideas based on theoretical approximations, computer simulation and numerical evaluation of population size probabilities are clearly universal, and the technique of using local linear approximations is also generally applicable provided that an equilibrium situation exists. In the general epidemic the number of susceptibles can never increase, and so its relevance is restricted to rare or isolated diseases for which any single outbreak can be regarded as an isolated phenomenon. However, more common diseases like measles, chicken-pox and influenza exhibit periodic flare-ups with infection being sustained at a low level in between times by a gradual spread to new susceptibles. Here the issue is one of persistence and critical community size. Some kind of long-term structure is therefore anticipated including the possibility of chaos (Bolker & Grenfell 1993).

To model such situations, suppose that new susceptibles are introduced into the population at rate α. Then we need to modify Equations 2.35 to

$$\mathrm{d}x(t)/\mathrm{d}t = -\beta x(t)y(t) + \alpha \quad \text{and} \quad \mathrm{d}y(t)/\mathrm{d}t = \beta x(t)y(t) - \gamma y(t) \tag{2.41}$$

These equations are similar to the Lotka–Volterra Equations 2.33—prey birth is replaced by prey immigration—and they possess equilibrium values $x^* = \gamma/\beta$ and $y^* = \alpha/\gamma$. On writing $x(t) = x^*[1 + u(t)]$ and $y(t) = y^*[1 + v(t)]$ for small $u(t)$ and $v(t)$, we have the linearized approximations

$$\mathrm{d}u/\mathrm{d}t \simeq -(\alpha\beta/\gamma)(u+v) \quad \text{and} \quad \mathrm{d}v/\mathrm{d}t \simeq \gamma u \tag{2.42}$$

Solving these equations (see Renshaw 1991) shows that both infective and susceptible populations undergo damped oscillations with period $2\pi/\xi$, phase difference θ where $\cos\theta = 1/\sqrt{(4\gamma\sigma)}$, and amplitude damping factor $\exp(-t/2\sigma)$. Here $\sigma = \gamma/\alpha\beta$ and $\xi = \sqrt{\{(\gamma/\sigma) - (1/4\sigma^2)\}}$. Note that unlike the earlier Lotka–Volterra process, $\theta \neq \pi/2$.

Given that there exists a vast number of potential nonlinear models which will give rise to the same linear Equations 2.42, it is clearly unrealistic to expect the same solution to provide an accurate quantitative prediction in all cases. All it can do is to predict some kind of oscillatory behaviour together with rough estimates of period and phase. For example, in his early deterministic work on the recurrent measles epidemics Soper (1929) took $1/\gamma$ to be 2 weeks (the approximate incubation period) and estimated σ from London data to be 68 weeks. This gives rise to a period of 74 weeks, which is roughly right, but produces a peak-to-peak damping factor of 0.58. This swift damping of the infective cycles towards a steady endemic state contradicts the epidemic facts, and demonstrates that the combination of deterministic mathematics with local linearization may well produce misleading results.

Simulation follows exactly as for the general epidemic process, except that there are now three types of event. So we break the unit interval (0,1) at the points $\beta XY/(\alpha + \gamma Y + \beta XY)$ and $(\gamma Y + \beta XY)/(\alpha + \gamma Y + \beta XY)$, and select the events infection, removal or immigration according to whether W is in the first, second or third segment. Studies show that initial values $(X(0), Y(0))$ that are sited well away from the equilibrium value (x^*, y^*) generally lead to a huge upsurge in the number of infectives and a correspondingly large drop in the number of susceptibles, followed by a severe crash in the infective population causing it to be wiped out. Conversely, simulations starting near to (x^*, y^*) initially follow a deterministic-type path before wandering about (x^*, y^*) thereafter. Thus if this model is to produce sustained flare-ups then we need $(X(0), Y(0))$ to be substantially different from (x^*, y^*), plus the occasional immigrant infective to ensure that the epidemic restarts should the infective population die out.

Figure 2.4 shows a simulation for $x^* = 3400$, $y^* = 100$, a susceptible immigration rate of $\alpha = 50$, and a relatively tiny infective immigration rate of $\delta = 0.1$. The start point lies well away from equilibrium, and seven flare-ups are seen to occur in the first 14 years. Although the mean cycle period of 97 weeks is considerably higher than the deterministic prediction of 68 weeks, this is because the process takes some time to restart whenever the infective population dies out. Indeed, the intercycle time is dominated by $1/\delta$ which becomes increasingly large as $\delta \to 0$. Note the great difference in cyclic behaviour between Fig. 2.3(b) for the stochastic Holling–Tanner process, which reaches zero only when the process terminates, and Fig. 2.4, which relies on extensive periods with $Y(t) \simeq 0$ to enable the susceptible population to restock to a sufficiently high level to enable a major outbreak to develop. Even then, careful examination of Fig. 2.4 shows that several very small outbreaks can occur before each major one. After 14 years the simulation undergoes a dramatic change, because during the next infective trough there are now just enough infectives remaining to keep the susceptible population in check. This causes a substantial reduction in the next infective peak, and so $\{X(t), Y(t)\}$ passes near enough to (x^*, y^*) for it to become trapped into undergoing damped cycles towards an endemic state. Although the peak-to-peak damping factor 0.73 is higher than the deterministic prediction 0.58, it is nevertheless qualitatively similar. So as the epidemic

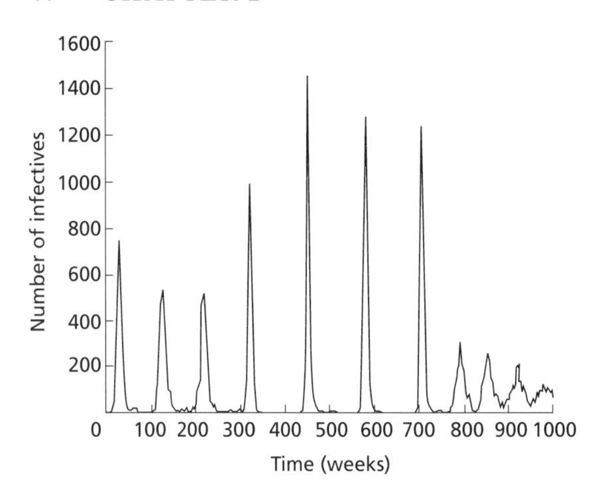

Fig. 2.4 A simulated stochastic recurrent epidemic corresponding to the deterministic representation $dx/dt = -0.000147xy + 50$ and $dy/dt = 0.000147xy - 0.5y + 0.1$ (reproduced from Renshaw, 1991, by permission of Cambridge University Press).

switches from the 'flare-up' to the 'endemic' state the deterministic local approximation does provide some useful information. The key to using deterministic results successfully clearly lies in careful matching with their simulated stochastic counterparts.

Is the move to the endemic state permanent? To reverse it the infective population will have to wander down close to the time axis, and to assess the probability of this event we employ the birth–death approximation. Suppose that $X(t) \simeq x^*$, and treat $\{Y(t)\}$ as a simple birth–death process with parameters $\lambda = \beta x^*$ and $\mu = \gamma$. On disregarding the occasional infected immigrant, use of the result in Equation 2.9 shows that the probability of infective extinction for $Y(0) = y^* = 100$ is given by $p_0(t) = [1 + (2/t)]^{-100}$. Thus the median time to extinction is determined from $p_0(t) = 0.5$ as 288 weeks, and so on a human time-scale this particular process has the potential for flip-flopping between the epidemic and endemic states. Another approach is to extend the single-species logistic result of Equation 2.29 to determine the variance–covariances of small susceptible and infective swings around the (endemic) equilibrium value (Bailey 1975; Renshaw 1991). Here we find that x^* and y^* lie 10 and <2 standard deviations away from zero, respectively, and so whilst early short-term extinction of susceptibles is virtually impossible, that for infectives is highly likely. This ties in with our simulation experience.

Although I have developed my main ideas in the context of epidemic theory, it is important to remember the strong paradigm that exists between ecological and epidemiological processes. For example, if we wish to obtain undamped deterministic epidemic cycles, then we might consider disease which is not only lethal to all those who contract it but which is also sufficiently virulent to suppress any live births among circulating infectives. Let us therefore suppose that susceptibles, but not infectives, give birth. Then on replacing the immigration term α by the birth term λx,

Equations 2.41 become the Lotka–Volterra predator–prey Equations 2.33, namely

$$dx(t)/dt = -\beta x(t)y(t) + \lambda x(t) \quad \text{and} \quad dy(t)/dt = \beta x(t)y(t) - \gamma y(t) \qquad (2.43)$$

However, we must reiterate that such deterministic equivalence does not extend to stochastic solutions. Indeed, it does not automatically follow that stochastic simulations of this epidemic process will emulate Lotka–Volterra realizations which dive onto an axis; indeed, it is quite possible to obtain several full cycles before the infectives become extinct. Such realizations certainly qualify as genuine recurrent epidemics on a human time-scale; whilst introducing the occasional immigrant infective, as before, enables the process to persist indefinitely.

Current developments

In recent years the great upsurge of interest in modelling population dynamics has been driven by concern over the AIDS epidemic, possibly to the detriment of other equally worthwhile issues. Much of the work on building mathematical models for the transmission dynamics of HIV infection is deterministic, although allowance for many sources of variability can be made in the models (see Isham 1988; Anderson *et al.* 1989). Although it will often be difficult or impossible to obtain explicit expressions for the properties of the corresponding stochastic models, the underlying epidemic process *is* stochastic and so considerable future effort *must* be devoted to deriving approximate theoretical, and exact numerical, stochastic results. Note that to use the general epidemic model as a simple descriptor for AIDS, we must interpret 'infected' as being infected with HIV but not yet having full AIDS, and 'recovered' as having progressed to a full AIDS diagnosis. So we must assume that those who have been infected with HIV cease to transmit the infection once they have been diagnosed as having AIDS (Isham 1993).

The idea enshrined in Equations 2.35, that each of the $y(t)$ infectives/predators is equally likely to infect/kill each of the $x(t)$ susceptibles/prey, is clearly unrealistic, and in both the epidemic and predator–prey scenarios a more sensible approach is to replace the $\beta x(t)y(t)$ component by say $\beta y(t)\min[x(t),k]$. Thus k reflects the maximum number of partners an HIV individual may have, or the maximum number of prey a predator will eat before becoming satiated. Realism is, of course, not the only goal, as we want a simple parsimonious model which can be interpreted easily. Nevertheless, it is vital to include sources of variation which are known to have an important influence on the course of an epidemiological or biological process. Both Isham (1993) and Mollison *et al.* (1994) detail a range of such possibilities, whilst Shaw (1995) considers specific issues that arise in plant pathology and describes problems which seem likely to be especially illuminating. Some important modelling extensions that need comprehensive examination are as follows.

Incubation/digestive periods

In the general stochastic epidemic, infected individuals are assumed to have a constant period of recovery, whilst predators are assumed to be constantly looking for new prey to kill irrespective of how much they have just eaten. In the AIDS context this means that the incubation periods between infection with HIV and the progression to full AIDS are exponentially distributed; a similar implication is involved for the progression of food through the intestinal tract. However, it is known that short incubation periods (of say up to 3 years) are very unlikely, as is meteoric digestion of food! To accommodate this the constant parameter β needs to be replaced by a function of the time since infection/eating. Whilst this can be any appropriate function, for the purpose of stochastic simulation, or numerical solution of the probability equations, it is best to assume a gamma distribution with integer index k. We can then represent the incubation/digestion period as a series of k stages, the durations of stay in each stage being independent, identically distributed exponential variables. A wider class of distributions can be considered by letting the ith stage have rate v_i.

Time-lag models

In practice there is often a time-lag between the inception of an action and the resulting change: seeds take time to develop into plants, whilst newborn animals take time to reach the age when they are capable of reproduction. For example, modifying the logistic Equation 2.21 to

$$dN(t)/dt = rN(t-t_G)[1-N(t-t_D)/K] \qquad (2.44)$$

allows for both a reproduction time-lag (t_G) and an interaction time-lag (t_D). In general, if the duration of the delay in the feedback loop is longer than the 'natural period' of the system, then large amplitude oscillations will result (Nisbet & Gurney 1982; Renshaw 1991). This is clearly another scenario for generating persistent cyclic behaviour. Stochastic simulation requires only a minor change to that for the basic logistic process; whilst the extension to multispecies situations should produce a rich diversity of dynamic behaviour.

Time-dependent parameters

Another way of allowing for time-dependency is through the parameter values. For example, seasonality may induce a yearly periodic effect into the contact rate β for certain mild diseases; the removal rate γ might increase with time with growing public awareness; whilst the predation rates s_{12} and s_{21} in the Lotka–Volterra process Equation 2.33 may alter as the killing and avoidance strategies of the two species evolve. At a deeper level, temporal nonstationarity can also be induced by the process itself. A classic example is the coevolution of myxomatosis and its rabbit host (Fenner and Myers 1978). The rapid emergence of drug-resistant parasitic strains makes this general issue one of high priority for population modellers.

Noisy parameters

So far we have introduced stochastic effects into a system by developing the full probability equations, as in Equation 2.15, and adding noise to the deterministic system, as in Equation 2.26. Related techniques involve constructing quasi-equilibrium probabilities, as in Equation 2.23, and using the Isham–Whittle Normal approximation method. A fundamentally different approach is to admit random variation through the parameters themselves. For example, the parameter β in Equation 2.35 might be driven by a stochastic logistic process, thereby reflecting random environmental oscillations around some equilibrium contact value.

Heterogeneous subgroups

Another important source of population structure is recognition that individual plants, animals, people, etc. may be classified as falling into various subgroups each with its own set of activity rates. So if β_{ij} denotes the infection rate between a susceptible in subgroup i and an infective in subgroup j, then the x_i susceptibles in the ith subgroup become infected at rate $\Sigma_j \beta_{ij} y_j$ where y_j denotes the number of infectives in subgroup j.

Demographic factors

Models can also be made more realistic by stratifying the population into various age-groups. This can either be viewed as being equivalent to (the previous case), or else it can be used in addition to it, though in this latter case the total number of resulting subgroups is likely to become unwieldy.

Change in behaviour

Further complexity can be introduced by allowing individuals to move from one subgroup to another. For example, predatory animals might change from being introverted to dominant, prey from strong to weak, and susceptibles from high-risk to low-risk.

The number of interesting situations open for investigation is clearly immense, and many of these will involve a high order of complexity. For example, wild animals will generally not be infected with a single parasitic species, but with a mix of species, some of which may have interacting population dynamics (Dobson 1985). Methods are needed that determine which species in a community will dominate, which can persist, which are doomed to extinction and under what circumstances alien species can invade. Such problems are particularly important not only in ecosystem dynamics but also in the conservation of rare species, and determining the ultimate outcome of a control measure (e.g. to a community of parasites) (Mollison *et al.* 1994).

Two specific examples

Until recently most areas of 'applicable mathematics' essentially used linear mathematics, in the sense that both the deterministic birth–death Equation 2.2 and its stochastic counterpart Equation 2.6 just involve N. As soon as terms in N^2 appear, as happens in the logistic, epidemic and Lotka–Volterra scenarios, stochastic mathematics becomes difficult. Once even greater nonlinearities are introduced, such as ratios in the Holling–Tanner process Equation 2.34, then theoretical progress generally becomes very demanding. However, theoreticians are now rising to the challenge. For example, Ball & O'Neill (1993) consider an epidemic in which the interaction term βxy in Equation 2.35 is replaced by $\beta xy/(x + y)$, which is clearly a variation on the Holling–Tanner scheme Equation 2.34. The justification for modifying this term is that AIDS spreads by individuals changing sexual partners. So if removed individuals are no longer available as sexual partners, and new partners are chosen at random from the population of possible partners, then the probability that a new partner of a given susceptible is infected is $y/(x + y)$. Surprisingly, the deterministic equation admits a complete closed-form solution. Moreover, the stochastic equations can be solved to a considerable degree, yielding total size distributions and threshold theorems, together with results relating to the splitting of the susceptibles into m subgroups, each of which experiences a different rate of infection. So considerable mathematical progress is still possible in spite of having the combined complexity of substantial nonlinearity with population heterogeneity. 'Practical' modellers who make an immediate dash for their favourite numerical sledge-hammer miss out on the deep insight that such a theoretical investigation can bring. The formation of mixed groups of mathematicians and biologists/epidemiologists who can communicate with each other is clearly essential if we are to raise our strike rate of both posing and asking the right questions.

A biological illustration relates to the population dynamics of nematode parasites of ruminants. These parasites are among the most studied in terms of empirical investigations of seasonal microclimatic influences on transmission. This is due to both the economic importance of these infections and the relative ease of performing experimental manipulations to estimate relevant parameters. Roberts & Grenfell (1992) demonstrate that the annual introduction of host animals, that have not previously been exposed to infection, on to infected pasture is sufficient to generate the qualitative annual patterns of parasitism that are observed. The parameters that determine the development rates of the free-living stages may be functions of time. Denote the mean density of infective free-living stages by L, the mean intensity of adult parasites by A, and the level of acquired immunity by r. Then the dynamics of parasite transmission and the hosts' immune response are enshrined in Equation 2.45.

$$\begin{aligned} \mathrm{d}L(t)/\mathrm{d}t &= -[\rho(t)+\beta(t)]L(t)+q(t)\lambda(r)A(t) \\ \mathrm{d}A(t)/\mathrm{d}t &= \beta(t)p(r)L(t)-\mu(r)A(t) \\ \mathrm{d}r(t)/\mathrm{d}t &= \beta(t)L(t)-\sigma r(t) \end{aligned} \qquad (2.45)$$

Here the function $p(r)$ denotes the probability of parasite establishment, $\lambda(r)$ the rate of egg production of adult parasites, $\mu(r)$ the effect of parasite mortality on acquired immunity in the host, σ the rate of loss of immunological memory, $q(t)$ the probability that the egg develops to the L_3 stage, $\rho(t)$ the loss rate of free-living stages from the environment, and $\beta(t)$ the contact rate. Allowing dependence on time enables us to capture the seasonal variation in the free-living part of the cycle.

Equations 2.45 clearly involve a much higher order of complexity than those of say the recurrent epidemic Equation 2.41. For whilst nonlinearity in the latter just enters through the product $x(t)y(t)$, in the former the parameters λ, ρ and μ are functions of the vastly more awkward form

$$r(t) = \text{constant} - \exp\left\{\int \beta(s)L(s)e^{\sigma(s-t)}\mathrm{d}s\right\} \qquad (2.46)$$

Grenfell and Roberts use approximation techniques to examine the stability of these deterministic equations, and further, more exact, procedures which exploit the full nonlinearity of this process will doubtless be undertaken. Numerical solutions may be obtained regardless of the complexity of the parameter functions. Theoretical stochastic investigations are under way, but they will almost certainly involve substantial simplification to the underlying process if progress is to be made, and will therefore not provide exact information on underlying behavioural characteristics. This can only be acquired from a simulation-based approach (Marion et al 1997).

On taking the individual components of Equation 2.45 to be non-negative, we can immediately extract the birth and death rates

$$\text{birth } B_L = q(t)\lambda(r)A(t) \qquad B_A = \beta(t)p(r)L(t) \quad B_r = \beta(t)L(t)$$

$$\text{death } D_L = [\rho(t)+\beta(t)]L(t) \quad D_A = \mu(r)A(t) \qquad D_r = \sigma r$$

Here we have treated r as a stochastic variable, though an alternative approach would be to evaluate and update the deterministic solution Equation 2.46 at each of the event times $t_1, t_2, \ldots$. Provided that $\rho(t)$, $\beta(t)$ and $q(t)$ change slowly relative to the inter-event times $(t_i - t_{i-1})$, they can be held constant in-between events, being updated at each $\{t_i\}$. So to simulate the process we:

1 update birth and death rates;
2 evaluate $Z = B_L + D_L + B_A + D_A + B_r + D_r$;
3 choose two random $U(0,1)$ numbers U_1 and U_2;
4 determine the event-type via;
if $0 \le U_1 < B_L/Z$ then $L \to L+1$,
if not but $B_L/Z \le U_1 < [B_L + D_L]/Z$ then $L \to L-1$, etc.;
5 evaluate the inter-event time $[-\log_e(U_2)]/Z$;
6 update, t, L, A and r, together with any time- or r-varying parameter values;
7 go to **1**.

Figure 2.5 shows a simulation run over $0 \le t \le 200$ for initial values which lie near to the endemic equilibrium of the deterministic system Equation 2.45. The broad picture of cyclic behaviour predicted by deterministic theory is clearly evident, and the added information on the variability of amplitude and period is highly instructive. The average stochastic period of $\simeq 15$ years is very close to that obtained from both deterministic prediction and stochastic approximation. Moreover, the simulated estimates for the standard deviations, namely $\sigma_L = 37\,000$, $\sigma_A = 350$ and $\sigma_r = 31\,000$, are similar in value to those obtained from both the local variance–covariance calculations and the Normal approximation. Figure 2.6 shows a corresponding stochastic and deterministic plot of $L(t)$ for $\mu(r) \equiv 30$, and we see that stochastic infection becomes extinct at time $t \simeq 31$ despite the fact that deterministic theory places the process firmly within the oscillatory domain. This discrepancy parallels that observed for the simple birth–death process, and highlights yet again that considerable care is required when using deterministic theory to infer properties relating to long-term persistence.

Spatial processes

The final approach I advocate is recognition that the environment has a spatial dimension, since individual population members rarely mix homo-

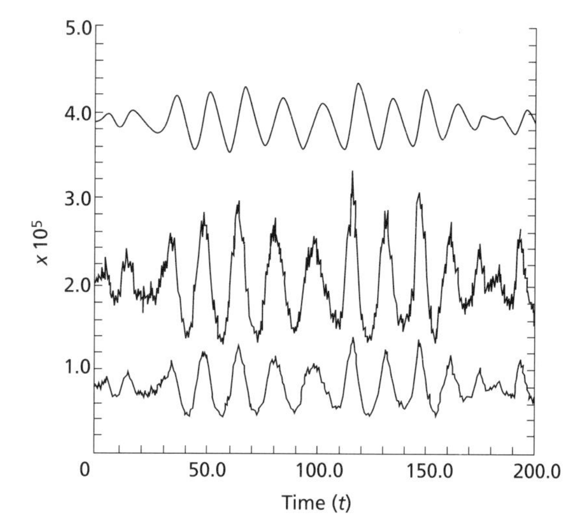

Fig. 2.5 Stochastic realization of the deterministic Grenfell and Roberts process Equation 2.45 with $\rho = 7$, $\beta = 0.365$, $q = 0.35$, $\sigma = 0.01$, $\mu(r) = 25$, $p(r) = 0.65$ and $\lambda(r) = 39\,420\exp\{10^{-6}r\}$ (taken from Marion *et al.* 1997) showing: $L(t)$ (lower curve), $200A(t) + 5 \times 10^4$ (middle curve) and $r(t) - 2.54 \times 10^6$ (upper curve).

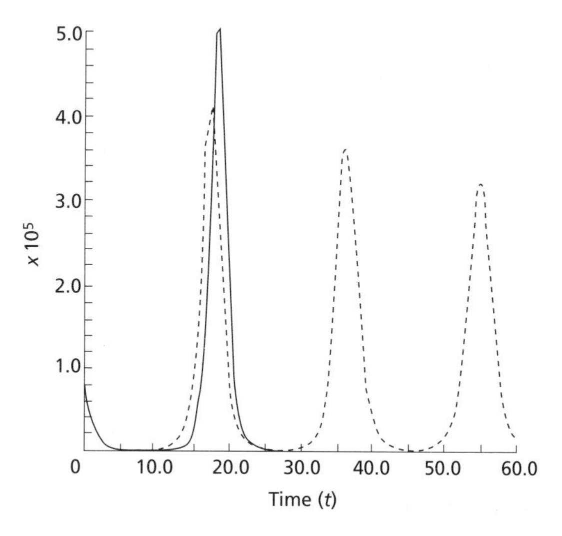

Fig. 2.6 Stochastic (full curve) and deterministic (broken curve) realization of $L(t)$ for the process shown in Fig. 2.5 (but with $\mu(r) \equiv 30$).

geneously over the territory available to them but develop instead within separate subregions. Such dispersal can be truly local (e.g. young oaks growing around a parent tree), mid-range (e.g. seals catching fish), long-range (e.g. aerosol dispersal of plant disease), or global (e.g. spread of human disease through intercontinental travel). All too often, modellers disregard this fundamentally important aspect of population development on the grounds that the resulting mathematics may be 'too difficult'. There is no point working with 'easy mathematics' if the biological process it relates to is fatally flawed. Fortunately, fairly simple models can be developed which highlight the effects that geographic restriction and species mobility may have on population development. These models can provide vital knowledge about the dispersal and control of many natural populations, not just of animals, insects and plants, but also of diseases such as malaria and rabies.

Most species attempt to migrate for a variety of both individual and population reasons, including: search for food; territorial extension for increasing population needs; widening the gene pool; and minimizing the probability of extinction. Migration can occur either between distinct sites, such as neighbouring valleys or islands in an archipelago, or within continuous media such as the air or sea. There is clearly an intrinsic difference between the 'diffusion model', in which individuals can move anywhere in space, and the 'stepping-stone model', in which they have to live at specific sites. Since the type and extent of the spatial environment can have a major effect on the way that populations develop (see Renshaw, 1991 for examples), it is important to ensure that any simplified spatial model structure accurately reflects the real-life process under study.

Given that many populations develop within reasonably well-defined subregions, the stepping-stone approach is a sensible one to consider first. We envisage the process as being spatially distributed amongst n sites or colonies, with migration or cross-infection being allowed between: nearest neighbours; all sites with transition rate independent of intersite distance; or, all sites but with transition rate decreasing as intersite distance increases (called the contact distribution). This scenario was first posed in a genetics context (Kimura 1953), then later in a much easier birth–death–migration situation (Bailey 1968). Here the population comprises an infinite number of colonies situated along a line, with each colony i being subject to a simple stochastic birth–death–migration process with birth and death rates λ and μ, respectively, and with migration rates v to each of the two nearest neighbours $i-1$ and $i+1$. Although the equations for the population size probabilities are intractable to direct solution, it is possible to develop fairly neat expressions for the mean number of individuals in colony i at time t. In practice, of course, we have to take just N sites with $i = 1, \ldots, N$; if birth is not present then provided that neither extinction nor population explosion can occur it is possible to write down the full equilibrium probability distribution (Bartlett 1949; Renshaw 1986, 1991). Note that here we effectively have a closed system since death or emigration are easily covered by introducing a further state from which there is no return. Unfortunately, Bartlett's result is rather opaque, and has therefore not been widely proclaimed. This is a pity, since its general nature has wide applicability.

Two-colony processes

As soon as birth is introduced the system is no longer closed since new individuals are produced. This means that even the simple stochastic two-colony birth–death–migration process is (currently) intractable to full solution, though means and variance–covariances can be obtained (Renshaw 1973a,b). Theoretical progress can be made by employing a variety of approximation techniques (see Renshaw 1986, 1991), and understanding of the process is substantially enhanced by viewing these results together with output from simulation runs. For example, Renshaw (1991) shows a realization for $\lambda_1 = \lambda_2 = 1$, $\nu_1 = \nu_2 = 0.05$, where the ν_i are migration rates, and $\mu_1 = 1.1, \mu_2 = 0.9$ starting from $X_1(0) = 10, X_2 = 0$. In spite of the net growth rate of colony 1 being negative ($\lambda_1 - \mu_1 - \nu_1 = -0.15$), it experiences an initial surge in population growth and is able to send migrants to colony 2 to start the process there. Colony 1 then crashes, closely followed by the collapse of colony 2, even though $\lambda_2 - \mu_2 - \nu_2 = 0.05$ is positive. Thus the stochastic behaviour is the opposite of that predicted by the deterministic analysis. Rough agreement between stochastic and deterministic realizations can occur, but the problem is clearly one of lack of consistency.

This difference becomes even more marked when we place interacting populations within a spatial domain. For example, Renshaw (1991) demonstrates that provided parameter values are carefully chosen, the stochastic Lotka–Volterra model Equation 2.33 can generate cyclic behaviour within just a 3-colony model (the 1-colony process always crashes on the first cycle). Trial results show that not only must predators be more mobile than prey, but the predator migration rate must be high enough to prevent a prey explosion occurring in a colony before predators arrive to control it. However, migration rates must not be so large that they synchronize the spatial structure out of the system. So whereas the cyclic behaviour of the nonspatial Holling–Tanner process Equation 2.34 is essentially induced through a stable deterministic limit cycle, here persistence occurs through a *stochastic dynamic*. It is precisely the ability of prey to be constantly on the move recolonizing empty sites, and predators to chase after them, that keeps the whole process alive.

N-colony process

In general, obtaining theoretical deterministic solutions of spatially interacting processes is either very difficult or impossible, and deriving stochastic solutions is even worse, so the birth–death–migration process represents the limit of our current mathematical expertise. However, provided we maintain some flexibility in outlook it still provides us with a reasonable array of possible model structures. For example, in genetics migration is equivalent to mutation (Armitage 1952), and models for phage reproduction within a bacterium also fall under this scheme (Gani 1965). If we restrict attention to the opening stages of a process, then the birth–death–migration scenario is clearly a reasonable approximation to spatial logistic and epidemic type processes.

For the two-colony process, expressions are available for means and variance–covariances along with expressions for approximate probabilities, conditions for ultimate extinction, and stochastic results for 'slightly connected colonies' for which migration occurs relatively rarely (see Renshaw 1973a,b). In particular, on denoting the net growth rates for colony i by $\xi_i = \lambda_i - \mu_i - \nu_i$, it may be shown (deterministically) that if $\xi_1\xi_2 > \nu_1\nu_2$ and $\xi_1,\xi_2 < 0$ then the total mean population size decreases, if $\xi_1\xi_2 = \nu_1\nu_2$ and $\xi_1,\xi_2 < 0$ it remains constant, otherwise it increases. Thus we can assess the impact that migration will have on overall population development; for example, the 2-colony version of the Kermack and McKendrick nonspatial threshold theorem follows immediately. However, although general solutions for the N-colony case in which individuals can migrate between any pair of colonies i and j at rate ν_{ij} are available (Renshaw 1972), they are in an opaque vector-matrix form. Some spatial structure must therefore be provided if more transparent solutions are to be obtained, and the simplest approach is to preclude large jumps so that individuals may migrate between nearest neighbours only.

In a linear setting this means that we have to consider the problem of 'edge-effects' at the boundary colonies $i = 1$ and $i = N$. Renshaw (1972) shows that the boundary conditions there play a major role in determining the form of the mean population values *throughout* the system; Renshaw (1980), for example, targets these specifically when modelling the classic spatial *Tribolium confusum* data of Neyman *et al.* (1956). Here flour beetles develop in a $10 \times 10 \times 10$-inch cubic container over a 4-month period, and both male and female populations show an increase in density away from the centre towards the edges, and along the edges towards the corners. Zoning the volume into 1000 one-inch subcubes with equal migration between them, but letting the edges and corners have different birth and death rates, produces a spatial distribution similar to the one observed. Usher and Williamson (1970) take a contrasting approach in which a proportion p of the beetles are 'movers' and $1 - p$ are 'stayers', with movers and stayers having different birth and death rates. As far as the deterministic equations are concerned these two processes are identical, which highlights the fact that it is not possible to infer model structure from final 'snapshot' data. Proper inference of process from pattern requires further information on the stochastic dynamics of the system.

Although one might object to the arbitrary subdivision of space into one-inch cubes, it turns out that no matter how finely the region is divided (say L^3 subcubes), the total number in each one-inch cube remains unchanged. Thus the deterministic spatial distribution is independent of the chosen scale of migration. One important consequence is that this equivalence holds true even when $L \to \infty$ and the process turns into Brownian motion. So in the steady state the steppingstone and Brownian models yield the same deterministic population sizes. Note, however, that it does not follow that the corresponding stochastic motions are compatible.

Nevertheless, this equivalence does not necessarily hold in nonsteady state situations. For example, suppose we return to Bailey's (1968) linear habitat process with individual migration rates of v_1 and v_2 to the right and left, respectively. If the initial population is located at a single clump of sites, and $\lambda > \mu$, then the population will develop as a travelling wave. The velocities of the left and right wavefronts are the solutions to the equations

$$v_1 + v_2 + \mu - \lambda = \sqrt{(c^2 + 4v_1v_2)} - c\log_e\{[c + \sqrt{(c^2 + 4v_1v_2)}]/(2v_1)\} \quad (2.47)$$

(Renshaw 1977), and if $v_1 > v_2$ the left wavefront moves right or left depending on whether $\lambda - \mu < (\sqrt{v_1} - \sqrt{v_2})^2$ or $\lambda - \mu > (\sqrt{v_1} - \sqrt{v_2})^2$, respectively. The equivalent diffusion velocities are

$$c_{\text{diff}} = (v_1 - v_2) \pm \sqrt{\{2(\lambda - \mu)(v_1 + v_2)\}} \quad (2.48)$$

and the left wavefront moves right only if $\lambda - \mu < (v_1 - v_2)^2/2(v_1 + v_2)$. A little algebra shows that these two sets of results are compatible only if the net growth rate $\lambda - \mu$ is small in comparison to the sum of the migration rates $v_1 + v_2$. Otherwise, similar migration patterns can give rise to substantially different velocities in the two cases.

In practice, birth, death and even migration rates are density-dependent, and we might therefore wish to replace the birth–death component by a logistic process. Far more interesting, however, is to consider a multispecies process developing within a spatial domain. Consider, for example, placing the Ball and O'Neill (1993) model within a nearest-neighbour setting, with $x_i(t)$, $y_i(t)$ and $z_i(t)$ denoting the number of susceptibles, HIV-infected and removed (i.e. full-blown AIDS or dead) individuals at site $i = 1, \ldots, N$. Then allowing for the migration of infectives gives rise to the (deterministic) representation

$$\begin{aligned} dx_i(t)/dt &= -\beta x_i(t)y_i(t)/[x_i(t) + y_i(t)] \\ dy_i(t)/dt &= \beta x_i(t)y_i(t)/[x_i(t) + y_i(t)] - (v_1 + v_2)y_i(t) + v_1 y_{i-1}(t) + v_2 y_{i+1}(t) \\ dz_i(t)/dt &= \gamma y_i(t) \end{aligned} \quad (2.49)$$

In contrast, allowing spatial *cross-infection* in the general epidemic Equation 2.35 yields

$$\begin{aligned} dx_i(t)/dt &= -x_i(t)[\beta y_i(t)+\beta_1 y_{i-1}(t)+\beta_2 y_{i+1}(t)] \\ dy_i(t)/dt &= x_i(t)[\beta y_i(t)+\beta_1 y_{i-1}(t)+\beta_2 y_{i+1}(t)]-\gamma y_i(t) \\ dz_i(t)/dt &= \gamma y_i(t) \end{aligned} \tag{2.50}$$

Spatial extensions to other types of process, such as predator–prey, competition, or the parasite–host model Equation 2.45, follow along similar lines.

An obvious spatial arrangement to consider first is a line of colonies $i = 1, \ldots, N$, with $y_{N-1}(t) \equiv y_{N+1}(t) \equiv 0$. This may produce substantial edge-effects, and to remove them arrange the colonies on a circle so that sites $i = 0$ and N and $i = 1$ and $N + 1$ are synonymous (i.e. $y_0(t) \equiv y_N(t)$, etc.). If N is large this may result in quite considerable differences in population density around the circle, and so to make the process more uniform one might allow interaction between all colonies and a central 'hub' site ($i = 0$). Thus infectives/predators, etc. may migrate directly from say site 3 to 867 by passing through 0, rather than by having to move through all the intervening sites around the 'rim'. Far more attention needs to be paid to investigating how the behavioural characteristics of deterministic and stochastic models change when they are placed within different spatial environments.

Turing ring processes

Although the mathematics surrounding spatial stochastic models is notoriously difficult, in a brilliant pioneering paper Turing (1952) developed elegant linear mathematical deterministic solutions which predict the types of behaviour likely to be encountered when sites lie on a ring (Murray 1989). Turing's analyses relate to a system of chemical substances, called morphogens, which react together and diffuse through tissue. Let the ring contain N 'cells', with the cell-to-cell diffusion constant for morphogens X and Y being μ and ν, respectively. The most general assumption one can make about the rates of chemical reaction is that X and Y increase at rates $f(X,Y)$ and $g(X,Y)$, respectively. The deterministic behaviour of the system is then defined by the $2N$ differential equations

$$\begin{aligned} dX_i/dt &= f(X_i, Y_i)+\mu(X_{i+1}-2X_i+X_{i-1}) \\ dY_i/dt &= g(X_i, Y_i)+\nu(Y_{i+1}-2Y_i+Y_{i-1}) \end{aligned} \tag{2.51}$$

If there exist positive values X^* and Y^* such that $f(X^*,Y^*) = g(X^*,Y^*) = 0$, then this ring system of equations possesses an equilibrium in which each $X_i = X^*$ and $Y_i = Y^*$.

Turing's aim was to examine whether it is feasible to generate *spatially* stable waves. Though his key area of concern was morphogenesis, namely the development of the pattern and form of embryo from fertilization until birth, his approach is totally general in its applicability. For example, Equations 2.51 include the spatial Ball and O'Neill Equations 2.49 as a special case. Though the potential number of choices for $f(X,Y)$ and $g(X,Y)$ is unlimited, only a relatively small number will represent realistic scenarios. Murray (1989) presents an excellent account of reaction-diffusion mechanisms that have been proposed as possible pattern formation processes in a wide variety of situations, including animal coats, butterfly wings and shells. Turing's idea is simple but profound. If in the absence of diffusion (i.e. $\mu = \nu = 0$) X_i and Y_i tend to a *linearly stable* uniform state, then under certain conditions spatially inhomogeneous patterns can evolve by *diffusion-driven instability*. Since diffusion is usually considered to be a stabilizing process we can see why this is such a novel concept.

Given the mathematical intractability of general nonlinear differential equations, Turing considered only local departures from equilibrium by writing $X_i(t) = X^* + x_i(t)$ and $Y_i(t) = Y^* + y_i(t)$. So for small $x_i(t)$ and $y_i(t)$ the functions f and g may be approximated by $f(X_i,Y_i) \simeq ax_i + by_i$ and $g(X_i,Y_i) \simeq cx_i + dy_i$ for local constants a, b, c and d. Note that *any* reaction functions which have the same linearized form will yield the same *initial development*. The open question, of course, is what happens when this local approximation is no longer valid and the nonlinear components of f and g begin to exert their authority? Turing analysed a variety of different possible types of population development, but the one of greatest biological interest involves stationary waves of finite length. For example, if for some I we take $a = I - 2$, $b = 2.5$, $c = -1.25$, $d = I + 1.5$ and $\mu = 2\nu$, then if $8\mu\sin^2(\pi s/N) = 1$ the system generates s waves around the ring of N cells. So for any given N, we can simulate the growth of an s-pointed 'starfish' by choosing an appropriate μ.

Initial analysis of any reaction-diffusion ring process can be carried out using this local deterministic approach; Renshaw (1994) presents a sharpened version involving a numerically superior third-order technique. He also shows how to reduce the ring size N, without disturbing the underlying structure of the process, by modifying the diffusion parameters. This is particularly valuable when conducting a large number of computer experiments for which short run-times are paramount. Two simulation procedures are open to us (Renshaw 1994). The exact strategy is to cycle through the colonies to determine the next event, and then to determine the time to this event. The problem is that the total number of possible events, namely $6N$, may be prohibitively large to check through even when using a fairly powerful workstation. An alternative, though approximate, strategy is to choose an 'appropriately small' time increment h, and for each time point $t = h, 2h, \ldots$ to examine each colony in turn to see whether a birth, death or migration occurs there. The difficulty is that it may be necessary to choose h so small that it can cope with any outrageously large X_i or Y_i *hot-spot* that develops by chance. At all other times (and indeed for all other sites) the process is highly inefficient since 'no change' will mostly be recorded. Whilst in principle the extension from one to two or even three dimensions is perfectly straightforward, if the total number of colonies is very large compute time can easily become prohibitive. To improve the second strategy we could make frequent changes to h to ensure that $h\max_i(X_i(t),Y_i(t)) < 0.01$ (say); whilst in the first we could check the largest $X_i(t)$- and $Y_i(t)$-values first. However, such modifications will result in only relatively modest gains, and for large simulation exercises it may well be necessary to use massive computing facilities. Unfortunately, efficient implementation on even parallel computers can be difficult, and success may depend on using dynamic load-balancing in which computational effort is kept high across all parts of the machine by continually reallocating colonies amongst the computer processors as the $\{X_i(t), Y_i(t)\}$ values change (Smith 1993).

Figure 2.7 shows a stochastic realization of a system on $N = 50$ sites starting in equilibrium at $X_i(0) \equiv Y_i(0) \equiv 10$, obtained by using the exact strategy. Two fundamental differences between this and the associated deterministic solution are immediately apparent.

1 Since $I = -0.2 < 0$ the deterministic solution remains unchanged at the equilibrium value, in total contrast to the explosive stochastic build-up into a wave structure with violently oscillating amplitudes.

2 A deterministic solution starting from nonequilibrium soon generates a smooth, symmetric five-wave structure during its decay to zero. This is clearly radically different from the 'Alpine arêtes' generated by the stochastic realizations, with their steeply sloping sides and irregular wave-spacing. Moreover, given that the number of waves counted depends on declaring a minimum threshold height, the question immediately arises as to 'What is a stochastic wave?' Taking $X_i = 100$ as a crude threshold value shows that for $t \geq 16$ about five such stochastic waves do exist, in line with the deterministic prediction. Examination of the space-time autocorrelations would provide a rough estimate of the time-scale over which individual waves exist during their period of waxing and waning.

Thus although the deterministic solution provides useful insight into the underlying morphology of the stochastic process, the presence of nonlinearity allows within-site stochastic population explosions to override the spatial diffusion elements of the process by causing nonuniformly spaced population centres to be self-sustaining.

Conclusions

Let us once again stress that *both* deterministic and stochastic analyses should always be undertaken. If stochastic realizations provide no useful additional information then by all means concentrate on the deterministic

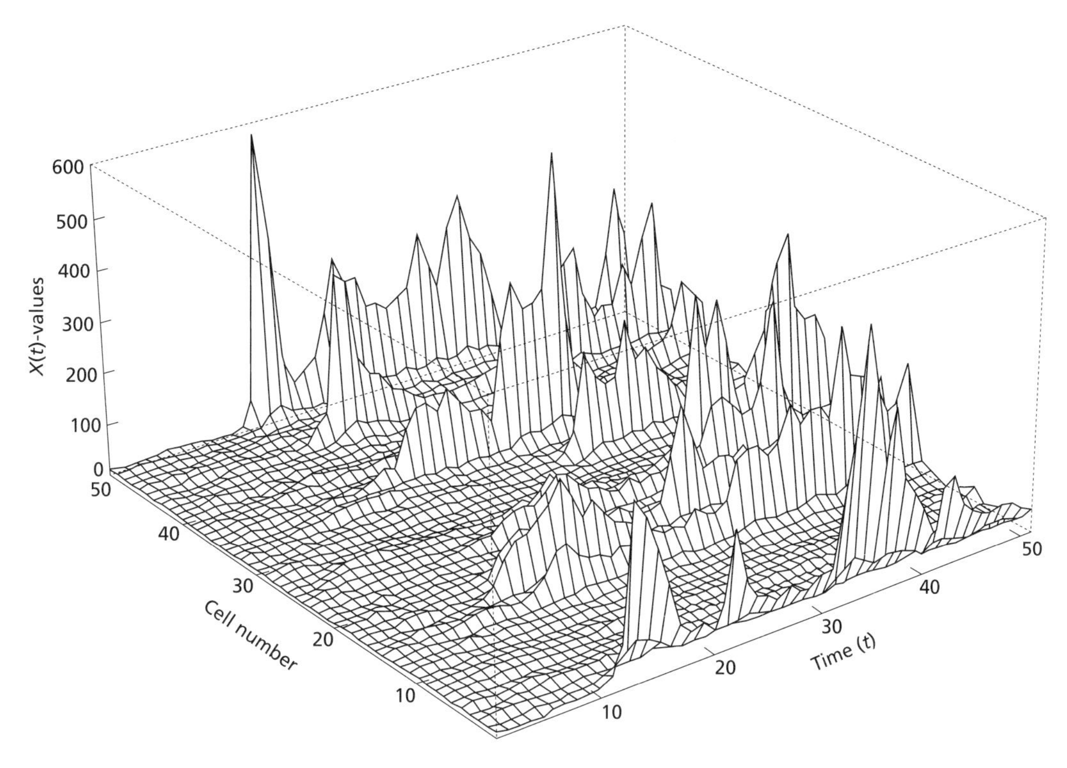

Fig. 2.7 Stochastic realization of the deterministic Turing process

$$dX_i/dt = 0.25X_iY_i - X_i(0.3 + 0.22X_i) + 1.3090(X_{i+1} - 2X_i + X_{i-1})$$

$$dY_i/dt = 0.13Y_i^2 - Y_i(0.05 + 0.125X_i) + 0.6545(Y_{i+1} - 2Y_i + Y_{i-1})$$

with $I = -0.2$, showing $\{X_i(t)\}$ over $i = 1, \dots, 50$ for $t = 0(1)50$ (reproduced from Renshaw, 1994, by permission of the Applied Probability Trust).

solution. However, this will not always be the case, and the difference can be fairly dramatic as the above example demonstrates. Indeed, blind obedience to deterministic space-time processes can lead to severe problems at the outset: with Turing ring systems slight changes in initial conditions can produce a variety of different spatial configurations, so setting a nasty trap for the unwary. Höfer & Maini (1996) report on the difficulties involved in generating realistic deterministic patterns in fish skin, and fresh insight into such situations could well be provided by injecting stochastic variation. Care is also needed when selecting the number of sites N; Renshaw (1994) illustrates a system in which a ring of just 20 sites provides a much better approximation to the 500-site case than does one with 40 sites. This disparity is especially important when a genuine diffusion scenario is being approximated by a stepping-stone process.

Whilst Turing rings are an excellent medium for investigating pattern on closed systems, such as the fetus of a zebra or giraffe, patterns developing on regions possessing a definite boundary, such as a butterfly wing, may well be heavily influenced by assumptions at the boundary itself. These can totally dominate population development, as illustrated by the flour beetle experiment, and considerable thought needs to be exercised in their construction.

As well as being useful in the interpretation of pattern, space-time population processes have a vital role to play in the analysis of extremely complicated biological processes that often occur in many marine and land-based scenarios. Given the lack of full information relating to population development in such situations, it is understandable that many modellers limit their concern to the construction of a set of deterministic equations that attempt to mimic the broad characteristics of the process under study. The key point is that unless models are *robust to stochastic variation*, conclusions drawn from them may well be incorrect. This fact holds true no matter how complicated the model appears to be, and the development of stochastic techniques that are appropriate in many-variable situations is clearly a matter of top priority. Ross *et al.* (1994), for example, present a convincing and highly detailed 13-variable deterministic model in order to examine the hypothesis that differences in annual cycle amongst four sea-lochs are products of their distinct environment and hydrography. Studying what happens when such models are placed in a fully stochastic setting presents a timely and worthwhile challenge.

If incorrect deterministic predictions result in wrong actions being taken, such as decisions relating to population control, then using deterministic models can be worse than not conducting an analysis at all. Mollison (1991) shows a striking example which highlights the inability of nonlinear deterministic epidemic models to handle the transition to endemicity. In their examination of how fox rabies might invade a new country, Murray *et al.* (1986) predict a roughly circular expanding wave of advance, followed after a quiet phase of about 7 years by another wave originating from the same starting point. Mollison challenges these findings on two counts. As regards the first advancing wave, the weight of evidence from fox rabies in Europe suggests that after a short while it could 'break back' across the devastated territory immediately behind it and induce an endemic equilibrium there. The deterministic model's inability to describe this is not surprising because the process is an essentially stochastic phenomenon. As to the second wave, close inspection reveals that this is an artefact of the modelling of population size as continuous rather than discrete and its associated inability to let population values reach zero. Because the density of infective foxes at the place of origin declines not to zero, but to 10^{-18} of a fox per square kilometre, the model allows this 'atto-fox' to start the second wave as soon as the susceptible population has grown sufficiently. Such numerical nonsense may, of course, be eliminated by placing any population variable equal to zero as soon as it declines to 0.5. Nevertheless, the discrepancies between the overall deterministic predictions and reality are a cause for serious concern.

Given that we can immediately write down an associated stochastic process for any deterministic system, there can be no excuse for not investigating the effect of taking a stochastic approach. Even if no theoretical studies are undertaken, early examination of simulated stochastic realizations is clearly vital if conclusions drawn from underlying deterministic analyses are to retain any degree of credibility.

References

Anderson, R.M., Blythe, S.P., Gupta, S. & Konings, E. (1989) The transmission dynamics of the human immunodeficiency virus type 1 in the male homosexual community in the United Kingdom: the influence of changes in sexual behaviour. *Philosophical Transactions of the Royal Society of London*, **B325**, 45–98.

Armitage, P. (1952) The statistical theory of bacterial populations subject to mutation. *Journal of the Royal Statistical Society*, **B14**, 1–33.

Bailey, N.T.J. (1964) *The Elements of Stochastic Processes with Applications to the Natural Sciences*. Wiley, New York.

Bailey, N.T.J. (1968) Stochastic birth, death and migration processes for spatially distributed populations. *Biometrika*, **55**, 189–198.

Bailey, N.T.J. (1975) *The Mathematical Theory of Infectious Disease and its Applications*, 2nd edn. Griffin, London.

Ball, F. & O'Neill, P. (1993) A modification of the general stochastic epidemic motivated by AIDS modelling. *Advances in Applied Probability*, **25**, 39–62.

Bartlett, M.S. (1949) Some evolutionary stochastic processes. *Journal of the Royal Statistical Society*, **B11**, 211–229.

Bolker, B.M. & Grenfell, B.T. (1993) Chaos and biological complexity in measles dynamics. *Proceedings of the Royal Society of London*, **B251**, 75–81.

Carlson, T. (1913) Über Geschwindigkeit und Grösse der Hefevermehrung in Würze. *Biochemische Zeitschrift*, **57**, 313–334.

Cox, D.R. & Miller, H.D. (1965) *The Theory of Stochastic Processes*. Methuen, London.

Davidson, J. (1938a) On the ecology of the growth of the sheep population in South Australia. *Transactions of the Royal Society of South Australia*, **62**, 141–148.

Davidson, J. (1938b) On the growth of the sheep population in Tasmania. *Transactions of the Royal Society of South Australia*, **62**, 342–346.

Dobson, A.P. (1985) The population dynamics of competition between parasites. *Parasitology*, **91**, 317–347.

Feller, W. (1939) Die Grundlagen der Volterraschen Theorie des Kampfes ums Dasein in wahrscheinlichkeits theoretischen Behandlung. *Acta Biotheoretica*, **5**, 1–40.

Fenner, F. & Myers, K. (1978) Myxoma virus and myxomatosis in retrospect: the first quarter century of a new disease. In: *Viruses and the Environment* (eds E. Kurstak & K. Maramorosch), pp. 539–570. Academic Press, New York.

Gani, J. (1965) Stochastic models for bacteriophage. *Journal of Applied Probability*, **2**, 225–268.

Grenfell, B.T., Bolker, B.M. & Kleczkowski, A. (1995) Seasonality and extinction in chaotic metapopulations. *Proceedings of the Royal Society of London*, **B259**, 97–103.

Höfer, T. & Maini, P.K. (1996) Turing patterns in fish skin? *Nature*, **380**, 678.

Holling, C.S. (1965) The functional response of predators to prey density and its role in mimicry and population regulation. *Memoirs of the Entomological Society of Canada*, **45**, 3–60.

Isham, V. (1988) Mathematical modelling of the transmission dynamics of HIV infection and AIDS: a review. *Journal of the Royal Statistical Society*, **A151**, 5–30.

Isham, V. (1991) Assessing the variability of stochastic epidemics. *Mathematical Biosciences*, **107**, 209–224.

Isham, V. (1993) Stochastic models for epidemics with special reference to AIDS. *Annals of Applied Probability*, **3**, 1–27.

Kermack, W.O. & McKendrick, A.G. (1927) Contributions to the mathematical theory of epidemics. *Proceedings of the Royal Society of London*, **A115**, 700–721.

Kimura, M. (1953) 'Stepping stone' model of population. *Annual Report of the National Institute of Genetics, Japan*, **3**, 62–63.

Lotka, A.J. (1925) *Elements of Physical Biology*. Williams and Wilkins, Baltimore.

Marion, G., Renshaw, E. & Gibson, G. (1997) Stochastic effects in a model of nematode infection in ruminants. *IMA Journal of Mathematics Applied in Medicine and Biology*, **15**, 97–116.

Mollison, D. (1986) Modelling biological invasions: chance, explanation, prediction. *Philosophical Transactions of the Royal Society*, **B314**, 675–693.

Mollison, D. (1991) Dependence of epidemic and population velocities on basic parameters. *Mathematical Biosciences*, **107**, 255–287.

Mollison, D., Isham V. & Grenfell, B. (1994) Epidemics: models and data (with discussion). *Journal of the Royal Statistical Society*, **A157**, 115–149.

Morgan, B.J.T. (1984) *Elements of Simulation*. Chapman and Hall, London.

Murray, J.D. (1989) *Mathematical Biology*. Springer-Verlag, Berlin.

Murray, J.D., Stanley, E.A. & Brown, D.L. (1986) On the spatial spread of rabies among foxes. *Proceedings of the Royal Society of London*, **B229**, 111–150.

Neyman, J., Park, T. & Scott, E.L. (1956) Struggle for existence. The *Tribolium* model. *Proceedings of the Third Berkeley Symposium on Mathematical Statistics and Probability*, **3**, 41–79. University of California Press, Berkeley and Los Angeles.

Nisbet, R.M. & Gurney, W.S.C. (1982) *Modelling Fluctuating Populations*. Wiley, New York.

Pearl, R. (1927) The growth of populations. *Quarterly Review of Biology*, **2**, 532–548.

Pearl, R. (1930) *Introduction to Medical Biometry and Statistics*. Saunders, Philadelphia.

Pearl, R. & Reed, L.J. (1920) On the rate of growth of the population of the United States since 1790 and its mathematical representation. *Proceedings of the National Academy of Sciences*, **6**, 275–288.

Renshaw, E. (1972) Birth, death and migration processes. *Biometrika*, **59**, 49–60.

Renshaw, E. (1973a) Interconnected population processes. *Journal of Applied Probability*, **10**, 1–14.

Renshaw, E. (1973b) The effect of migration between two developing populations. *Proceedings of the 39th Session of the International Statistical Institute*, **2**, 294–298.

Renshaw, E. (1977) Velocities of propagation for stepping-stone models of population growth. *Journal of Applied Probability*, **14**, 591–597.

Renshaw, E. (1980) The spatial distribution of *Tribolium confusum*. *Journal of Applied Probability*, **17**, 895–911.

Renshaw, E. (1986) A survey of stepping-stone models in population dynamics. *Advances in Applied Probability*, **18**, 581–627.

Renshaw, E. (1991) *Modelling Biological Populations in Space and Time*. University Press, Cambridge.

Renshaw, E. (1994) Non-linear waves on the Turing ring. *Mathematical Scientist*, **19**, 22–46.

Ripley, B.D. (1987) *Stochastic Simulation*. Wiley, New York.

Roberts, M.G. & Grenfell, B.T. (1992) The population dynamics of nematode infections of ruminants: the effect of seasonality in the free-living stages. *IMA Journal of Mathematics Applied in Medicine and Biology*, **9**, 29–41.

Ross, A.H., Gurney, W.S.C. & Heath, M.R. (1994) A comparative study of the ecosystem dynamics of four fjords. *Limnology and Oceanography*, **39**, 318–343.

Sang, J.H. (1950) Population growth in *Drosophila* cultures. *Biological Reviews*, **25**, 188–219.

Shaw, M.W. (1995) Simulation of population expansion and spatial pattern when individual dispersal distributions do not decline exponentially with distance. *Proceedings of the Royal Society of London*, **B259**, 243–248.

Smith, M. (1993) Dynamic load-balancing strategies for data parallel implementations of reaction–evolution–migration systems. *International Journal of Modern Physics*, **C4**, 107–119.

Soper, H.E. (1929) Interpretation of periodicity in disease prevalence. *Journal of the Royal Statistical Society*, **92**, 34–73.

Tanner, J.T. (1975) The stability and the intrinsic growth rates of prey and predator populations. *Ecology*, **56**, 855–867.

Turing, A.M. (1952) The chemical basis of morphogenesis. *Philosophical Transactions of the London Mathematical Society*, **B237**, 37–72.

Usher, M.B. & Williamson, M.H. (1970) A deterministic matrix model for handling the birth, death and migration processes of spatially distributed populations. *Biometrics*, **26**, 1–12.

Verhulst, P.F. (1838) Notice sur la loi que la population suit dans son accroissement. *Corr. Math. et Phys. publ. par A. Quetelet*, T.X (also numbered T.II of the third series), 113–121.

Volterra, V. (1926) Fluctuations in the abundance of a species considered mathematically. *Nature*, **118**, 558–560.

Volterra, V. (1931) *Leçons sur la Théorie Mathématique de la Lutte pour la Vie*. Gauthier-Villars, Paris.

Whittle, P. (1955) The outcome of a stochastic epidemic – a note on Bailey's paper. *Biometrika*, **42**, 116–122.

Whittle, P. (1957) On the use of the normal approximation in the treatment of stochastic processes. *Journal of the Royal Statistical Society*, **B19**, 268–281.

3: Spatial models of interacting populations

M. Keeling

Introduction

The quantitative study of the effects resulting from the spatial distribution of species within their habitat has recently become the focus of much ecological research (Durrett 1988; Hastings 1990; DeAngelis & Gross 1992; Kareiva 1994; Durrett & Levin 1994a; May 1995; Tilman & Kareiva 1996; Diekmann *et al.* 1999). The position of individuals in an environment is well known to have a dramatic influence on their behaviour, survival and reproductive success. In this chapter I will examine the importance of understanding and modelling biological interactions in a spatial environment as well as high-lighting some of the new problems and phenomena that this creates. The discussion will be broken into four parts: in the first part I will describe examples of situations where the spatial distribution of organisms has been found to be important; in the second part I will discuss some of the distinctive features and problems which are generally encountered whenever a spatial framework is introduced; and in the third section I will show the main types of spatial models that are used and highlight their applicability and drawbacks. The fourth and final section is devoted to problems and ideas arising from current research.

Why and when is spatial modelling necessary?

Spatial modelling is any mathematical description of a system that accounts for the positions or distribution of the various components as well as their quantities. Rather than the densities of each population simply being a function of time (e.g. see the effects of stochasticity in a nonspatial approach in Chapter 2), they are also a function of space. Just as the increase from a single-species to a two-species system leads to the possibility of far richer behaviour, then the increase from one spatial site to two interacting sites can do the same. This idea follows through as the number of spatial sites increases until, when modelling a real system, a vast array of different behaviours become possible.

A brief history of spatial modelling

It has only been with the advent and accessibility of powerful computers that a quantitative understanding of the effects of space has begun to be reached (Huston *et al.* 1988; McGlade 1993; Levin *et al.* 1997). Previously, variations in the density of a population over its range were only considered in qualitative terms and variations were assumed mainly to be due to changes in the underlying habitat—the fact that the system had a spatial component was widely overlooked. Some early research was done on the spatial distribution of plants (Blackmann 1935; Stewart & Keller 1936; Singh & Das 1938) but this was predominantly from a statistical viewpoint. Little consideration was given to attempting to modelling the distributions observed, or predicting how spatial patterns could arise.

The idea that the presence of a spatial component in a model could lead to the formation of spatial patterns was proposed in 1952 by Alan Turing (Turing 1952). He demonstrated that for a particular class of two-species systems, minute fluctuations in the starting environment could be amplified and large-scale structures could emerge. This type of pattern formation can be seen in the well-known Belousov–Zhabotinsky reaction (Roux & Swinney 1981); more recently Turing structures (as these patterns are known) have been cited as a likely means of limb development in embryos and creation of marking on animals' fur and skin (see review of Murray 1988 and Cohen & Stewart 1993). This area of work will be discussed later.

Heterogeneity

Spatial modelling allows heterogeneities, or local deviations away from the global mean, to be introduced and these phenomena can drastically affect the global dynamics. Examples of spatial heterogeneities exist at all scales in the environment, e.g. from huge planktonic blooms in the oceans (Levin & Segel 1976) to the patterns made by slime-moulds on a Petri dish (Hofer *et al.* 1995). Some of the most well-studied spatial heterogeneities in ecology are the waves of infection which travel through populations (Mollison 1977; Durrett 1993; Rand *et al.* 1995) and patterns found in vegetation (Pacala 1987; Czaran & Bartha 1992; Hendry & McGlade 1995). If we imagine examining a satellite photograph of northern Europe, then a patchwork of rural and urban areas can be discerned. Zooming in on the scene, more heterogeneities are visible in the rural environment; first isolated woodland habitats, then heterogeneities within the patches of woodland and then individual trees. Finally heterogeneities are found within each tree, with species that have evolved to exploit each microhabitat from the highest canopy to the roots in the ground. Many of these species contain parasites which are again adapted for life in or on a particular region of their host. The existence and importance of heterogeneity at all scales can be demonstrated; which of these need to be considered depends on the species being modelled and the range over which they can interact.

Spatial patterns are often found where the movement, or range of influence, of individuals is small compared with the habitat they occupy. A vast number of ecosystems are composed of such individuals, the obvious exceptions being many aquatic environments where strong currents cause long-range mixing. Short-scale behaviour often leads to the formation of strong correlations at a local level, for example in the case of disease spread one infected organism is most likely to infect the surrounding organisms and hence strong local correlations arise. These correlations in turn lead to sweeping waves of infection that are characteristic of diseases in many spatially distributed hosts, from fungal disease in trees (Maddison *et al.* 1996) to rabies in wildlife populations (Murray *et al.* 1986). Plants in particular fall into this category, as during their lifetime they can only exert an influence over a very limited area, and even the effects of long-range seed dispersal do little to reduce the local correlations that can build up (Hendry *et al.* 1996). Local correlations and many other features of spatial systems are discussed in more detail in the next section.

Population structures are also a result of other forms of heterogeneity derived from social, demographic or genetic processes, but unfortunately there are limited data or understanding of exactly how these act. For example, although we understand how a parent's genotype is reflected in its offspring and how variation in the phenotype may affect the survival and reproductive fitness of an organism, the lack of understanding of exactly how the genotype determines a phenotype limits the modelling to coarse caricatures (Hartvigsen & Starmer 1995; Keeling & Rand 1995). Both age and social heterogeneity have also been included in some epidemiological models, but these tend to suffer from a lack of good data for parameterization (Yorke *et al.* 1978; Castillo-Chavez *et al.* 1989; Bolker & Grenfell 1993).

Causes of heterogeneity

The standard models which do not take space or individuals into account are often termed *homogeneous* or *mean-field*, and rely on the following two assumptions.

1 The populations are infinite, which means that all of the variables can be continuous (for real populations the values must be integers) and it is only the relative size or density of the species that is important. Thus very small population densities (often as low as 10^{-20}—termed nanopopulations) are allowable, so extinctions are far less likely. The use of infinite populations precludes any form of probabilistic uncertainty in the models.

2 The populations are fully mixed, meaning that every individual experiences the presence of all other individuals equally. In an ecosystem a given individual is far more likely to interact with others around it than with a random individual which is some distance away.

These two assumptions are often called the Law of Mass Action and are based on principles from physics; but while they lead to an accurate description of the random movement of gas particles they have little to do with many real ecosystems. Ecology is the study of organisms and their environment, which means that where an individual is located is as important as its behaviour and interactions. In this chapter, I will concentrate on spatial models which demonstrate one or more of the following properties.

1 Limited interactions, in which a given organism can only interact with others within a small range (see Chapter 4 for a detailed description of interactions in viscous populations).

2 Finite populations, so that individuals are important and local demographic extinctions are possible.

3 Probabilistic interactions, so that random events can also play a role (see Chapter 2).

Very few models possess all of these properties and which of them are necessary, or useful, entirely depends on the problem being studied; each brings its own set of associated difficulties and constraints (Durrett & Levin 1994b). However, it is the presence of limited-range interactions (item 1 above) that allows the consideration of space together with the formation of correlations and heterogeneities, and hence this property will be present in all the models discussed.

There are two main forms of heterogeneity seen in the natural world, self-induced heterogeneities (**Type i**) and underlying heterogeneities (**Type ii**); these interact with each other to produce the rich, complex spatial patterns that are observed in real ecosystems. Self-induced heterogeneities are features such as Turing structures or aggregations of organisms which simply arise due to the amplification of initial random fluctuations. Underlying heterogeneities are reflections of physical changes (such as gradients in temperature, moisture, soil or altitude) producing patterns in the species present. To see how these two features overlap and interact consider a simple human example. Local geography defines likely places for a settlement to be started (**Type ii**), and the presence of a human settlement attracts more people so a pattern of urban and rural areas forms (**Type i**); a new disease introduced to this area spreads through the population in a wave-like manner (**Type i**) but is distorted by the pattern of human density (**Type ii**). Although both of these forms of heterogeneity are important, most studies of spatial models have concentrated on the formation of structures from a uniform environment (**Type i**). This is mainly due to a lack of good spatial data to initialize a nonuniform model and due to the mathematical simplicity of considering a homogeneous background. In the following sections, I will be guided by the trends in current research and will be mainly concerned with self-induced spatial patterns although some examples of nonuniform starting conditions will also be described (Green 1990; Clayton *et al.* 1997).

Features of spatial models

In this section I describe in more detail some of the features seen in many generic spatial models. As with real ecosystems each model with have its own specificity and properties, but there are some features which tend to recur. These types of behaviour are observed in all models from the very simplest patch models to highly complex interacting particle systems, thus supporting the hypothesis that these features are generic and should be the driving force behind many complex natural interactions.

Aggregation and local correlations

The main differences between spatially structured and homogeneous systems can be explained by the presence of local correlations and the aggregating effects that cause heterogeneities. There is a positive spatial correlation when the presence of an organism of one species increases the likelihood of finding an organism of the same or another species in the surrounding area. A species is said to be aggregated if there is a positive spatial correlation with itself. A good ecological example is the many plants that propagate by producing rhizomes; the presence of one plant in an area means that there are very likely to be others in the immediate vicinity. Many types of fish are clear cases of species which aggregate due to their behaviour rather than as an outcome of some physical limitation; they tend to clump together into large schools rather than spread themselves evenly over their feeding or spawning grounds (Niwa 1994). Negative correlations can also be found; for example many animals are territorial, so the presence of one will reduce the chance of finding another within a certain distance (Sibley & Smith 1985).

The phenomenon of aggregation or local spatial correlations can have a profound effect on the dynamics of any system (Hassell & May 1974; Hassell & Pacala 1990; Bolker *et al.* 1999). Suppose there exists an aggregated population with a carrying capacity of K. Although the average density of the population may be far from the carrying capacity, the fact that the individuals are highly aggregated may mean that in many areas the density is close to K and thus density-dependent effects will be felt (Hassell & May 1985). When two or more species are involved then the effects of spatial correlations can be even more dramatic: two examples are described below.

Refuges and stabilizing behaviour

A refuge is a site where organisms are sheltered from the effects of another species. This is usually thought of as an area which is inaccessible to predators or parasites due to some physical or geographic barrier (Begon *et al.* 1996). For example the cold temperatures at high altitude may protect a host from various forms of parasite, or a species that exists high up in the canopy could escape many of its predators by virtue of the fact that the branches cannot support the weight of larger organisms. However, in spatial models where species have a limited range, ephemeral refuges may arise simply due to aggregation of the organisms. For example, some prey may escape predation if all the predators are aggregated in one area for the purposes of reproduction or the prey may become aggregated into clumps thereby limiting some forms of predation, when predators avoid crossing the large barren areas between clumps of prey (DeRoos *et al.* 1991; McCauley *et al.* 1993). The effect of refuges in these two examples is to reduce the predator population in comparison to the global prey population; a similar feature is observed in host–parasite systems where a significantly lower number of parasites are predicted by a spatial model compared with a homogeneous form (Rand *et al.* 1995).

One of the main effects of refuges is in stabilizing the global behaviour of the system. If the prey can be driven to local extinction by predation and yet refuges exist, then there are always a small number of prey remaining which can recolonize the system. Often the effect is not quite so dramatic; in populations that cycle the decrease in predation afforded by one part of the cycle leads to an abundance of prey which can spread to neighbouring sites damping down their oscillations and hence stabilizing the global system (Hastings 1977; McLaughlin & Roughgarden 1991; Keeling 1997; Lynch *et al.* 1998). Thus although at a local scale the dynamics would seem to be complex, at larger scales with the inclusion of localized interactions, the system converges to constant population levels.

Spatial patterns

Most spatial patterns can be considered as being generated by one of four basic processes: aggregation, spatial instabilities, multiple stable solutions or complex local dynamics. The simplest source of spatial patterning is the aggregation and local correlations due to the directed movement of individuals. These are the cause of many common spatial phenomena such as shoals of fish and insect swarms (Levin 1992; Gueron & Levin 1993; Reuter & Breckling 1994).

Turing structures arise in partial differential equations and are one of the best understood forms of spatial heterogeneity due to spatial instability (see later). In general these partial differential equations must have two (or more) species, the *activator* and the *inhibitor*. The presence of the activators leads to an increase in the number of inhibitors and the presence of the inhibitors reduces the number of activators. All that remains is that both species should move randomly through the system and that the inhibitor should move appreciably faster than the activator. Although the nonspatial dynamics produce a stable, fixed population, the inclusion of random movement leads to destabilization and the appearance of permanent spatial structures. This type of behaviour is often observed with predator–prey (and parasitoid–host) models where an increase in predators (inhibitors) reduces the numbers of prey (activators) and more prey increases the number of predators. Hence even though the homogeneous equations are often stable, Turing-type structures and their associated spatial patterns in general should be expected in a vast number of simple two-species models.

Spatial patterns can also arise for other reasons; if these are multiple, stable, fixed points in the homogeneous dynamics then a spatial model may have large uniform patches of the population levels described by each of the fixed points. This theme of spatial modelling, causing increased coexistence, has been dealt with by a number of authors (Hanski 1983; Hassel *et al.* 1994) but can be clearly seen if a very simple spatial model with genetics is examined. Let us imagine a population whose fitness depends upon a single diallelic locus, with alleles labelled **a** and **A**. If heterozygote inferiority is now assumed, then standard theory predicts that the system should eventually arrive at a population which is either all **AA** or all **aa**; but on the inclusion of a spatial element this theory fails. Instead, if organisms can breed only with others in their immediate vicinity then large patches of homozygote individuals are obtained, separated by narrow heterozygote bands (Fig. 3.1). A more detailed account can be found in Keeling *et al.* (1997a); this model, by necessity, is highly simplified and such striking genetic patterns have not been observed in nature.

Another source of spatial structure is due to the local dynamics; if the homogeneous equations lead to an oscillatory system then when spatial coupling is included, a gradual change in the phase of the orbit may be observed on moving through space. This gradual change together with the oscillations will appear as waves moving through the system. (It should be noted that the action of local coupling will usually reduce the amplitude of the oscillations.) This type of behaviour is observed in the coupled map lattice system introduced below (Hassell *et al.* 1991a), in the spread of a disease (Johansen 1994, 1996) and due to the cycles in Lotka–Volterra type systems (Solé & Valls 1991).

A profusion of parameters

One of the main problems with spatial modelling is the increase in the

Fig. 3.1 The spatial pattern for a spatially extended genetic model with heterozygote inferiority. Large areas of all **AA** (black) and all **aa** (white) are clearly visible. This model is a coupled map lattice.

number of parameters and possible situations. For example, there is now a vast of choice choice over initial conditions. Previously, only the global average of each species needed to be specified, but including space means that the distribution of each species also has to be known. This distribution can be made as complex and varied as necessary, but most often the initial value at each point in space is chosen as being randomly distributed about the global mean.

Given the starting point, the next consideration is the form and magnitude of the spatial interactions. Unfortunately, ecological data usually only give the global averages for a population; to calculate the amount of spatial coupling requires detailed behavioural information as well. It is very rare to find a case where there is detailed population dynamics data together with an understanding of the behaviour at an individual level which is simple enough to be modelled. Attempting to balance individual behavioural effects and global population level dynamics is still one of the most difficult yet fascinating areas of current research in ecology.

The inclusion of space also increases the number of situations that need to be studied; this can be illustrated by a simple example. Imagine the invasion of a system by a particular species; the standard homogeneous approach is to introduce the species at low density and look at the long-term solutions of the model. With spatial modelling this causes problems as to exactly where the new species should be introduced. Should the invader be placed at a high density in one point, a low density over the entire space or some combination of the two? In spatial modelling it is perfectly possible to have the success or failure of an invasion dependent on the initial distribution of the invading species (McGlade 1995), as multiple attractors often exist (Solé *et al.* 1992). It is situations such as these that demonstrate the extra complexity and greater realism involved in even the simplest of ecological questions when space is included.

Ideally, the information about any given ecological system should be gathered with the problems of spatial modelling in mind. This would mean that every global measurement would need to be accompanied by a measure of its spatial distribution. Very often this is simply impractical due to limitations on both time and resources; however one source of data which would provide much of the necessary information is that of satellite photographs, and an increasing amount of research is making use of these (see Chapter 11).

A question of scale

One of the most important questions that emerges from the consideration of space is the scale at which the modelling should take place (Wiens 1989; Menge & Olson 1990; DeRoos *et al.* 1991; Levin 1992; Vail 1993). This can be broken into two independent problems: (i) which is the smallest scale of heterogeneity that should be considered; and (ii) how large a system is it necessary to study. I will consider the second question in the final section as it represents a significant problem in current research. As has already been shown, heterogeneity exists at all scales, from nanometres to thousands of kilometres; but many of these can be ignored as being too small to influence the organisms in question or so large that the organism views its habitat as uniform. This type of qualitative argument is appropriate if the model treats space as a continuum; however in many systems space is broken up into a lattice of sites and so a quantitative value for the area of a site is required.

In individual-based systems (such as the interacting particle systems outlined below and in Chapter 1) each site should correspond to the area covered by or under direct control of an organism. With plants this is obviously the area they occupy, but for a predator–prey or host–parasite system the choice is less clear and will depend on the speed with which the organisms move and the density at which they live. In nonindividual models there is the implicit assumption that the populations of each spatial unit obey a set of homogeneous equations together with external coupling. Therefore a balance must be struck between too fine a scale in which case the model will be computationally intensive, and too large a scale in which case too many heterogeneities may be ignored.

The question of the size of each site is inevitably linked to the range over which spatial interactions take place. For speed and simplicity of modelling it is best to only allow interactions in the local neighbourhoods, but this may be in direct conflict with taking a small enough scale, so these two factors need to be balanced. Which type of spatial model is used will be heavily dependent on the choice of scales. Planktonic blooms in the oceans where there is gradual spatial variation and large numbers of organisms can be successfully modelled by a coupled map lattice or set of partial differential equations (see below and Levin 1992), but systems governed by the interactions of a few large territorial predators would be better suited by an individual-based approach.

Types of spatial model

In this section four types of spatial model are described, in increasing order of complexity (summarised in Table 3.1). To facilitate a greater understanding, I shall apply the different modelling techniques to a simple host–parasite system, adding the spatial interactions to a simple homogeneous framework. Unfortunately, due to the nature of the different models they are designed to act upon, different types of homogeneous equations, largely depending on whether the model uses discrete or continuous time. Nevertheless the conclusions are generic.

For discrete time models, the Nicholson–Bailey host–parasitoid equations will be used as the homogeneous model (Nicholson & Bailey 1935; Begon *et al.* 1996). These equations were formulated from a purely theoretical approach and yet have been shown to be a good approximation to laboratory experiments. Let H_t and P_t be the number of hosts and parasitoids respectively at time t. If the hosts reproduce at rate λ and the parasitoids randomly search out an area a looking for hosts then the familiar equations are obtained:

$$\begin{aligned} H_{t+1} &= \lambda H_t e^{-aP_t} \\ P_{t+1} &= H_t(1-e^{-aP_t}) \end{aligned} \tag{3.1}$$

In other words, parasitized host produces parasitoids in the next generation, and healthy hosts produce hosts. Equation 3.1 has an equilibrium point (where $H_{t+1} = H_t$ and $P_{t+1} = P_t$):

$$H^* = \frac{P^*}{\lambda - 1} \quad P^* = \frac{\log(\lambda)}{a}$$

In all the calculations that follow it is assumed that $\lambda = 2$ and $a = 1$. Although accurate for small laboratory populations, the dynamics of Equation 3.1 are intrinsically unstable, with larger and larger oscillations of the hosts and parasitoids occurring until one or other of the species becomes extinct. Spatial modelling and, in particular, aggregating effects have long been suggested as a means of stabilizing these dynamics so that a more realistic behaviour is obtained (Hassell & May 1974, 1988; Hassell *et al.* 1991b).

For continuous systems I will examine the following host–parasite/predator system (see May 1981):

$$\frac{dH}{dt} = H - \frac{H}{1+H}P$$
$$\frac{dP}{dt} = b\frac{H}{1+H}P - P^2 \qquad (3.2)$$

This has a stable fixed point (where the rate of change of H and P is zero):

$$H^* = P^* - 1 \quad P^* = \frac{b - \sqrt{b^2 - 4b}}{2}$$

Again, there is a fixed host birth rate and the birth rate of parasites increases with the number of hosts, but there is also a density dependence for both the parasites' death rate and the number of hosts they can parasitize. The slightly unusual form of the homogeneous equations is necessary if interesting spatial effects are to be seen.

These two fairly simple models can be used to demonstrate the range of behaviours possible, the different techniques used and problems encountered when the spatial component is also taken into account (Table 3.1).

Patch models

Consider a landscape which can be represented by a series of habitats or patches labelled 1 up to N. Each habitat contains its own population of hosts

Table 3.1 Summary of the different types of spatial model.

Type	Space	Time	Population	How behaviour is expressed	Uses
Patch or metapopulation model	Several homogeneous patches	Any	Continuous	A small number of coupled equations	Isolated, uniform habitats
Coupled map lattices	Discrete lattice	Discrete	Continuous	Large number of coupled equations	Large number of individuals
Partial differential equations	Continuous	Continuous	Continuous	Partial differential equations	Large number of individuals with random movement
Interacting particle systems (cellular automata and artificial ecologies)	Discrete lattice	Any	Discrete	Set of probabilistic rules	Small numbers or low densities of organisms

and parasitoids, with the number of these in habitat i being given by H^i and P^i respectively (i.e. the Nicholson–Bailey model). If the patches are completely isolated then each set of H and P would perform its own expanding oscillations about the fixed point irrespective of the other patches and soon the entire system would go to extinction. Now suppose there exists a small amount of movement between the patches; at first assume that individuals move between patches completely at random. This means that each patch loses a fixed proportion of its inhabitants (at a rate μ) and these are divided equally amongst the remaining $N-1$ habitats. Such a set of coupled subpopulations is known as a patch or metapopulation model (Taylor 1990; Hanski & Gilpin 1991; Levin *et al.* 1993; Grenfell & Harwood 1997). For each patch there are two equations:

$$\begin{aligned} H^i_{t+1} &= \lambda H^i_t \mathrm{e}^{-P^i_t} - \mu\lambda H^i_t \mathrm{e}^{-P^i_t} + \frac{\mu}{N-1}\sum_{j\neq i} \lambda H^j_t \mathrm{e}^{-P^j_t} \\ P^i_{t+1} &= H^i_t(1-e^{-P^j_t}) - \mu H^i_t(1-e^{-P^i_t}) + \frac{\mu}{N-1}\sum_{j\neq i} H^j_t(1-e^{-P^j_t}) \end{aligned} \qquad (3.3)$$

For simplicity this can be both written and calculated in two parts; the first is the standard homogeneous equations (compare with Equation 3.1), the second is the coupling:

$$\begin{aligned} h^i_{t+1} &= \lambda H^i_t \mathrm{e}^{-P^i_t} \\ p^i_{t+1} &= H^i_t(1-\mathrm{e}^{-P^i_t}) \\ H^i_{t+1} &= (1-\mu)h^i_{t+1} + \frac{\mu}{N-1}\sum_{j\neq i} h^j_{t+1} \\ P^i_{t+1} &= (1-\mu)p^i_{t+1} + \frac{\mu}{N-1}\sum_{j\neq i} p^j_{t+1} \end{aligned} \qquad (3.4)$$

For high values of coupling μ there is sufficient mixing between the patches that eventually they all act as one giant, homogeneous patch and share the same unstable dynamics. As the coupling is reduced then the orbits of each pair (P^i,H^i) stabilize, and although qualitatively similar behaviour is seen in each patch there is enough variation to maintain a presence of each species. The dynamics of one patch and of the global system are shown in Fig. 3.2. This quasi-periodic orbit is only one example of the type of behaviour that can be observed; for other initial conditions but the same parameter values simple periodic orbits and other quasi-periodic orbits can also be found. It is interesting to note that only a minute amount of coupling is necessary to stabilize the system, and that for μ as low as 0.01 and with 10 patches there is still sufficient coupling to homogenize the system and lead to parasite extinction. This result would suggest that in the real world only host–parasitoid systems where there is very weak coupling should be observed. However, true spatial structure (as developed in the next section)

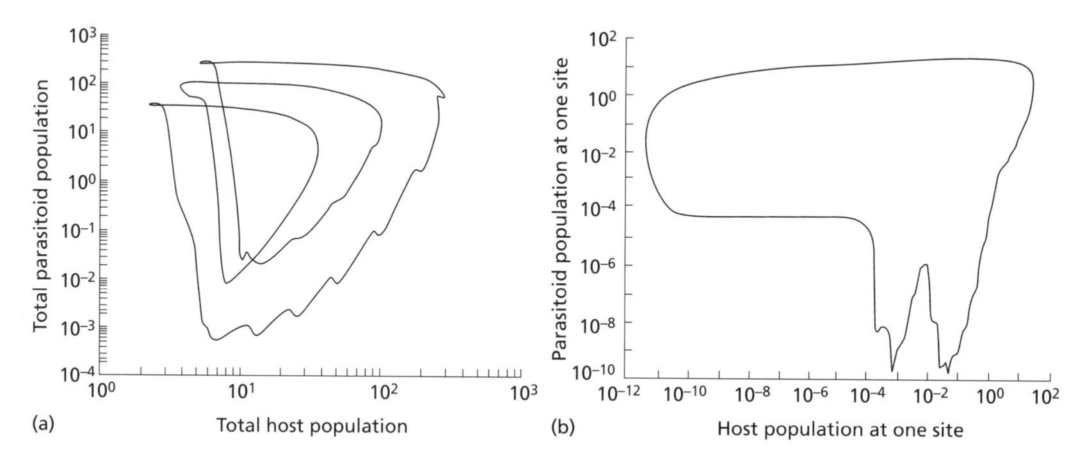

Fig. 3.2 (a) The total number of parasitoids against the total number of hosts in all the patches. The parameters used were $\mu = 0.0001$ and $N = 10$. (b) The orbit for the number of parasitoids in one patch against the number of hosts; the dynamics in all the other patches are qualitatively similar.

and demographic stochasticity will reduce the homogenizing effect and hence extend the conditions under which host–parasitoid systems occur.

Examples and extensions

There are a wide variety of extensions that could be applied to this model to increase its realism and also complexity. The first would be to have variation in the amount of coupling. For example, the coupling between two patches could be dependent on the patches involved so that some habitats are closely linked whereas other hardly ever interact; this leads in turn to coupling only to local neighbours which is the basis of the next section (coupled map lattices). Another modification to the model would be to give the parasitoids some degree of selection, so that they tend to move into patches that have an abundance of hosts (van Baalen & Sabelis 1993). So far only identical patches have been considered, but there is no reason why the parameters a and λ cannot vary between patches. Including all these complications would lead to a very complex model indeed, which would require a large number of parameters to be measured from the system being modelled, but would be capable of producing a vast array of behaviours.

Much theoretical work has been done on the behaviour of patch or metapopulation models (Levin *et al.* 1993) which can be very complex even when dealing with just two patches (Hastings 1993; Lloyd 1995). One scenario ideally suited to be modelled by homogeneous patches, together with coupling between them, is the transmission of disease in a developed country. Each city and large town would act as a uniform patch and the coupling would be a measure of the amount of traffic between any two sites. This technique has been successfully applied to the modelling of measles (Grenfell *et al.* 1995a,b; Keeling 1997b) where weak coupling (0.1%) between 10 identical towns is shown to increase the persistence of the disease. Unfortunately it is difficult to estimate the coupling parameters from the global data available. The lower extinction rate is due to rescue effects from populations which have become decorrelated. This phenomenon has been thought to play an important role in the survival of many ecological systems (Hilborn 1975; Fahrig & Merriam 1985; Nee & May 1992; Ruxton 1994; Hess 1996).

Coupled map lattices

Coupled map lattices (CML) are a more advanced form of patch model. Instead of abstract patches with arbitrary coupling between them, space is represented by a lattice of homogeneous cells, and only interactions between neighbouring cells are allowed. Again these interactions are most commonly taken as simple couplings, but more complicated directed movement is also possible. A simple conceptual example is to imagine a forest plantation; each tree provides a habitat for a community of organisms with their own dynamics and only limited movement between trees is possible. Thus a tree corresponds to a single site on the CML, with internal population dynamics and local coupling (see Fig. 3.3).

CML have been extensively used in the physics literature and have proved popular with ecologists due to their relatively simple formulation and the vast range of complex behaviour and spatial patterns that can arise. Much research has been done on basic ecological equations with small amounts of coupling to neighbouring sites, for example the logistic map (Kaneko 1985, 1990), the Lotka–Volterra model (Solé & Valls 1991), the Nicholson–Bailey map (Hassell *et al.* 1991a) and other simple single-species models (Bascompte & Solé 1994). These models demonstrate interesting spatial patterns and complex dynamics with very long transient behaviour (often over millions of generations). As well as spatial extensions of long-standing problems, CML have also been used in novel situations including plant growth (Hendry *et al.* 1996), evolution (Cocho & Martinez-Mekler 1991) and gene interactions (Bignone 1993).

The lattice is usually a two-dimensional square grid, which is the simplest form to implement and means that each cell can be labelled by a pair of co-ordinates (x,y). This square grid can occasionally impose an orientation and structure onto the results but if the coupling is small this is rarely a problem. Hexagonal grids and irregular spacing have been used to attempt to more

Fig. 3.3 Representation of the neighbourhood structure of a coupled map lattice. Each tree in a plantation is a habitat for beetles, which have their own population dynamics with local coupling (Kaneko 1990; Bascompte & Solé 1994).

accurately capture true space, but these approaches are highly computationally intensive. The neighbourhood structure used is most commonly either the four-cell (North, South, East and West) von Neumann type or the eight-cell (including the diagonals) Moore type, although larger and more complex neighbourhoods are possible.

When the Nicholson–Bailey equations are spatially extended using the CML framework, then using the same notation as was used in the patch model the equations become:

$$\begin{aligned} H_{t+1}(x,y) &= (1-\mu)h_{t+1}(x,y) + \frac{\mu}{N}\sum_{(a,b)\in Nbd(x,y)} h_{t+1}(a,b) \\ P_{t+1}(x,y) &= (1-\mu)p_{t+1}(x,y) + \frac{\mu}{N}\sum_{(a,b)\in Nbd(x,y)} p_{t+1}(a,b) \end{aligned} \tag{3.5}$$

where N is the size of the neighbourhood, $Nbd(x,y)$ is the set of cells which are neighbours of (x,y) and, as in the patch model, lower case denotes the action of the standard homogeneous equations.

As a square lattice is being considered rather than abstract patches, two new problems arise: firstly how big should the grid be, and secondly how should the edges be treated. Both these problems arise for almost all spatial models; the question of grid size will be dealt with later, but for now it will be assumed that a sufficiently large grid has been used so that it does not affect the dynamics. The boundaries are commonly taken as periodic (or toroidal) which means that the top of the grid meets the bottom and the left-hand side meets the right. An alternative to this is zero boundary conditions, which corresponds to an absence of species (e.g. because of a hostile environment) outside the lattice. There is usually very little qualitative difference between the results obtained with these two approaches.

The spatial host–parasitoid model

A CML version of the Nicholson–Bailey equations was first studied by Hassell *et al.* (1991a) and since then various modifications have been addressed (Boerlijst *et al.* 1993; Hassell *et al.* 1994; Rohani *et al.* 1994; Rohani & Miramontes 1995; Ruxton & Rohani 1996). However the main results can be summarized as follows: the CML generally produces bounded global dynamics. Whereas in the patch model the heterogeneities were randomly distributed, with the inclusion of a more realistic space, the CML can demonstrate a variety of spatial patterns. For small couplings ($\mu < 0.18$), the homogeneous dynamics of each cell are less controlled by the action of their neighbours and so the system exhibits chaos. This means that long-term predictions cannot be made about the state of the system, and at any one time the arrangement on the spatial lattice appears random. As the coupling increases, the lattice becomes more ordered, as there is a higher correlation between adjacent sites, until spiral waves of hosts form, which are being chased by waves of parasites (Fig. 3.4). In the original paper (Hassell *et al.* 1991a) the situation studied was more complex, as the coupling was assumed to be different for hosts and parasitoids, and this produces even more types of spatial behaviour. When the coupling for the parasitoids is much greater

Fig. 3.4 A snapshot of the spatial pattern of parasitoids in the coupled map lattice version of the Nicholson–Bailey model, Equations 3.5. This is from a 100×100 lattice with coupling $\mu = 0.5$.

than the coupling for the hosts, a stationary 'crystal lattice' arrangement forms; this may be connected to the Turing patterns which are found when host and parasitoid couplings are widely differing.

Figure 3.4 shows the spiral patterns of parasitoids which are formed at $\mu = 0.5$; darker regions represent a higher density of parasitoids. The host population demonstrates very similar patterns. In keeping with the work of Hassell *et al.* (1991a) an eight-cell Moore neighbourhood has been used, although this has little effect on the dynamics. Although the production of spiral waves in an ecological model is very interesting, it is questionable how realistic these patterns are and whether such phenomena will ever be observed in the natural world where long-range effects and stochastic fluctuations in the finite population would destroy such structures (Rohani *et al.* 1997).

The move from abstract patches to a spatial lattice has brought a far richer range of behaviour; this however has been achieved at the expense of a great increase in computation. A typical patch model may possess five to ten patches; a typical CML may have in excess of 2500 (50×50) homogeneous cells. As with the patch models there are problems with the assumption of full mixing (homogeneity) within the cell and yet only partial mixing with adjacent cells; this is especially acute with very low coupling. Another difficulty associated with this is the fact that interactions are generally limited to the surrounding four or eight cells; this means that initially correlations can only form on the most local of scales. The effect of increasing the range over which cells can interact is discussed in more detail in the final section.

The final problem which is inherent in all models with continuous populations is the ability of the system to recover from very low population levels (often termed *nanopopulations*) and the lack of extinction at a local level. In both a patch model and a CML it is often useful to consider integer populations and stochastic behaviour (Grenfell *et al.* 1995b; see also Chapter 2). At any one time-step, all the fundamental processes such as birth, death and interactions have a probability of occurring. At any one time there is always a chance that a species will become locally extinct; this chance will be higher for smaller populations. The phenomenon of local extinction forces rescue effects from neighbouring populations to play a vital role in stabilizing the global dynamics.

Partial differential equations

Partial differential equations and reaction-diffusion equations in particular are mathematically well understood (e.g. Okubo 1980; Hagan 1982; Murray 1982; Vickers *et al.* 1993; Holmes *et al.* 1994; Hutson 1999). Whereas ordinary differential equations (ODEs) give the temporal changes in terms of the global population, partial differential equations (PDEs) give the temporal changes at each point in space in terms of the local densities and the spatial gradients. PDEs and the simpler reaction-diffusion equations have in general attracted large amounts of attention due to their mathematical tractability rather than their ecological relevance. However, the existence of strong theoretical results in this area can help to direct research in other fields of spatial modelling.

Reaction-diffusion equations are a subset of PDEs and are formed of two distinct parts: the reaction terms come from the homogeneous equations and space is incorporated in terms of diffusion. This method treats space as a continuum, although if any numerical simulations are required then of course the spatial elements are treated discretely. Given a set of differential equations, such as Equation 3.2, and the assumption that the species in the system move randomly (i.e. they diffuse through space) then the new equations become:

$$\begin{aligned} \dot{H} &= \frac{\partial H}{\partial t} = H - \frac{H}{1+H}P + \mu_H \nabla^2 H \\ \dot{P} &= \frac{\partial P}{\partial t} = b\frac{H}{1+H}P - P^2 + \mu_P \nabla^2 P \end{aligned} \qquad (3.6)$$

where H and P are functions of both time and space and ∇^2 is mathematical shorthand for the spatial derivative

$$\frac{\partial^2}{\partial x^2} + \frac{\partial^2}{\partial y^2}$$

Theoretical criteria

For a given set of equations, it can be determined exactly whether or not Turing structures (spatial instabilities) will be found in such a system. For a two-species system, the three following conditions need to be satisfied at the homogeneous fixed point for this type of spatial pattern to be observed:

$$\frac{\partial \dot{H}}{\partial H}\frac{\partial \dot{P}}{\partial P} < 0 \qquad \frac{\partial \dot{H}}{\partial H} + \frac{\partial \dot{P}}{\partial P} < 0$$

$$\frac{\partial \dot{H}}{\partial H}\frac{\partial \dot{P}}{\partial P} + \frac{\partial \dot{H}}{\partial P}\frac{\partial \dot{P}}{\partial H} > 0$$

All these conditions are met by Equation 3.6; all that remains is for there to be a sufficient difference between the diffusion parameters μ_H and μ_P. For the given system it is also necessary that μ_P is much bigger than μ_H; the phenomenon of predators moving through faster than the prey appears to be common in biological systems, so Turing structures may well be seen. Figure 3.5 shows the density of prey in the system. In contrast to the patch model

Fig. 3.5 A solution to the partial differential Equation 3.6; the dark areas show where there is a large density of prey. The area shown is 5×5.

and CML, where spatial heterogeneity acts to stabilize the local and global dynamics, this host–parasite pattern emerges due to spatial instabilities about a fixed point which would be stable in a homogeneous system.

Biological uses

These reaction-diffusion equations have been primarily used in developmental biology to model the interactions of morphogens. The patterns formed can then be used to explain many features, from the positioning of limbs to the markings on an animal's skin (Murray 1981, 1988; Tsonis *et al.* 1989; Savic 1995; Hofer & Maini 1996). Modelling this type of chemical reaction has several advantages over modelling ecosystems: firstly there is usually a vast number of chemicals, so the assumption of continuous variables is valid, and secondly chemicals move in a random manner so that they can be modelled by diffusion. Despite the problems that are encountered, PDEs have been successfully used in a variety of ecological applications including plankton dynamics (Malchow 1993), the spread of insect pathogens (Dwyer 1992), predator–prey dynamics (Mimura & Murray 1978; Dunbar 1983), hypercycle models (Boerlijst & Hogeweg 1995) and territory formation (White *et al.* 1996).

Interacting particle systems

By far the most recent and least understood type of spatial model is an interacting particle system (IPS) (Hogeweg 1988; DeAngelis & Gross 1992;

DeAngelis & Rose 1992; Ermentrout & Edelstein-Keshet 1992; Bullock 1994; Judson 1994; Durrett & Levin 1994a). Interacting particle systems (IPS) will be used as a general term for any individual-based model on a lattice, including probabilistic cellular automaton and artificial ecologies. (In much of the statistical literature, the term IPS is reserved exclusively for probabilistic cellular automaton.) An IPS uses a discrete lattice of cells (like the CML) but in this case each cell can only be in one of a finite set of states and the behaviour of each cell is specified by a set of probabilistic rules. It is assumed throughout this section that the systems are updated synchronously, that is all the cells are updated at once; the alternative, asynchronous updating is where cells are updated one at a time, chosen randomly weighted by the rates of change. This type of updating is more natural but far slower, and in most ecological applications there is usually little difference in the final results. In general, asynchronous updating can be thought of as being comparable with differential, continuous time systems and synchronous updating is the analogue of difference, discrete time systems. Whether a model is classified as a cellular automaton or artificial ecology depends on the exact nature of the updating rules; these two classes are dealt with separately.

Probabilistic cellular automata

A probabilistic cellular automaton is defined by the set of possible states for a cell and the probabilistic updating rules; these give the probability of a cell being in a particular state at the next iteration and depends on the state of the cell and its surrounding neighbourhood in this iteration. Probabilistic cellular automata are of most use when considering slow-moving or sessile organisms (Karlson & Jackson 1981; Karlson & Buss 1984) or in spatial situations where stochasticity and integer populations are essential (Durrett & Levin 1994a; Bascompte & Solé 1996). This type of spatial model can be best illustrated by a simple example. Keeping with the theme of hosts and parasites, each cell can be in only one of three states: vacant; occupied by a healthy host; or occupied by a parasitized host. The probabilistic rules will be as follows.

1 A healthy host can spread into a vacant site in its local neighbourhood with probability g.

2 A parasite can be transmitted from a parasitized host to a healthy host in its neighbourhood with probability T.

3 A parasitized host dies with probability V.

The neighbourhood is taken to be of the von Neumann four-cell type. Further details of this system can be found in Rand *et al.* (1995). In brief, the rules correspond to the direct transmission of a virulent disease through a slow growing sessile population. This formulation is well suited for modelling the spread of a disease through plants where slow vegetative spread and local parasite or pathogen dispersal is likely (Maddison *et al.* 1996). It can also be applied to human and other populations where the net movement of each individual is small in comparison to the speed the infection travels (Smith & Harris 1991; Rhodes & Anderson 1996).

This model is related to the Nicholson–Bailey host–parasitoid model with limited interaction; each organism can only experience a small fraction (four cells) of the global environment. There is also density-dependent growth of hosts as there can only be at most one host per cell. The local rules therefore correspond to the following Nicholson–Bailey type equations:

$$\begin{aligned} H_{t+1} &= (1 - H_t - P_t)\mathrm{e}^{-4gH_t} + H_t\mathrm{e}^{-4TP_t} \\ P_{t+1} &= H_t(1 - \mathrm{e}^{-4TP_t}) + (1 - V)P_t \end{aligned} \tag{3.7}$$

Even with such a simple set of rules this type of model has several advantages over more conventional models. Firstly it deals with discrete populations (Durrett & Levin 1994b), so that the behaviour of individuals counts; for the first time the model truly deals with the number of individuals at a site and not just the expected densities. The probabilistic nature of the model and the integer-based populations can cause dramatic fluctuations on a local scale; these random demographic effects are frequently seen in the natural world and can profoundly affect the global dynamics (Keeling & Rand 1995, 1999). The fact that each site can only experience a few neighbouring sites means that often the local environment is forced to lie far from the global mean. For example, with the four-cell von Neumann neighbourhood, the proportions in the local environment must be multiples of a quarter.

An example of the spatial distribution of the model is shown in Fig. 3.6. Two phenomena can be noticed: the existence of spatial correlations and temporary refuges. The correlations are clearly visible; the parasites appear to spread out in wavefronts through the hosts, leaving only vacant sites behind, so parasites tend to be aggregated and there is a negative correlation between parasites and hosts. The refuges are any large areas of healthy hosts which are far from any parasites; a refuge is longer lasting if the hosts are separated from the parasites by a barrier of vacant cells.

Figure 3.7 shows the typical dynamics of the system. It is clear that the values fluctuate about some long-term average; however there is a large amount of variance in the system. This high degree of stochasticity in the data is due to the probabilistic rules and the fact that there are sudden bursts of infection whenever parasites locate a large healthy host population. The presence of strong spatial correlations and the associated large patches of healthy hosts means that the number of parasites in the cellular automaton model is far lower than predicted by the homogeneous mean-field equations. Obviously, if this lower parasite level is a consistent feature of spatial

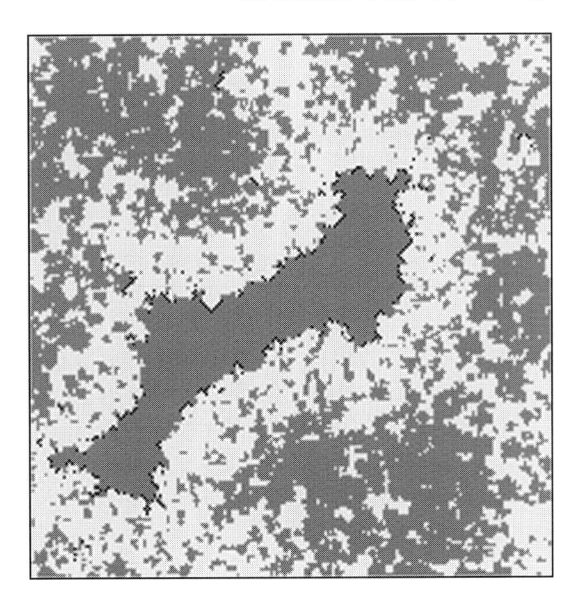

Fig. 3.6 A snapshot of the spatial distribution of vacant sites (pale grey), hosts (grey) and parasites (black) from the cellular automaton model. The results are from a 200×200 lattice with $g = 0.05$, $T = 0.6$ and $V = 1$.

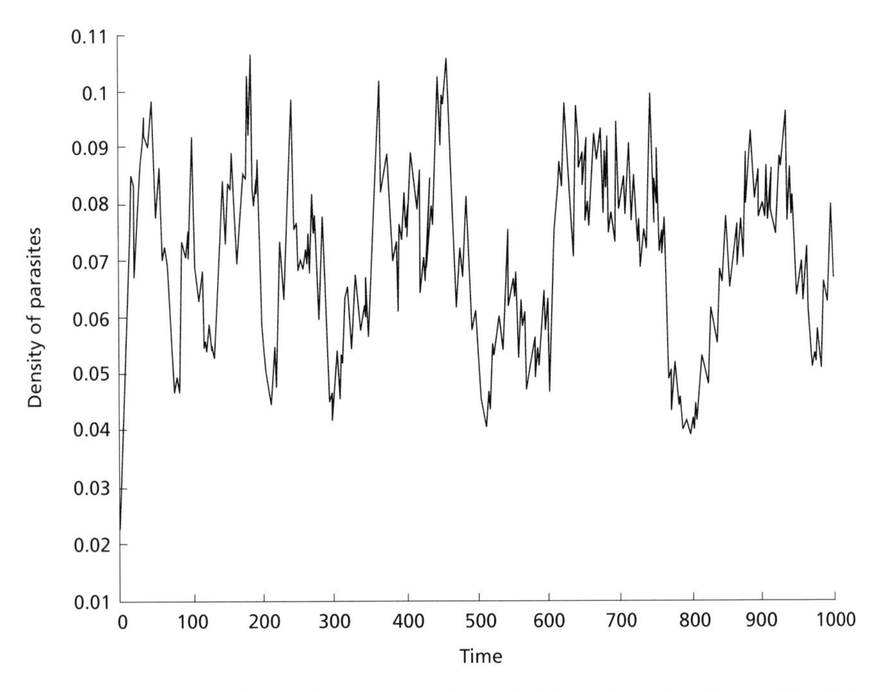

Fig. 3.7 An example of the density of parasites in a typical simulation, taken from a 100×100 lattice with $g = 0.05$, $T = 0.5$ and $V = 1$.

systems then it could have very wide-ranging epidemiological consequences (Mollison 1997).

While using probabilistic rules adds to the realism of models, it introduces many new problems. The most notable is the need for multiple simulations (McGlade & Price 1993; Rand *et al.* 1995); this is especially apparent when invasions are considered. In standard deterministic models an invasion is either always successful or always fails; in stochastic models this is no longer the case and even a species which should be very successful may fail due to demographic fluctuations, causing extinction for the initially small population (see Chapter 2). This is observed all the time in the natural world where even the most carefully planned introductions still fail (Pimm 1991) and invasion is ruled by chance events (Mollison 1986; Kornberg & Williamson 1987). It would appear that as the models include greater realism, and hence become more complex, then more of the problems that are found with obtaining results from real systems will occur.

Artificial ecologies

These are similar to cellular automata, in that they are defined by a set of probabilistic rules acting on a lattice of discrete states, but with this type of model the probability of a cell being in a particular state in the next iteration depends on the states of the neighbourhood in the next iteration as well as on the current state of the cell and its neighbourhood. This means that whereas cellular automata can only model the spread of organisms, artificial ecologies allow the organisms to *move*. This movement of organisms between sites (as opposed to random transmission and death) means that the quantities of each species are far better conserved, and in general stochastic fluctuations are lower and the global behaviour richer. Thus, these types of system are the ideal tool for studying spatial populations and ecological systems if the behaviour of individuals is understood in detail (McGlade 1993). The host–parasitoid system can be formulated as an artificial ecology (compare to Wilson *et al.* 1993 for Lotka–Volterra systems; and McGlade 1995; Keeling *et al.* 1997(a) for a four-species system). Each host lays two eggs each year within a short range (10 cells) of where it was born and both hosts and parasitoids can only survive for one year. The parasites move randomly to a neighbouring cell at each step, searching for hosts with a limited efficiency. If the parasitoid is in the same cell as a host, it can locate and lay a single egg with a given probability. Finally, the parasitized host is killed when the parasitoid larvae emerge, but only a fraction of these larvae survive to adulthood. Forty movements of the parasitoid constitute a year, after which the cycle begins again. The numbers of hosts and parasitoids are recorded at the start of each year.

The values used in the artificial ecology would give rise to the following set of Nicholson–Bailey type equations:

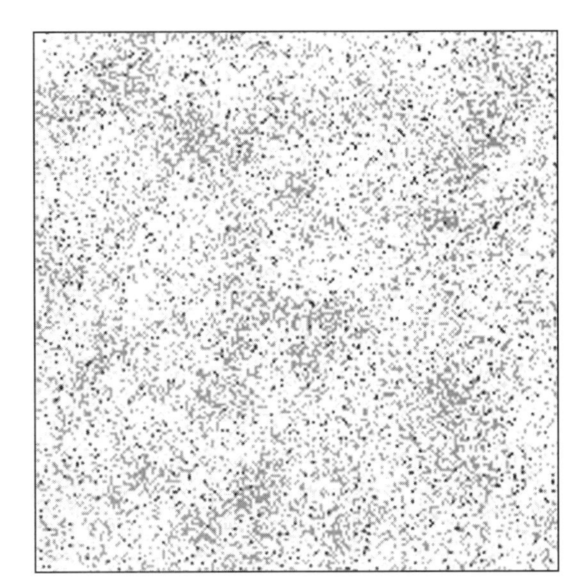

Fig. 3.8 Example of the artificial ecology of 300 × 300 cells. Parasitoids are shown in black, hosts in grey, parasitized hosts in pale grey.

$$\begin{aligned} H_{t+1} &= 2H_t \mathrm{e}^{-20P_t} \\ P_{t+1} &= 0.2H_t(1-\mathrm{e}^{-20P_t}) \end{aligned} \tag{3.8}$$

which are identical to Equation 3.1 under a change of variables.

Figure 3.8 shows an example of the spatial patterns observed and Fig. 3.9 shows the dynamics of this system. Large oscillations are observed in which the numbers of parasitoids and hosts slowly cycle over a period of approximately 14 years. Due to the movement of individuals, there is less stochastic fluctuation than observed in the host–parasite cellular automaton, and this is reflected in the smoother behaviour that is produced. It should be noted that once again the spatial model produces a lower density of parasites (centred about 0.021) compared with the theoretical prediction of 0.0347.

Further research and open questions

This final section will be devoted to topics that are either fields of ongoing research or are open questions with no definite answers. The list of topics is by no means exhaustive, but is intended to give the reader a flavour of current research and indicate the directions in which it might move.

Caricature models or accurate simulations

The models used in the CML, reaction-diffusion systems and cellular automata all come under the classification of caricature or generic models. They attempt to capture the essential features of the system without concen-

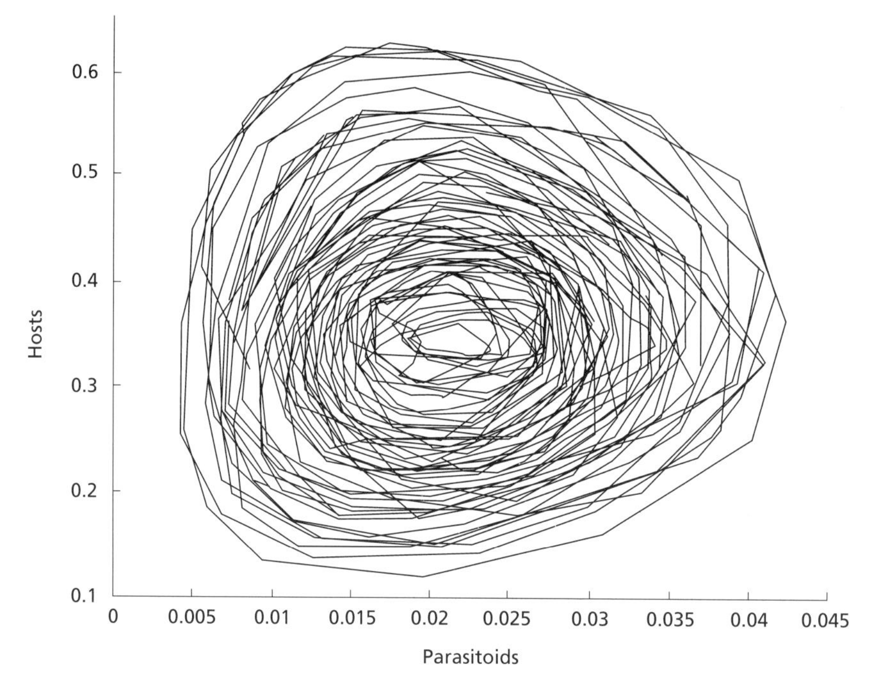

Fig. 3.9 This graph shows the dynamics of the host and parasitoid populations. Although affected by stochastic fluctuations the populations are clearly cycling.

trating on specific details. The artificial ecology is closer to a simulation, although more biological realism could be added. Whether a caricature model or an accurate simulation is used depends very much on the system being studied and the detail to which the interactions are understood (Mangel 1985; Walters 1986; Breckling 1992; Durrett & Levin 1994a).

When large, intensive computer simulations are considered, the prime example is weather prediction. Here an attempt is made to create as accurate a simulation as possible, taking every detail, feature and reaction into account. This is plausible because very accurate meteorological data exist, the individual physical processes are fairly simple and well understood, stochastic effects rarely produce drastic changes in the weather, and most important highly accurate forecasts are usually required. For many ecological systems none of these hold, and apart from the lack of understanding and detailed parameters there is rarely a need for accurate predictions. Most often modelling is used to increase the understanding of a system, so that experiments can be performed with the model that would be difficult, time-consuming or potentially devastating using the real system. The general statements that come from a caricature model give more widely applicable results than predictions from a simulation, where the number of complex interactions often makes an intuitive understanding impossible.

Obviously exceptions do exist to this general rule. Forest-based simulators such as FORET and SORTIE (Levin *et al.* 1997) attempt to describe accurately the composition of forests. The computer models the nutrients in the soil, the amount of light available including the effects of shading by other trees and the dispersal of seeds; together with detailed behavioural rules of individual trees this allows accurate prediction of the densities of species. In conversation ecology when the number of individuals is often known in detail and a precise answer is required, then explicit simulations may again be necessary (Dublin *et al.* 1990; Swart & Lawes 1996; Wu *et al.* 1996; Clayton *et al.* 1997).

Length scales

As was discussed previously when dealing with lattice-based models, there is the question of how small to make each cell so that homogeneous dynamics still hold. At the other end of the spectrum is the question of how large to make the lattice or what size of lattice should the data be sampled from. When trying to simulate an actual ecosystem then obviously the lattice should be taken to be of a comparable size to the real environment. If, however, a caricature model is being used then the decision is not so simple.

If too small a lattice is used then the true dynamics of the system are not observed, often with dramatic results. For example in the host–parasite cellular automata if a grid size of less than 70×70 is used then the parasites soon die out; rapid extinctions on small grids is a very common scenario for spatial models and is observed for both the CML and the artificial ecology previously described. In stochastic models a small grid implies a relatively large amount of random noise which will make data analysis far more difficult. On the other hand if the lattice is too large then not only are needless calculations being performed, but also any interesting (nonstationary) dynamics may be averaged out; therefore it is important to view the system at the correct scale. Many authors have already looked at this problem (Wiens 1989; Levin 1992; Rand & Wilson 1995), but here the simple framework of a method which relies on the scaling of the standard deviation is given (further details can be found in Keeling *et al.* 1997a and Wilson & Keeling 1999).

Let us assume sampling of the ecosystem occurs in a window of size $L \times L$, and for every window size it is possible to calculate the standard deviation in the density of one species σ_L. Hopefully, as the sampling size becomes large, the standard deviation should tend to zero. In fact for almost all realistic models the standard deviation should be proportional to $1/L$. To find the length at which this scaling begins, plot X_L against L (where $X_L = L\sigma_L$) and look for the point where the curve asymptotes to a constant. Figure 3.10 shows the curve X_L for the host-parasitoid artificial ecology; this graph

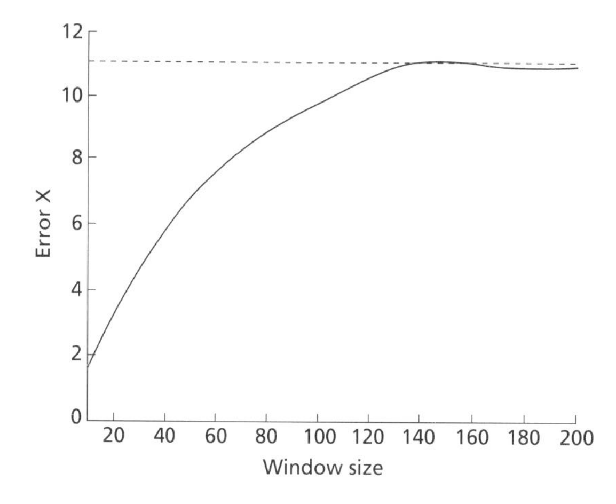

Fig. 3.10 The graph of the error X_L against the lattice size L clearly asymptotes around 140, demonstrating the scaling behaviour that was predicted above this size.

clearly demonstrates a length scale of around 140. Further analysis of the results from various nonstationary spatial models has shown that at this length-scale, the ratio of the dynamics to the amount of noise is minimal and hence this is the optimal sampling size.

Changing the range of interactions

In all of the above true spatial models, the range of interaction was limited to either the neighbouring four or eight sites. It is sometimes necessary to involve species which interact over far larger ranges, and the problems this causes and the results that can be expected are dealt with now. Three main features arise from using larger neighbourhoods: the correlations formed are weaker as the centre site spends less time interacting with each neighbour; in stochastic models the fluctuations experienced by each site are less as they are affected by an average over more sites; and finally the numerical computation is far slower. With larger neighbourhoods it is usual to decrease many of the parameters accordingly, for example in the simple host–parasite cellular automaton if the neighbourhood was increased from the four-cell von Neumann to the eight-cell Moore model, then the growth rate and transmission rate should be halved. An alternative to large neighbourhoods is to consider the same number of neighbours, but allow these to be chosen randomly from within a given range. This larger-range behaviour generally just reduces the spatial correlations.

Normally with extended neighbourhoods the centre site is assumed to interact equally with all its neighbours. However, a far more realistic approach would be to ensure that the level of interaction decreases with distance. For example if the amount of interaction at a given distance was normally distributed then this would correspond to a random movement of individuals. If the principle of interaction decreasing with distance is applied to the standard neighbourhoods, then it is clear that the centre site of the eight-cell Moore neighbourhood should be affected less by the corner sites than the edge sites; however this has been rarely included in any models.

The decrease in local correlations and stochastic effects means that the system starts to behave more and more like the homogeneous equations (Keeling & Rand 1999). This is reassuring to know, as it shows that the standard models used throughout ecology are not totally redundant in a spatial context. With long-range interactions, the simple host–parasite cellular automaton discussed above can be approximated by the Nicholson–Bailey equations with logistic growth of hosts and, with a range of three cells or more, the fit is fairly accurate (Keeling & Rand 1999). This supports the assumption that most systems with fairly long-range interactions can be well approximated by homogeneous equations.

Data analysis and data creation

One problem that is very evident with spatial modelling is how the data should be sorted and analysed. There is a vast amount of reported research, both technical and nontechnical, on a variety of methods which can be used on time series (Diggle 1983; Sugihara *et al.* 1990; Tong 1990; Yao & Tong 1994; Ellner & Turchin 1995). This subsection is not designed to introduce the field of time-series analysis, but rather illustrate where spatial models fit into this subject. One element of this work however must be introduced, namely that of *embedding* (Broomhead & King 1986). Imagine a host–parasite model where the state of the system at one time is perfectly predictable from the state of the system at a previous time; all nonstochastic models fall into this category. Embedding theory states that although ideally the numbers of hosts and parasites at one time are required to predict the future of the system, prediction is possible only if the numbers of one species are known, but that these data are available at several different times.

This idea can be extended for spatial systems; suppose that only the global densities of hosts and parasites were required to predict the global densities in the future, then it is quite likely that predictions could be made using only the global density of hosts and some aspects of the hosts' spatial pattern. The advantages of this idea are clear when it is realized that there is often a vast difference in the ease of measuring the numbers of the species in the ecosystem. Embedding theory allows prediction from information about the species which is the most readily measured. This becomes a huge bonus if the species can be measured from satellite photographs, in which case vast amounts of data become readily available, and detection and prediction of species numbers can become automated.

The output from spatial models is an ideal test-bed for any new ideas or analysis techniques which are designed for satellite or other spatial data.

Spatial models provide an 'unlimited' supply of data which is free from measurement error and easily manipulated. Also the study of data from spatial models can lead to a better understanding of the way measurements are made in the environment and how heterogeneities are dealt with.

Simple alternatives to spatial models

One of the most exciting areas of research in spatial models is the attempt to formulate simple equations which take space into account without explicitly modelling it. For many years there have been simple modifications which can be used if a species is normally distributed rather than uniform over space. However it would be useful to be able to model systems with arbitrary spatial correlations and even to be able to predict these correlations from the model itself. Various forms of moment–closure models have been proposed to approximate spatial heterogeneities in terms of a few key variables.

One of the most promising techniques is that of pairwise modelling (Matsuda *et al.* 1987; Sato *et al.* 1994; Altmann 1995; Keeling *et al.* 1997b; see also Chapter 4). This technique is most easily described for interacting particle systems although continuous space analogues exist (Bolker *et al.* 1999; Duarte *et al.* 1998). A homogeneous approach would simply calculate the densities, that is the proportion of sites occupied by each species, whereas a pairwise approach models the densities for adjacent pairs of sites (Fig. 3.11). For example in the homogeneous formulation the proportions of host, parasites and vacant cells would be modelled, but in the pairwise approach nine different variables are used, such as the proportion of host–host pairs, host–parasite pairs, etc. In this way local correlations can easily arise and be incorporated into the dynamics.

This type of pairwise model usually gives a far better prediction of a spatial system than the associated homogeneous counterparts, and has several advantages over true spatial models. The first is speed and simplicity. In the host–parasite example given above, spatial correlations can be captured by just nine variables rather than thousands of sites. Also these equations are written as either differential or difference equations and hence a vast number of advanced mathematical techniques can be used. Their deterministic nature is often another big advantage as multiple simulations are no longer necessary, although as discussed previously stochastic fluctuations can play an important role in many ecosystems. Probably the greatest advantage possessed by pairwise models is the ease with which they can be generalized to include details such as more complex or larger neighbourhoods.

Obviously a greater degree of accuracy could be gained by simulating the proportions of each neighbourhood type rather than each pair. In going

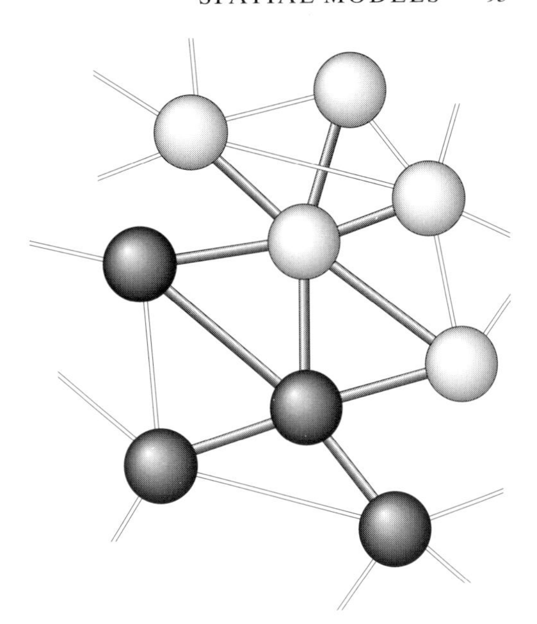

Fig. 3.11 A representation of a pairwise approach to modelling, showing a centre black–white pair together with its local neighbourhood. In this example there is aggregation of black sites and white sites.

from modelling a pair of adjacent sites to a four-cell neighbourhood, the number of variables in the host–parasite system rises from 9 to 243 (ignoring symmetries), and it is questionable as to whether this dramatic explosion in the number of variables can be justified by the slight increase in accuracy.

Yet again whether to use a true spatial model or a pairwise simulation depends very much on the application concerned. A pairwise model often leads to a greater understanding of the role played by local spatial correlations, whereas an interacting particle system may provide greater realism. However, it is clear that pairwise models and all of the models outlined above have an important role to play in increasing our awareness of the spatial environment and the effect it can have on the dynamics of interacting populations.

References

Altmann, M. (1995) Susceptible-infectious-recovered epidemic models with dynamic partnerships. *Journal of Mathematical Biology*, **33**, 661–675.

Bascompte, J. & Solé, R.V. (1994) Spatially-induced bifurcations in single-species population dynamics. *Journal of Animal Ecology*, **63**, 256–264.

Bascompte, J. & Solé, R.V. (1996) Habitat fragmentation and extinction thresholds in spatially explicit models. *Journal of Animal Ecology*, **65**, 465–473.

Begon, M., Harper, J.L. & Townsend, C.R. (1996) *Ecology. Individuals, Populations and Communities, 3rd*. Blackwell Science, Oxford.

Bignone, F.A. (1993) Cells–gene interactions—simulations on a coupled map lattice. *Journal of Theoretical Biology*, **161**, 231–249.

Blackmann, G.E. (1935) A study by statistical methods of the distribution of species in grassland associations. *Annals of Botany*, **49**, 749–777.

Boerlijst, M.C. & Hogeweg, P. (1995) Attractors and spatial patterns in hypercycles with negative interactions. *Journal of Theoretical Biology*, **176**, 199–210.

Boerlijst, M.C., Lamers, M.E. & Hogeweg, P. (1993) Evolutionary consequences of spiral waves in a host–parasitoid system. *Proceedings of the Royal Society of London*, **B253**, 15–18.

Bolker, B.M. & Grenfell, B.T. (1993) Chaos and biological complexity in measles dynamics. *Proceedings of the Royal Society of London*, **B251**, pp. 75–81.

Bolker, B. & Pacala, S.W. (1997) Using moment equations to understand stochastically driven spatial pattern formation in ecological systems. *Theoretical Population Biology*, **52**, 179–197.

Bolker, B.M., Pacala, S.W. & Levin, S.A. (1999) Moment methods for stochastic processes in continuous space and time. In: *The Geometry of Ecological Interactions: Simplifying Spatial Complexity* (eds U. Dieckmann, R. Law & J.A.J. Metz). Cambridge University Press, Cambridge.

Breckling, B. (1992) Uniqueness of ecosystems versus generalizability and predictability in ecology. *Ecological Modelling*, **63**, 13–27.

Broomhead, D.S. & King, G.P. (1986) Extracting qualitative dynamics from experimental data. *Physica*, **D20**, 217–236.

Bullock, J. (1994) Individual-based models. *Trends in Evolution and Ecology*, **9**, 299.

Castillo-Chavez, C., Hethcote, H.W., Andreasen, V., Levin, S.A. & Liu, W.M. (1989) Epidemiological models with age structure, proportionate mixing and cross immunity. *Journal of Mathematical Biology*, **27**, 233–258.

Clayton, L., Keeling, M.J. & Milner-Gulland, E.J. (1997) Bringing home the bacon: a spatial model of wild pig hunting in Sulawesi, Indonesia. *Ecological Applications*, **7**, 642–652.

Cocho, G. & Martinez-Mekler, G. (1991) On a coupled map lattice formulation of the evolution of genetic sequences. *Physica*, **D51**, 119–130.

Cohen, J. & Stewart, I. (1993) Let T equal tiger. *New Scientist*, **140**, 40–44.

Czaran, T. & Bartha, S. (1992) Spatio temporal dynamic models of plant population and communities. *Trends in Evolution and Ecology*, **7**, 38–42.

DeAngelis, D.L. & Gross, L.J. (eds) (1992) *Individual-based Models and Approaches in Ecology. Populations, Communities and Ecosystems*. Chapman & Hall, New York.

DeAngelis, D.L. & Rose, K.A. (1992) Which individual-based approach is most appropriate for a given problem? In: *Individual-based Models and Approaches in Ecology. Populations, Communities and Ecosystems* (eds D.L. DeAngelis & L.J. Gross). Chapman and Hall, New York.

DeRoos, A.M., McCauley, E. & Wilson, W.G. (1991) Mobility versus density-limited predator–prey dynamics on different spatial scales. *Proceedings of the Royal Society of London*, **B246**, 117–122.

Diekmann, U., Law, R. & Metz, J.A.J. (eds) (1999) *The Geometry of Ecological Interactions: Simplifying Spatial Complexity* Cambridge University Press.

Diggle, P. (1983) *Statistical Analysis of Spatial Point Patterns*. Academic Press, London.

Duarte, L.C., Boldrini, J.L. & dos Reis, S.F. (1998) Scaling phenomena and ecological interactions in space: cutting to the core. *Trends in Evolution and Ecology* , **13**, 176–17.

Dublin, H., Sinclair, A. & McGlade, J.M. (1992) Elephants and fire as the causes of multiple stable states in the Serengeti Mara woodlands. *Journal of Animal Ecology*, **59**, 1147–1164.

Dunbar, S.R. (1983) Travelling wave solutions of diffusive Lotka–Volterra equations. *Journal of Mathematical Biology*, **17**, 11–32.

Durrett, R. (1988) Crabgrass, measles and gypsy moths: an introduction to interacting particle systems. *The Mathematical Intelligencer*, **10**, 37–47.

Durrett, R. (1993) Stochastic models of growth and competition. In: *Patch Dynamics*. Springer Lecture Notes in Biomathematics, Vol. 96. Springer, Berlin.

Durrett, R. & Levin, S.A. (1994a) Stochastic spatial models: a user's guide to ecological applications. *Philosophical Transactions of the Royal Society of London*, **343**, 329–350.

Durrett, R. & Levin, S.A. (1994b) The importance of being discrete and spatial. *Theoretical Population Biology*, **46**, 363–394.

Dwyer, G. (1992) On the spatial spread of insect pathogens—theory and experiment. *Ecology*, **73**, 479–494.

Ellner, S. & Turchin, P. (1995) Chaos in a noisy world: new methods and evidence from time series analysis. *American Naturalist*, **145**, 343–375.

Ermentrout, G.B. & Edelstein-Keshet, L. (1992) Cellular automata approaches to biological modelling. *Journal of Theoretical Biology*, **160**, 97–133.

Fahrig, L. & Merriam, G. (1985) Habitat patch connectivity and population survival. *Ecology*, **66**, 1762–1768.

Gonzales-Andujar, J.L. & Perry, J.N. (1993) Chaos, metapopulations and dispersal. *Ecological Modelling*, **65**, 255–263.

Green, D.G. (1990) Cellular automata models of crown-of-thorns outbreaks. In: *Acanthaster and the Coral Reef: A Theoretical Perspective* (ed. R. Bradbury). Lectures in Biomathematics, vol 88, pp. 157–166. Springer Verlag, Berlin.

Grenfell, B. & Harwood, J. (1997) (Meta) population dynamics of infectious diseases. *Trends in Evolution and Ecology*, **12**, 395–399.

Grenfell, B.T., Kleczkowski, A., Gilligan, C.A. & Bolker, B.M. (1995a) Spatial heterogeneity, nonlinear dynamics and chaos in infectious diseases. *Statistical Methods in Medical Research*, **4**, 160–183.

Grenfell, B.T., Bolker, B.M. & Kleczkowski, A. (1995b) Seasonality and extinction in chaotic metapopulations. *Proceedings of the Royal Society of London*, **B259**, 97–103.

Gueron, S. & Levin, S.A. (1993) Self-organization of front patterns in large wildebeest herds. *Journal of Theoretical Biology*, **165**, 541–552.

Hagan, P.S. (1982) Spiral waves in reaction-diffusion equations. *SIAM Journal of Applied Mathematics*, **42**, 762–786.

Hanski, I. (1983) Coexistence of competitors in patchy environments. *Ecology*, **64**, 493–500.

Hanski, I. & Gilpin, M. (1991) Metapopulation dynamics: brief history and conceptual domain. *Biological Journal of the Linnean Society*, **42**, 3–16.

Hartvigsen, G. & Starmer, W.T. (1995) Plant–herbivore coevolution in a spatially and genetically explicit model. *Artificial Life*, **2**, 239–259.

Hassell, M.P. & May, R.M. (1974) Aggregation in predators and insect parasites and its effects on stability. *Journal of Animal Ecology*, **43**, 567–594.

Hassell, M.P. & May, R.M. (1985) From individual behaviour to population dynamics. In: *Behavioural Ecology: Ecological Consequences of Adaptive Behaviour* (eds R.M. Sibley & R.H. Smith). Blackwell Scientific Publications, Oxford.

Hassell, M.P. & May, R.M. (1988) Spatial heterogeneity and the dynamics of parasitoid–host systems. *Annales Zoologici Fennici*, **25**, 55–61.

Hassell, M.P. & Pacala, S.W. (1990) Heterogeneity and the dynamics of host parasitoid interactions. *Philosophical Transactions of the Royal Society of London*, **B330**, 203–220.

Hassell, M.P., Comins, H. & May, R.M. (1991a) Spatial structure and chaos in insect population dynamics. *Nature*, **353**, 255–258.

Hassell, M.P., May, R.M., Pacala, S.W. & Chesson, P.L. (1991b) The persistence of host–parasitoid associations in patchy environments. 1. A general criterion. *American Naturalist*, **138**, 568–583.

Hassell, M.P., Comins, H. & May, R.M. (1994) Species coexistence and self organizing spatial dynamics. *Nature*, **370**, 290–292.

Hastings, A. (1977) Spatial heterogeneity and the stability for predator–prey systems. *Theoretical Population Biology*, **12**, 37–48.

Hastings, A. (1990) Spatial heterogeneity and ecological models. *Ecology*, **71**, 426–428.

Hastings, A. (1993) Complex interactions between dispersal and dynamics: lessons from coupled logistic equations. *Ecology*, **74**, 1362–1372.

Hendry, R.J. & McGlade, J.M. (1995) The role of memory in ecological systems. *Proceedings of the Royal Society of London*, **B259**, 153–159.
Hendry, R.J., McGlade, J.M. & Weiner, J. (1996) A coupled map lattice model of the growth of plant monocultures. *Ecological Modelling*, **84**, 81–90.
Hess, G. (1996) Disease in metapopulation models—implications for conservation *Ecology*, **5**, 1617–1632.
Hilborn, R. (1975) The effect of spatial heterogeneity on the persistence of predator prey interactions. *Theoretical Population Biology*, **8**, 346–355.
Hofer, T. & Maini, P.K. (1996) Turing patterns in fish skin. *Nature*, **380**, 678.
Hofer, T., Sherratt, J.A. & Maini, P.K. (1995) *Dictyostelium discoideum*. Cellular self-organization in an excitable biological medium. *Proceedings of the Royal Society of London*, **B259**, 249–257.
Hogeweg, P. (1988) Cellular automata as a paradigm for ecological modelling. *Applied Mathematics and Computation*, **27**, 81–100.
Holmes, E.E., Lewis, M.A., Banks, J.E. & Veit, R.R. (1994) Partial differential equations in ecology: spatial interactions and population dynamics. *Ecology*, **75**, 17–29.
Huston, M., DeAngelis, D. & Post, W. (1988) New computer models unify ecological theory. *BioScience*, **38**, 682–691.
Hutson, V. (1999) Techniques for reaction-diffusion equations. In: *The Geometry of Ecological Interactions: Simplifying Spatial Complexity.* (eds U. Dieckmann, R. Law and J.A.J. Metz).
Johansen, A. (1994) Spatiotemporal self-organization in a model of disease spreading. *Physica*, **D78**, 186–193.
Johansen, A. (1996) A simple-model of recurrent epidemics *Journal of Theoretical Biology*, **178**, 45–51.
Judson, O.P. (1994) The rise of the individual-based model in ecology. *Trends in Evolution and Ecology*, **9**, 9–14.
Kaneko, K. (1985) Spatiotemporal intermittency in coupled map lattices. *Progress in Theoretical Physics*, **74**, 1033–1044.
Kaneko, K. (1990) Supertransients, spatiotemporal intermittency and stability of fully developed spatiotemporal chaos. *Physics Letters*, **A149**, 105–112.
Kareiva, P. (1994) Space the final frontier for ecological theory. *Ecology*, **75**, 1.
Karlson, R.H. & Buss, L.W. (1984) Competition, disturbance and local diversity patterns of substratum-bound clonal organisms: a simulation. *Ecological Modelling*, **23**, 243–255.
Karlson, R.H. & Jackson, J.B.C. (1981) Competitive networks and community structure: a simulation study. *Ecology*, **62**, 670–678.
Keeling, M.J. (1999) Spatial models of interacting populations In: *The Geometry of Ecological Interactions: Simplifying Spatial Complexity*. (eds U. Dieckmann, R. Law & J.A.J. Metz).
Keeling, M.J. (1997) Modelling the persistence of measles. *Trends in Microbiology*, **5**, 513–518.
Keeling, M.J. & Rand, D.A. (1995) A spatial mechanism for the evolution and maintenance of sexual reproduction. *Oikos*, **74**, 414–424.
Keeling, M.J. & Rand, D.A. (1999) Space and fluctuations in the dynamics of infection. In: *From Finite to Infinite Dimensional Dynamical Systems* (ed. P. Glendinning). Kluwer, Amsterdam.
Keeling, M.J., Mezic, I., Hendry, R.J., McGlade, J. & Rand, D.A. (1997a) Characteristic length scales of spatial models in ecology. *Philosophical Transactions of the Royal Society of London*, **B352**, 1589–1601.
Keeling, M.J., Rand, D.A. & Morris, A.J. (1997b) Correlation models for childhood diseases. *Proceedings of the Royal Society of London*, **B264**, 1149–1156.
Kornberg, H. & Williamson, M.H. (1987) Quatitative aspects of the ecology of biological invasions. *Philosophical Transactions of the Royal Society of London*, **B314**, 501–724.
Levin, S.A. (1992) The problem of pattern and scale in ecology. *Ecology*, **73**, 1943–1967.

Levin, S.A. & Segel, L.A. (1976) Hypothesis for origin of planktonic patchiness. *Nature*, **259**, 659.

Levin, S.A., Powell, T.M. & Steele, J.H. (1993) *Patch Dynamics: Springer Lecture Notes in Biomathematics*. Springer, Berlin.

Levin, S.A., Grenfell, B., Hastings, A. & Perelson, A.S. (1997) Mathematical and computational challenges in population biology and ecosystems science. *Science*, **275**, 334–343.

Lloyd, A.L. (1995) The coupled logistic map: a simple model for the effects of spatial heterogeneity on population dynamics. *Journal of Theoretical Biology*, **173**, 217–230.

Lloyd, A.L. & May, R.M. (1996) Spatial heterogeneity in epidemic models. *Journal of Theoretical Biology*, **179**, 1–11.

Lynch, L.D., Bowers, R.G., Begon, M. & Thompson, D.J. (1998) A dynamic refuge model and population regulation by insect parasitoids. *Journal of Animal Ecology*, **67**, 270–279.

McCauley, E., Wilson, W.G. & DeRoos, A.M. (1993) Dynamics of age-structured and spatially structured predator–prey interactions: individual-based models and population-level formulations. *American Naturalist*, **142**, 412–442.

McGlade, J.M. (1993) Alternative ecologies. *New Scientist*, 14–16.

McGlade, J.M. (1995) Dynamics of complex ecologies. In: *Modelling the dynamics of biological systems* (eds E. Mosekilde & O.G. Mouritsen). Springer Series in Synergetics, **vol. 65**. Springer, Berlin.

McGlade, J.M. & Price, A.R.G. (1993) Multidisciplinary modelling—an overview and practical implications for the governance of the Gulf region. *Marine Pollution Bulletin*, **27**, 361–377.

McLauglin, J.F. & Roughgarden, J. (1991) Pattern and stability in predator–prey communities: how diffusion in spatially variable environments affects the Lotka–Volterra model. *Theoretical Population Biology*, **40**, 148–172.

Maddison, A.C., Holt, J. & Jeger, M.J. (1996) Spatial dynamics of a monocyclic disease in a perennial crop. *Ecological Modelling*, **88**, 45–52.

Malchow, H. (1993) Spatiotemporal pattern-formation in nonlinear nonequilibrium plankton dynamics *Proceedings of the Royal Society of London*, **B251**, 103–109.

Mangel, M. (1985) *Decision and control in uncertain resource systems*. Academic Press, Orlando.

Matsuda, H., Tamachi, N., Ogita, N. & Sasaki, A. (1987) A lattice model for pupulation biology. In: *Mathematical Topics in Biology: Springer Lecture Notes in Biomathematics*. Springer, Berlin.

May, R.M. (1981) *Theoretical Ecology*, 2nd edn. Blackwell Scientific Publications, Oxford.

May, R.M. (1995) Spatial chaos and its role in ecology and evolution. In: *Frontiers of Theoretical Biology: Springer Lecture Notes in Biomathematics*. Springer, Berlin.

Menge, B.A. & Olson, A.M. (1990) Role of scale and environmental factors in regulation of community structure. *Trends in Evolution and Ecology*, **5**, 52–57.

Mimura, M. & Murray, J.D. (1978) On a diffusive prey–predator model which exhibits patchiness. *Journal of Theoretical Biology*, **75**, 249–262.

Mollison, D. (1977) Spatial contact models for ecological and epidemic spread. *Journal of the Royal Statistical Society*, **B39**, 283–326.

Mollison, D. (1986) Modelling biological invasion: chance, explanation, prediction. *Philosophical Transactions of the Royal Society of London*, **B314**, 675–693.

Mollison, D. (1997) *Epidemic Models: Their Structure and Relation to Data*. Cambridge University Press, Cambridge.

Murray, J.D. (1981) A pre-pattern formation mechanism for animal coat markings. *Journal of Theoretical Biology*, **88**, 161–199.

Murray, J.D. (1982) Parameter space for Turing instability in reaction diffusion mechanisms: a comparison of models. *Journal of Theoretical Biology*, **98**, 143–163.

Murray, J.D. (1988) How the leopard gets its spots. *Scientific American*, **258**, 62–69.

Murray, J.D., Stanley, E.A. & Brown, D.L. (1986) On the spatial spread of rabies by foxes. *Proceedings of the Royal Society of London*, **B229**, 111–150.

Nee, S. & May, R.M. (1992) Dynamics of metapopulations: habitat destruction and competitive coexistence. *Journal of Animal Ecology*, **61**, 37–40.

Nicholson, A.J. & Bailey, V.A. (1935) The balance of animal populations, part I. *Proceedings of the Zoological Society of London*, **3**, 551–598.

Niwa, H.S. (1994) Self-organizing dynamic-model of fish schooling *Journal of Theoretical Biology*, **171**, 123–136.

Okubo, A. (1980) *Diffusion and Ecological Problems: Mathematical Models*. Springer Verlag, Berlin.

Pacala, S.W. (1987) Neighbourhood models of plant population dynamics. 3. Models with spatial heterogeneity in the physical environment. *Theoretical Population Biology*, **31**, 59–392.

Pimm, S.L. (1991) *The Balance of Nature*? University of Chicago Press, Chicago.

Rand, D.A. & Wilson, H.B. (1995) Using spatio-temporal chaos and intermediate-scale determinism to quantify spatially extended ecosystems *Proceedings of the Royal Society of London*, **B259**, 111–117.

Rand, D.A., Keeling, M.J. & Wilson, H.B. (1995) Invasion, stability and evolution to criticality in spatially extended, artificial host–pathogen ecologies. *Proceedings of the Royal Society of London*, **B259**, 55–63.

Reuter, H. & Breckling, B. (1994) Self organisation of fish schools—an object oriented model. *Ecological Modelling*, **75**, 147–159.

Rhodes, C.J. & Anderson, R.M. (1996) Persistence and dynamics in lattice models of epidemic spread. *Journal of Theoretical Biology*, **180**, 125–133.

Rohani, P. & Miramontes, O. (1995) Host–parasitoid metapopulations: the consequence of parasitoid aggregation on spatial dynamics and searching efficiency. *Proceedings of the Royal Society of London*, **B260**, 335–342.

Rohani, P., Godfray, H.C.J. & Hassell, M.P. (1994) Aggregation and the dynamics of host–parasitoid systems—a discrete-generation model with within-generation redistribution. *American Naturalist*, **144**, 491–509.

Rohani, P., Lewis, T.J., Grunbaum, D. & Ruxton, G.D. (1997) Spatial self-organisation in ecology: pretty patterns or robust reality? *Trends in Evolution and Ecology*, **12**, 70–74.

Roux, J.C. & Swinney, H.L. (1981) In: *Nonlinear Phenomena in Chemical Dynamics* (eds C. Vidal & A. Pacault). Springer, Berlin.

Ruxton, G.D. (1994) Low levels of immigration between chaotic populations can reduce system extinctions by inducing asynchronous regular cycles. *Proceedings of the Royal Society of London*, **B256**, 189–193.

Ruxton, G.D. & Rohani, P. (1996) The consequences of stochasticity for self-organised spatial dynamics, persistence and coexistence in spatially extended host–parasitoid communities. *Proceedings of the Royal Society of London*, **B263**, 625–631.

Sato, K., Matsada, H. & Sasaki, A. (1994) Pathogen invasions and host extinction in lattice structured populations. *Journal of Mathematical Biology*, **32**, 251–268.

Savic, D. (1995) Model of pattern formation in animal coatings. *Journal of Theoretical Biology*, **172**, 299–303.

Sibley, R.M. & Smith, R.H. (1985) *Behavioural Ecology: Ecological Consequences of Adaptive Behaviour*. Blackwell Scientific Publications, Oxford.

Singh, B.N. & Das, K. (1938) Distribution of weed species on arable land. *Journal of Ecology*, **26**, 455–466.

Smith, G.C. & Harris, S. (1991) Rabies in urban foxes (*Vulpes vulpes*) in Britain—the use of a spatial stochastic simulation-model to examine the pattern of spread and evaluate the efficacy of different control regimes. *Philosophical Transactions of the Royal Society of London* **B334**, 459–479.

Solé, R.V. & Valls, J. (1991) Order and chaos in a 2D Lotka–Volterra coupled map lattice. *Physics Letters*, **A153**, 330–336.

Solé, R.V., Valls, J. & Bascompte, J. (1992) Spiral waves, chaos and multiple attractors in lattice models of interacting populations. *Physics Letters*, **A166**, 123–128.

Stewart, G. & Keller, W. (1936) A correlation method for ecology as exemplified by studies of native desert vegetation. *Ecology*, **17**, 500–514.

Sugihara, G., Grenfell, B.T. & May, R.M. (1990) Distinguishing error from chaos in ecological time series. *Philosophical Transactions of the Royal Society of London*, **B330**, 235–251.

Swart, J. & Lawes, M.J. (1996) The effects of habitat patch connectivity on samango monkey (*Cercopithecus mitis*) metapopulation persistence *Ecological Modelling*, **93**, 57–74.

Taylor, A.D. (1990) Metapopulations, dispersal, and predator–prey dynamics: an overview. *Ecology*, **71**, 429–433.

Tilman, D. & Kareiva, P. (1996) *Spatial Ecology: The Role of Space in Population Dynamics and Interspecific Interactions*. Princeton University Press, Princeton, NJ.

Tong, H. (1990) *Non-linear Time Series. A Dynamical System Approach*. Oxford Science Publications, Oxford.

Tsonis, A.A., Elsner, J.B. & Tsonis, P.A. (1989) On the dynamics of a forced reaction-diffusion model for biological pattern formation. *Proceedings of the National Academy of Sciences USA*, **86**, 4938–4942.

Turing, A. (1952) The chemical basis of morphogenesis. *Philosophical Transactions of the Royal Society of London*, **B237**, 37–72.

Vail, S.G. (1993) Scale-dependent responses to resource spatial pattern in simple-models of consumer movements. *American Naturalist*, **141**, 199–216.

van Baalen, M. & Sabelis, M.W. (1993) Coevolution of patch selection strategies of predator and prey and the consequences for ecological stability. *American Naturalist*, **142**, 646–670.

Vickers, G.T., Huston, V.C.L. & Budd, C.J. (1993) Spatial patterns in population conflicts. *Journal of Mathematical Biology*, **31**, 411–430.

Walters, C. (1986) *Adaptive Management of Renewable Resources*. Collier Macmillan, London.

White, K.A.J., Murray, J.D. & Lewis, M.A. (1996) A model for wolf pack territory formation and maintenance. *Journal of Theoretical Biology*, **178**, 29–43.

Wiens, J.A. (1989) Spatial scaling in ecology. *Functional Ecology*, **3**, 385–397.

Wilson, H.B. & Keeling, M.J. (1999) Stochastic models and low-dimensional, deterministic dynamics in spatially extended systems. In: *The Geometry of Ecological Interactions: Simplifying Spatial Complexity.* (eds U. Dieckmann, R. Law & J.A.J. Metz). Cambridge University Press, Cambridge.

Wilson, W.G., De Roos, A.M. & McCauley, E. (1993) Spatial instabilities within the diffusive Lotka–Volterra system: individual-based simulation results. *Theoretical Population Biology*, **43**, 91–127.

Wu, Y.G., Sklar, F.H., Gopu, K. & Rutchey, K. (1996) Fire simulations in the Everglades landscape using parallel programming. *Ecological Modelling*, **93**, 113–124.

Yao, Q.W. & Tong, H. (1994) On prediction and chaos in stochastic-systems. *Philosophical Transactions of the Royal Society of London*, **A348**, 357–369.

Yorke, J.A., Hethcote, H.W. & Nold, A. (1978) Dynamics and control of the transmission of gonorrhea. *Sexually Transmitted Diseases*, **5**, 51–56.

4: Correlation equations and pair approximations for spatial ecologies

D.A. Rand

Introduction

With the advent of readily accessible high-power computers it has become relatively easy to produce simulations of interesting spatial ecologies. Even at this early stage these have demonstrated that there is a range of new, interesting and important biological phenomena to be discovered in such systems (see Chapter 3 for references). These phenomena had previously been undiscovered because they are due to space, stochasticity and discreteness and because of the concentration on mean-field models. In such models fluctuations and correlations are ignored because it is assumed that not only does everyone behave like the mean individual, but also all individuals perceive the same environment.

With the benefit of hindsight, some of the new phenomena found in spatial individual-based systems are relatively intuitive: for example, coexistence and diversity are easier, epidemics are less violent and have more realistic persistence and critical community size (Keeling 1995; Keeling *et al.* 1997), evolutionary velocities are often much slower (Rand *et al.* 1995) and it is often much easier for cooperative and altruistic behaviour to invade (Matsuda *et al.* 1992; Nakamaru *et al.* 1997; van Baalen & Rand 1997). Others are more surprising or more difficult to understand: e.g. parasites can drive huge spatial genetic host diversity in sexual species (Rand & Keeling 1995), spiral waves can remove the parasites that destroy hypercycles (Boerlijst & Hogeweg 1991) and host–parasite systems can have a critical state towards which they evolve (Rand *et al.* 1995). What these phenomena have in common is that they cannot occur in mean-field systems. They all rely on the presence of space and most of them also depend on the fact that individuals are discrete and that there are relatively strong stochastic effects so that fluctuations or correlations or both are important.

Unfortunately, these explicit spatial population models, which usually are some form of probabilistic cellular automata, are notoriously difficult to analyse. Thus, while being excellent for developing intuition and formulating conjectures, as models they suffer from a number of deficiencies. Most importantly, we lack real mathematical understanding of them and, in most cases of interest, cannot say whether the behaviour we see is reasonable or not. Moreover, their formulation usually deviates considerably from biological realism and it is difficult to estimate the significance of this or to compare their structure to data. Finally, because of the absence of theory, one is often reduced to simulation alone which is not very satisfactory.

In this chapter I want to address this problem and introduce a class of general models which, while allowing us to address questions raised by space, discreteness and stochasticity, are nevertheless much more tractable and controllable and also can be more directly connected with biological data. They will enable an analytical approach. I refer to them as *correlation equations* because the basic idea is to derive stochastic differential equations for the time evolution of certain low-order correlations. This approach has its foundation in certain areas of physics (such as statistical mechanics and the theory of weak turbulence) where correlation equations are derived to describe the statistical structure of complex fields which are defined stochastically or by partial differential equations. It has also been used to model chemical reactions (ben-Avraham *et al.* 1990; Tretyakov *et al.* 1994). For applications in ecology, evolution and epidemiology, the derivation can be justified directly from biological hypotheses rather than being deduced from some more basic system or equations. This is because such biological dynamics are dominated by interactions between individuals which can be nicely captured by low-order correlations, and also because the stochastic background of biological systems even more effectively destroys high-order correlations. Moreover, such models are more robust to the assumptions underlying their derivation and these assumptions are more susceptible to experimental verification.

Many interesting biological systems are well described by what I call *pair approximations* which give the simplest correlation equations extending the mean-field equations. In these a closure is used which gives a system of stochastic differential equations for the second-order moments or pair numbers. These can work very well. One reason for this is that many biological systems are dominated by pairwise interactions which introduce important correlations and fluctuations. Two good examples are sexually transmitted diseases and evolutionary games played between individuals such as the Prisoner's Dilemma (see p. 138). Moreover, as we will see, in many systems, the higher-order correlations can be approximated or modelled as stochastic noise.

Such pair approximations have been developed for a range of systems such as simple host–parasite models (Satō *et al.* 1994; Keeling 1995; Keeling & Rand 1996); epidemics (Keeling 1995; Keeling & Rand 1996; Morris 1997); plant dynamics (Harada & Iwasa 1994; Satō & Konno 1995; Bolker & Pacala 1997); spatial games (Morris 1997; Nakamaru *et al.* 1997); and the evolution of altruism (Matsuda *et al.* 1992; Harada *et al.* 1995; van Baalen &

Rand 1997). They were pioneered by the Japanese school which includes Ezoe, Harada, Iwasa, Kubo, Matsuda, Nakamaru, Ogita, Sasaki, and Satō, and extensively studied by Keeling and Morris in their Warwick theses (Keeling 1995; Morris 1997). Bolker and Pacala also made an important contribution with a related study of some simple ecological systems in continuous space (Bolker & Pacala 1997).

My aim in this discussion is two-fold. Firstly, to survey some interesting examples and applications. Here I cannot claim to be exhaustive and must admit to a strong bias towards my own interests. Secondly, to provide a toolkit that will help others to apply these ideas to their own problems. This aim forces a more detailed approach to the mathematical underpinnings.

In this paper I proceed as follows. Firstly, I discuss the derivation of such equations from the underlying stochastic process, starting with the calculation of the master equation and then discussing its closure to obtain a differential equation. In particular, I give a careful discussion of the various pair approximation and closure procedures. This may seen unduly mathematical to the more biologically minded, but I believe that, firstly, it is important to understand the underlying principles and, secondly, on the practical side, this section lays down a clear procedure for deriving such equations that can then be applied to a very wide range of biological systems.

On p. 111 I start the applications by discussing a range of applications in infectious diseases and host–parasite systems. Including pair correlations can tell us a lot about previously unexplored phenomena. Firstly, I discuss the spatial SIR model and, in particular, consider how the establishment and maintenance of a disease depends upon spatial structure. Secondly, I use the pair approximation to analyse the evolution to criticality in the host–parasite systems of Rand *et al.* (1995) and also consider the host–parasite system of Satō*et al.* (1994). Finally, I consider a model for measles from Keeling *et al.* (1997) that gives a much improved fadeout structure and critical community size which compares well with data.

I then consider spatial games (p. 127). The discussion of this depends strongly upon Morris (1997). An important aspect of this section is the careful comparison between the pair approximation and the observed behaviour of stochastic simulations on regular and irregular, static and dynamic networks. Regarding cooperation and altruism (p. 132), I consider reciprocal altruism in the Prisoner's Dilemma game and discuss a model of Nakamaru *et al.* (1997) in which, in contrast to mean-field models, cooperative strategies can invade noncooperating resident populations. For unconditional altruism, the principle current approach concerns trait groups and relies upon fluctuations in local population structure to succeed. Following Matsuda *et al.* (1992) and van Baalen & Rand (1997), I explain a new and, I believe, more natural model in which correlations rather than fluctuations enable altruists to invade nonaltruistic populations.

Unfortunately, through lack of space, I am unable to treat all the applications I would have liked. Some ecological applications have suffered, notably the treatment of vegetative propagation versus seed production in plant systems by Harada & Iwasa (1994), the pair approximation Lotka–Volterra system (Matsuda *et al.* 1992; Nakamaru *et al.* 1997) and the treatment of Bolker & Pacala (1997) of a similar approach to pair approximation in a continuous two-dimensional space.

Finally, I would like to point out that many of the calculations in this paper were computed using Maple and I have made some of the worksheets available via my web page http://www.maths.warwick.ac.uk/~dar/. From this you can also get many of the differential equations that I have had to leave out because of lack of space.

Deriving the correlation equations

In the framework I consider, space is represented by a network of sites x. These sites represent either individuals or empty sites that individuals can occupy. The edges e join neighbouring sites. This mathematical structure is used to capture the idea that two individuals are neighbours if they regularly interact with each other. This relation may coincide with geographical proximity so that the network is essentially two-dimensional or it may represent some more complex interaction structure such as that seen in childhood diseases like measles or sexually transmitted diseases like HIV/AIDS. There, although the global structure is two-dimensional, the local structure can be higher-dimensional.

The states of each site are finite in number. Typically, they are either empty (ϕ) or correspond to occupation of the site by an individual in some state or of some type. In a simple host–parasite system the individuals might have two states, susceptible and infected, in a spatial game the states of the individuals will correspond to the strategies being played and in a simple predator–prey system the sites might either be empty or occupied by a predator or a prey.

The state of the system $\sigma = \{\sigma_x\}$, often called a *configuration*, is determined when we associate to each site x in the network a state σ_x.

Notation

Before I proceed to discuss how to calculate correlation equations from the event structure I need to introduce some basic notation and conventions that will be used throughout the chapter. Assume that we have a given configuration σ. I use the following notation:

σx and σe: denote respectively the state of the site x and the edge e. Thus $\sigma x = i$ and $\sigma e = ij$ mean that the state of x is i and one site of e is in state

i while the other is in state j. In this case e_i and e_j denote the sites of the edge e which are respectively in states i and j.
Q_x: denotes the number of neighbours of a site x. I use Q to denote the space-average of this. Very often, in particular applications, Q_x is assumed to be independent of x.
$Q_x(i) = Q_x^\sigma(i)$: denotes the number of neighbours of x which are in state i. Although $Q_x(i)$ depends upon the configuration σ, for notational convenience, I drop the reference to σ. The reader should keep this dependence in mind. In particular, since σ varies with time, so do the $Q_x(i)$ and I will derive equations of motion to describe this.
$[i]$, $[ij]$ and $[ijk]$: denote the number of sites, edges and triples in states i, ij and ijk respectively.
ρ_i and ρ_{ij}: denote the density of sites and edges in states i and ij respectively. Thus, if Q is the average number of neighbours and N the total population size, $\rho_i = [i]/N$ and $\rho_{ij} = [ij]/QN$.
$Q(i|j)$ and $Q(i|jk)$: denote the space-averaged values of the number of i neighbours of respectively a j site and a j site in a jk edge, i.e.

$$Q(i|j)=\frac{1}{[j]}\sum_{\sigma x=j} Q_x(i) \quad \text{and} \quad Q(i|jk)=\frac{1}{[jk]}\sum_{\sigma e=jk} Q_{e_j}(i)$$

Note that

$$Q(i|j)=\frac{[ij]}{[j]} \quad \text{and} \quad Q(i|jk)=\frac{[ijk]}{[jk]}+\delta_{ik} \tag{4.1}$$

$\eta_x(i|j)$ and $\eta e_j(i|jk)$: are the local fluctuations from these average values, i.e. $\eta_x(i|j) = Q_x(i) - Q(i|j)$ and $\eta e_j(i|jk) = Qe_j(i) - Q(i|jk)$. They satisfy

$$\sum_{\sigma x=j} \eta_x(i|jk)=0 \quad \text{and} \quad \sum_{\sigma e=jk} \eta_{ej}(i|jk)=0 \tag{4.2}$$

It is important to adopt a consistent convention about how to count edges. I will distinguish between the edge from x to y and the edge from y to x. This convention has the consequence that edges in state ii are counted twice. It is this that accounts for the fact that there are often factors of two in our equations.

Events and the master equations

I now discuss these various steps in the calculation of the correlation equations. The first step, the derivation of the master equation, is only quickly discussed in abstract here because our main aim is to discuss the various pair approximations. As a consequence, to really understand this, it may be helpful to read the discussion of on the various applications alongside this discussion.

All processes are assumed to be local: the immediate fate of an individual is affected only by its neighbours. Generally speaking, the state of a neighbourhood determines the rates at which certain events occur that transform the neighbourhood from its present state to a new one. This defines the dynamics. A crucial notion to get clear is that of an *event*, i.e. a change of state of some particular site or edge, typically a biological event such as birth, death or infection at that location.

The basic equation which allows us to do this is the following. Consider a function $f(\sigma)$ which gives us the expectation of some average quantity of the configuration σ. For example, f might be the expected total number $[i]$ of sites in some given state i or the total number of edges $[ij]$ in state ij. The rate of change of $f(\sigma)$ is then given by summing over all events ε the rate $r^{\sigma}(\varepsilon)$ at which that event occurs multiplied by the change $\delta f(\varepsilon)$ the event causes in f, i.e.

$$\frac{\mathrm{d}f}{\mathrm{d}t} = \sum_{\text{events } \varepsilon} r^{\sigma}(\varepsilon)\,\delta f(\varepsilon) \tag{4.3}$$

Both $r^{\sigma}(\varepsilon)$ and $\delta f(\varepsilon)$ will depend upon the configuration σ. I refer to Equation 4.3 as the *master equation* for f.

To derive the correlation equations for a particular biological system I will proceed as follows.

1 Calculate the contribution of each event type to the sum in the master Equation 4.3 for $f = [\alpha\beta]$.

2 Use this to get an exact expression for $\mathrm{d}[\alpha\beta]/\mathrm{d}t$ involving the pair numbers $[ij]$, the local densities $Q(i|jk)$ and the correction terms of the form below.

3 (*Pair closure.*) Determine an approximation (usually of the form $Q(i|jk) \approx \kappa Q(i|j)$) so that $Q(i|jk)$ can be replaced in the expression for $\mathrm{d}[\alpha\beta]/\mathrm{d}t$ by terms involving pair and singleton numbers only.

4 (*Stochastic closure.*) Incorporate any biases in the correction terms into the expression for $\mathrm{d}[\alpha\beta]/\mathrm{d}t$ trying to ensure that the remainder can be well-modelled by random noise. In this way the pair approximation for the system, which is a stochastic differential equation, is obtained.

5 If this equation does not capture the right behaviour consider also including differential equations for crucial higher-order correlations.

Steps 1 and 2: calculating the master equation for $f = [\alpha\beta]$

If the discussion that follows is not clear to the reader I suggest that he or she works through the simple derivation of the differential equations on p. 111 and Equations 4.13 and 4.14 on p. 129.

The basic events are either associated to a given site x or to an edge e connecting two neighbouring sites. The important thing about the site events is that for some states i and j they transform a site x in state i into state j at a rate $r(x)$ that is only dependent on the state of x's neighbourhood and therefore their contribution to df/dt can be written as a sum of the form $\Sigma_{\sigma x=i} r(x)\delta f(x)$ where $\delta f(x)$ is the change in f caused by this event at x. An example would be a constant rate death event where $r(x) = d$ is independent of x. Similarly, for an edge event which transforms ij edges into state $i'j'$, the contribution to df/dt can be written as $\Sigma_{\sigma e=ij} r(e)\delta f(e)$. An example of such an edge event would be a migration where the states of the vertices in an edge are interchanged.

Many events can be considered as either site or edge events. An example is birth of an i-individual. This can be associated with empty sites and then the rate of the births is given by summing rates over all the i neighbours able to give birth into the site. On the other hand, it can also be considered as associated with an edge e in state $i\phi$ with the birth event corresponding to the change $i\phi \to ii$. Then the rate is just determined by the neighbourhood of the i individual in the pair. The second approach regarding the event as an edge event is preferable in this case because the first event is composite, since several sites can be the cause of the birth, whereas the second is a simple birth only involving birth from the occupied site of the edge. If the edge approach is adopted the contribution of these birth events is written as a sum over edges in state $i\phi$. Of course, this choice is only a matter of convenience and does not affect the answer. In the applications in this chapter, I consider births, infections, migrations and replacement as edge events and the only site events are constant rate death and infectious recovery events.

If $f = [\alpha\beta]$ then typically for a site event both $r(x)$ and $\delta f(x)$ will be simply calculated functions of the $Q_x(k)$. While for an edge event as above $r(e)$ and $\delta f(e)$ may also involve the $Qe_i(k)$. If the resulting expression for the contribution to df/dt is linear in these then, because of Equation 4.2, one can replace them in the sum by their average values $Q(k|i)$ and $Q(k|ij)$ defined above on p. 103. Otherwise, in the nonlinear case, replacing the terms of form $Q_x(k)$ and $Qe_i(k)$ with $Q(k|i)$ and $Q(k|ij)$ introduces correction terms which are often correlations of the form

$$\Gamma(k|i|j) = \frac{1}{[i]} \sum_{\sigma x=i} \eta_x(k|i)\eta_x(l|i) \tag{4.4}$$

or

$$\Gamma(k|ij|l) = \frac{1}{[ij]} \sum_{\sigma e=ij} \eta_e(k|ij)\eta_e(l|ji)$$

where $\eta_x(k|i)$ and $\eta_x(k|ij)$ are as on p. 104. See the applications for examples.

Regular and irregular networks

There are two complications in the case of an irregular network where Q_x varies from site to site. The first is that one cannot use the relations $Q[i] = \Sigma_j[ij]$, which only hold for regular networks, to reduce the number of equations. One does have that $Q_i[i] = \Sigma_j[ij]$ where Q_i is the average of the Q_x over the i-sites, but this is obviously much less useful. The second complication is more important. If the Q_xs and $Q_x(i)$ values only enter the master equation in a linear way then it is straightforward and exact to replace them by the average values Q, $Q(i|j)$ and $Q(i|jk)$ in much the same way as for regular networks. Luckily, this is the case for quite a wide range of systems such as the epidemiological ones I am about to consider: irregularity does not introduce such nonlinearities and we can deal with both regular and irregular systems together. However, if they enter the master equation in a nonlinear way, as is the case for our models for games and for altruism and cooperation, then this cannot necessarily be done and there can then be genuine differences in the dynamics for regular and irregular networks. An example is given on p. 132 (see Fig. 4.5).

Pair closure

The expression for $d[\alpha]/dt$ and $d[\alpha\beta]/dt$ that are obtained in this way are of the form

$$\frac{d}{dt}[\alpha] = X_\alpha([i];[ij];Q(i|jk)) + \zeta_\alpha(t) \tag{4.5}$$

$$\frac{d}{dt}[\alpha\beta] = X_{\alpha\beta}([i];[ij];Q(i|jk)) + \zeta_{\alpha\beta}(t) \tag{4.6}$$

where the ξ_α and $\xi_{\alpha\beta}$ are the correction terms. Note that I have emphasized their dependence upon time t because they are functions of the spatial configuration σ^t at time t. Let us now consider how to close these equations so as to put them into the form of a differential equation or a stochastic differential equation involving only the $[i]$ and $[ij]$.

Our approach to this closure involves (a) approximating the $Q(i|jk)$ by functions of the $[i]$ and $[ij]$; (b) keeping track of all the correction terms; and (c) a strategy for dealing with the correction terms usually incorporating any biases into the deterministic part of the equation so that their mean is zero. Then they are replaced with random noise which approximates the statistical structure of their fluctuations due to the time-dependence of the configuration.

Approximating the $Q(i|jk)$

Note that $[ijk]$ is the sum of $Q_x(i)(Q_x(k) - \delta_{ik})$ over all x such that $\sigma x = j$ where δ_{ij} is 1 if $i = k$ and 0 otherwise. Thus $Q(i|jk) = Q(i|j) + Q(k|j)^{-1} \Gamma_x(i|j|k)$ where $\Gamma(i|j|k)$ is given in Equation 4.4. This calculation only makes sense if both $[ij]$ and $[jk]$ are nonzero, and is independent of whether or not $i = k$.

It is easy to see that in general the corrections $\Gamma(i|j|k)$ will have a nonzero mean. This occurs both because of the existence of nonzero triple correlations and also because knowing that the j site has a k neighbour constrains the number of i neighbours. I show an example of this in Fig. 4.8 for a simulation of a spatial game. It is therefore necessary to adopt a more sophisticated strategy to approximate the $\Gamma(i|j|k)$ than just putting them equal to zero.

In certain simulations one finds that the $Q_x(i)$, $\sigma x = j$, are either Poisson distributed or distributed as in Bernoulli trials. In these cases one gets good approximations as follows.

Poisson statistics

In approximating $Q(i|jk)$ consider separately the cases $i = k$ and $i \neq k$. In the first case, if it is assumed that the $Q_x(i)$ are Poisson distributed over j sites x then the mean $Q(i|j)$ equals the variance which is $\Gamma(i|j|i)$. In the second case, if it is assumed that $i \neq k$ and $Q_x^\sigma(i)$ and $Q_x^\sigma(i)$ are independently Poisson distributed over j sites with means $Q(i|j)$ and $Q(k|j)$. Then, by the independence, the covariance $\Gamma(i|j|k)$ is zero. In either case one deduces that the approximation to take is

$$[ijk] \approx \frac{[ij][jk]}{[j]} \quad \text{or equivalently} \quad Q(i|jk) \approx Q^P(i|jk) = Q(i|j) + \delta_{ik}$$

Bernoulli trial statistics

In this case I assume that the number of i-neighbours of a j at x is chosen from $Q_x \equiv Q$ repeated independent trials with the probability of choosing an i being given by $a = Q(i|j)/Q$. To approximate $Q(i|ji)$ note that the average number of successes is Qa and the variance $\Gamma(i|j|i)$ is $Qa(1 - a)$. A longer calculation using the multinomial distribution gives $\Gamma(i|j|k)$ when $i \neq k$. In both cases one deduces the following approximation:

$$[ijk] \approx \kappa \frac{[ij][jk]}{[j]} \quad \text{or equivalently} \quad Q(i|jk) \approx Q^B(i|jk) = \kappa Q(i|j) + \delta_{ik}$$

where $\kappa = (Q - 1)/Q$. This is the meaning of κ throughout the chapter. It is worth noting for future reference that this approximation is equivalent to the assumption

$$\Gamma(i|j|k) \approx Q(i|j)\left(\delta_{ik} - \frac{Q(k|j)}{Q}\right) \tag{4.7}$$

and that, with this approximation, the sums $\Sigma_{\sigma x=j} Q_x(i) Q_x(k)$ and $\Sigma_{\sigma e=jk} Q_{e_j}(i) Q_{e_k}(l)$ are approximated by $[ij]Q^B(k|ji)$ and $[jk]Q^B(i|jk)Q^B(l|kj)$.

We have studied three types of networks: lattices with a constant number of neighbours, random networks with a globally two-dimensional structure (typically obtained by connecting all points in a random distribution of points on the plane which are within a given distance of each other) and dynamic networks where births and deaths destroy and create sites. In a wide range of systems of the first two types, the Bernoulli approximation is good, while the Poisson distribution is more relevant for the dynamic networks.

Clumped network structures

Minus van Baalen and Andrew Morris (Morris 1996) have made the following very interesting observation. Note that if there is a high probability φ that two neighbours of a given site are themselves neighbours (i.e. there are lots of triangles in the network) then one should expect that the $Q(i|jk)$ will need further corrections because the site in state i is likely to be connected to that in state k. The approximation to take in this case is

$$[ijk] = \kappa \frac{[ij][jk]}{[j]} \left\{ (1-\varphi) + \frac{\varphi N}{Q} \frac{[ik]}{[i][k]} \right\} \tag{4.8}$$

Let me try and justify Equation 4.8. Firstly, let us consider the open triples, i.e. the triples xyz where there are edges from x to y and y to z but no edge from z to x. Let $[ijk]_{\text{open}}$ denote the number of such open triples in state ijk and $[i \cdot k]_{\text{open}}$ denote the number of those with $\sigma x = i$, $\sigma z = j$ and y in any state. Let $[ijk]_\Delta$ and $[i \cdot k]_\Delta$ denote the analogous numbers for triples that are in triangles. The basic assumption that I will make is that

$$\frac{[ijk]_\Delta}{[i \cdot k]_\Delta} \approx \frac{[ijk]_{\text{open}}}{[i \cdot k]_{\text{open}}} \tag{4.9}$$

The idea behind this is that these ratios will depend weakly on the absence or presence of an edge connecting the i-site to the k-site since this is not directly connected to the j-site. This approximation is supported by simulations.

Firstly, consider $[i \cdot k]_{\text{open}}$. Since the i-site is not directly connected to the

k-site and the intermediate site is free, it is reasonable to estimate $[i \cdot k]_{\text{open}}$ by

$$[i \cdot k]_{\text{open}} \approx \sum_{\text{open triples}} \rho_i \rho_k = (1-\varphi)\frac{Q(Q-1)}{N}[i][j]$$

Secondly, note that the arguments justifying the Bernoulli approximation above give us that one can approximate $[ijk]_{\text{open}}$ by $(1-\varphi)\kappa[ij][jk]/[j]$. Thirdly, note that $[i \cdot k]_{\Delta}$ is the number of triangles containing a ik pair. It is therefore reasonable that the ratio of $[i \cdot k]_{\Delta}$ to the total number of triangles is approximated by the proportion of all edges in state ik. The number of pairs is QN and the number of triangles is $\varphi Q(Q-1)N$. Putting all this together with Equation 4.9 gives the approximation in Equation 4.8.

One can regard φ and Q as parameters that can be varied. For a given fixed value of Q one can regard increasing φ as increasing the clumping of the sites in the network because if it is near one then the only way it can be compatible with the given value of Q is if there are clumps in which φ is large but which are only weakly connected to each other so as to get the right average number of neighbours. Thus using the substitution given in Equation 4.8 allows us to model different network structures. I will consider this again on pp. 116–118, in connection with epidemics and host–parasite systems. A nice example is measles (see pp. 123–127) where the change in structure between school vacations and term-time can be modelled by an increase in both φ and Q (Keeling *et al.* 1997).

Corrections from triple correlations

Suppose we have a situation where the Bernoulli trials approximation is expected to work. Then our estimate of $[ijk]$ is $E_{ijk} = \kappa[ij][jk]/[j]$. Although, in simulations one often finds that $[ijk] \approx E_{ijk}$ in some significant cases $T_{ijk} = [ijk]/E_{ijk}$ fluctuates about some mean which is significantly different from 1. It is not surprising that the largest divergences seem to occur when $i = k$. This is the case for example in the host–pathogen system discussed on pp. 118–122. In these cases it is necessary to take triples into account but only in the sense that $E_{ijk}/[ijk]$ is a time-independent constant different from 1. To obtain the correct behaviour from the pair model it may be crucial to allow for this. Thus another substitution scheme is $[ijk] \approx \kappa T_{ijk}[ij][jk]/[j]$ where the time-independent factors T_{ijk} are estimated from simulations or data.

Pair closure

Having chosen the appropriate substitutions scheme and substituting into Equation 4.5 to replace the terms of form $Q(i|jk)$, a set of equations of the following form is obtained:

$$\frac{\mathrm{d}}{\mathrm{d}t}[\alpha] = X_{\alpha}([i];[ij]) + \xi_{\alpha}(t) \tag{4.10}$$

$$\frac{\mathrm{d}}{\mathrm{d}t}[\alpha\beta] = X_{\alpha\beta}([i];[ij]) + \xi_{\alpha\beta}(t) \tag{4.11}$$

where, if the job has been done properly, the corrections $\xi_{\alpha}(t)$ and $\xi_{\alpha\beta}(t)$ can be regarded as low-amplitude random noise with zero mean. This is what I call the *pair approximation*. It is different from what some other authors call the pair approximation because they tend to restrict it to the equivalent to the approach described for Bernoulli trials.

We now forget that the correction terms come from the fluctuations in the configuration and replace them with random noise with approximating statistical structure. Thus one obtains a stochastic differential equation to investigate. If the deterministic part of the equation has as its attractor just a stable fixed point, then it is likely, provided the noise is of low amplitude, that the only effect of the noise is to cause the state to have small fluctuations around the deterministic equilibrium. Thus the noise terms can be ignored and set to zero. However, there are biological situations where the noise can play a very significant role and the behaviours with and without it are very different. An example of this occurs in the discussion of measles in the section below. It is therefore necessary to treat the question of how to handle the noise terms case by case.

Simple infection dynamics

I start the discussion of applications by considering some very simple infection dynamics because these illustrate most clearly the way in which pair approximations are derived and some very simple examples give rise to interesting new phenomena. Pair approximation is very natural here because the basic natural history of infection involves a process whereby infection is passed from one individual to another during an interaction. Moreover, the structure of the network describing the interactions will clearly play an important role in the transmission dynamics. For example, transmission in highly interconnected networks will be faster than in sparsely connected ones. Finally, and most importantly, it is clear that correlations will play an important role because an infection where the infected are clumped together will spread more slowly than one where they are scattered evenly throughout the susceptibles.

Throughout the discussion of this section I will assume that the networks are regular, i.e. $Q_x \equiv Q$. As explained on p. 107, this assumption does not affect the form of the equations (because the master equation depends linearly on the Qs for these systems), but does enable us to reduce the number of them.

The contact process

Although, because of lack of space, I will not discuss it in any detail, I must mention the so-called contact process (Durrett 1988; Bezuidenhout & Grimmett 1990). This is the simplest mathematical model of an infective process that one can think of. In it sites of a lattice represent hosts who are either uninfected and susceptible (S) or infected (I). Infected individuals infect their susceptible neighbours at a given rate β which is the transmissibility of the disease. They can also recover, becoming susceptible again at the rate v. The contact process can also be thought of as an ecological model in which infection corresponds to birth and recovery to death.

A most interesting fact about the contact process is that there is a critical transmissibility β_c below which the infection dies out (Bezuidenhout & Grimmett 1990). At the critical β_c, long-range correlations become important and the pair approximation is therefore poor. The contact process therefore gives insight into the way in which the approximations can fail because of important long-range correlations.

The contact model was analysed using a pair approximation in Matsuda *et al.* (1992). A comparison between the the mean-field and pair approximations gives that both their equilibria are greater than the true value for the contact process but that the pair approximation is substantially more accurate provided β is greater than the true critical value and is very accurate for relatively large values. Both the mean-field and pair models underestimate β_c but the pair model does substantially better than the mean-field model.

Standard SIR equation

In this system, which can be regarded as a simple extension of the contact process, recovery leads to an immune recovered state R instead of S. These immune individuals can die in which case they are replaced by a susceptible. Thus including infection and recovery the events are as shown symbolically below. I shall in future adopt a similar format for describing the models and represent them in the following form:

1 *infection*: at edge e, $SI \xrightarrow{\beta} II$ at rate β constant;
2 *recovery*: at site x, $I \xrightarrow{v} R$ at rate v constant;
3 *simultaneous death and birth*: at site x, I or $R \xrightarrow{\mu} S$ at rate μ constant.

I now calculate the pair approximation for this system. Since the number of singletons and pairs are respectively N and QN and $Q[i] = \Sigma_{j=S,I,R}[ij]$ it suffices to find the equations for the $[i]$ and $[ij]$ with i and j equal to S or I.

As an illustration let us calculate the SI term in some detail. We consider the contribution from the various events. The notation $\overset{\text{event-type}}{+=}$ used below means that this term is the contribution from the given event type to the master equation. Using this notation:

$$\frac{d}{dt}[SI] \overset{\text{recovery}}{+=} \sum_{\sigma x=1} -\nu Q_x(S) = -\nu[SI]$$

$$\overset{\text{infection}}{+=} \sum_{\sigma e=SI} \beta(Q_{e_S}(S) - Q_{e_S}(I)) = \beta([SSI]-[ISI]-[SI])$$

$$\overset{\text{death/birth}}{+=} \sum_{\sigma x=I,R} \mu Q_x(I) - \sum_{\sigma x=I} \mu Q_x(S) = \mu(Q[I]-2[SI])$$

Note that since all the Q's enter in a linear fashion the equations for this system are at this point exact, i.e. there are no correction terms. One of the pair approximations for the triples discussed on pp. 108–110 can now be used to close this equation.

One can similarly calculate the rest of the equations which, together with their Bernoulli pair approximations, are as follows:

$$\frac{d}{dt}[S] = \mu(N-[S]) - \beta[SI], \quad \frac{d}{dt}[I] = -(\mu+\nu)[I] + \beta[SI]$$

$$\frac{d}{dt}[SS] = 2\mu([SI]+[SR]) - 2\beta[SSI] \approx 2\mu([SI]+[SR]) - 2\beta\kappa Q(S|S)[SI]$$

$$\frac{d}{dt}[SI] = \mu(Q[I]-2[SI]) + \beta([SSI]-[ISI]-[SI]) - \nu[SI]$$

$$\approx \mu(Q[I]-2[SI]) + \beta\kappa[SI](Q(S|S)-Q(I|S)) - (\beta+\nu)[SI]$$

$$\frac{d}{dt}[II] = -2\mu[II] + 2\beta([ISI]+[SI]) - 2\nu[SI]$$

$$\approx -2\mu[II] + 2\beta[SI](\kappa Q(I|S)+1) - 2[II]$$

Establishment and invasion

One of the basic problems of epidemiology is to determine the criteria for establishment of a disease, i.e. when will a small number of infecteds give rise to long-term persistence of the disease (Anderson & May 1992). For the systems I consider this is equivalent to the condition that the trivial equilibrium with $[I] = 0$ is unstable.

The nontrivial equilibrium values for the pair approximation can be calculated exactly, for example by Maple (see the worksheets mentioned in the Introduction). The critical transmissibility is given by $\beta_c = \sigma/\tilde{Q}$ where $\tilde{Q} = Q - 2 - 2\mu/\sigma$. For $\beta > 0$ less than this, the trivial solution $[I] = 0$ is stable and

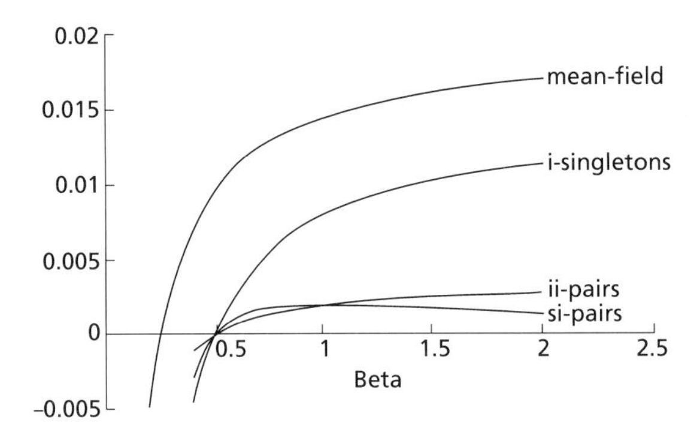

Fig. 4.1 A plot of the equilibrium proportions of singleton and pairs for the SIR Bernoulli pair approximation as β passes through the critical value. The parameter values are $\mu = 0.02$, $v = 1$ and $Q = 4$. Note how small $[SI]$ is compared with $\rho_S \rho_I$. For comparison, the equilibrium of the Q-neighbour mean-field system is also shown.

the nontrivial equilibrium above is negative and unstable. As β is increased past β_c, the nontrivial solution becomes positive and the trivial equilibrium exchanges its stability with it. In Fig. 4.1 I show how the singleton and pair numbers change as β passes through β_c and compare this with a Q-neighbour mean-field model.

In fact, the usual SIR model (see Section 6.1 of Anderson & May 1992) is the mean-field model derived using the assumption that the whole population is homogeneously mixed. Its critical transmissibility is $\beta_c = \sigma/N$ and for $\beta > \beta_c$ it gives the number of infectives as $[I] = \mu(N\sigma^{-1} - \beta^{-1})$. Thus the pair approximation gives radically different predictions from this. This, however, is not the mean-field model that is discussed in Fig. 4.1. In this I assume that each individual only interacts with only Q neighbours and all see the same environment so that $Q_x(I) = Q(S|I) = Q[S]/N$. Thus this model has built into it that each individual only interacts with a small subset of the population, but assumes this subset has the same statistical structure as the population as a whole. In this case $[SI]$ is approximated by $Q[S][I]/N$ to give a differential equation for $[S]$ and $[I]$ whose critical value is given by $\beta_c = \sigma/Q$ and whose equilibrium number of infectives is given by $[I] = \mu N(\sigma^{-1} - (Q\beta)^{-1})$. Thus in the pair approximation the onset of the disease as measured by the critical transmissibility is considerably delayed in comparison to the mean-field models and, above this, the levels of infection are generally lower. Below on pp. 117–119, I consider further how this depends upon network structure.

Vaccination and spatial structure of invasion

I have already calculated β_c above but now I show how also to calculate the correlation structure of the infecteds as they invade. Because of this one can start to ask sophisticated questions, about vaccination strategies for example, and below I will use a similar approach to show that the network structure can have a profound effect upon establishment.

Let $\vec{n}$ denote the vector $([I],[SI],[II])$. Then $\vec{n} = [I](1, Q[S|I], Q(I|I))$ and the equations above involving I together with the Bernoulli trials approximation can be written as

$$\frac{d\vec{n}}{dt} = M \cdot \vec{n}$$

where

$$M = \begin{bmatrix} -\sigma & \beta & 0 \\ \mu Q & \beta\kappa q - \beta - \sigma - \mu & 0 \\ 0 & 2\beta\kappa(Q(I|S)+1) & -2\sigma \end{bmatrix}$$

$\sigma = \mu + v$ and $q = Q(S|S) - Q(I|S)$

To study invasion of a purely susceptible population one would take $Q(S|S) = Q$. However, I would also like to consider the case where the population contains a proportion of immune individuals so as to study the effects, for example, of vaccinating a proportion of the population or the reinfection of a previously infected population in which the infection has died out. Thus I will assume that $Q(R|S)$ is externally fixed at invasion. Since initially $Q(I|S) = 0$ this means that $q = Q(S|S) = Q - Q(R|S)$ and so I shall regard q as being fixed in this way.

It is clear that the eigenvalues of M are the two eigenvalues of the 2×2 matrix M_0 given by the four upper-left entries of M and in addition -2σ (corresponding to the eigenvector $(0, 0, 1)$). Since the latter is negative I am only interested in the other two eigenvalues. These are given by the trace tr M_0 and the determinant det M_0 of M_0 and the dominant eigenvalue is zero when tr $M_0 \leq 0$ and det $M_0 = 0$. A simple calculation shows that the real parts of these are negative and hence that the uninfected equilibrium is stable provided $\beta < \beta_c(q)$ where

$$\beta_c(q) = \frac{\sigma(\sigma + \mu)}{\sigma\kappa q + \mu Q - \sigma}$$

If β is larger than this then the largest eigenvalue λ_+ has a positive real part and hence the infection will be established. This critical value is a decreasing function of q and its minimum value is the critical transmissibility $\beta_c = \sigma/\tilde{Q}$ given above, as one would expect.

As well as deriving the invasion criteria the correlation structure of the spatial population which invades can also be found. The invading population will have associated with it the exponentially growing vector $\vec{n}(t)$ but this will be approximately given by $\vec{n}_0 \exp(\lambda_+ t)$ where $\vec{n}_0$ is the eigenvector associated with λ_+. But as we saw above $\vec{n} = [I](1,Q(S|I),Q(I|I))$. Thus, $(1,Q(S|I), Q(I|I))$ is an eigenvector with eigenvalue λ_+. This eigenvector for our particular problem is easily calculated and it follows from this that at invasion

$$Q(S|I) = \frac{\alpha}{2\beta} \text{ and } Q(I|I) = \frac{\alpha\sigma - 2\mu\beta Q}{\sigma^2 + \sigma x - \mu\beta Q}$$

where $\alpha = \chi + \sqrt{(\chi^2 + 4\mu\beta Q)}$ and $\chi = (\beta(\kappa q - 1) - \mu - 2\sigma)$

SIR on clumped networks and critical transmissibilities

Now I want to consider the dynamics of such a disease on a network which has a clumped structure as described on pp. 109–110. We will see that lots of clumping in the network radically changes the transmission of the disease. To study this I use the clumped network approximation given by Equation 4.12 instead of the Bernoulli trials approximation. The resulting pair approximation which replaces that in Equation 4.12 is given in the worksheets mentioned in the Introduction.

Critical transmissibilities

For a fixed Q the clumping is increased by increasing φ as explained on pp. 109–110. We will find that there is a critical $\varphi = \varphi_c$ at which β_c diverges to infinity, i.e. the disease is unable to establish.

By standard structural stability theory, for small $\varphi > 0$ the onset of the disease is qualitatively like that for the standard pair approximation above except that onset of the disease is slightly delayed (i.e. β_c is larger; see Fig. 4.2a) and the number of infectives is reduced.

In particular as β passes through β_c the trivial solution exchanges stability with the nontrivial solution. This is characterized by the fact that one of the eigenvalues of the linearized system about the trivial solution becomes zero and hence by the fact that the determinant of the matrix M giving the linearized system is zero.

As in the previous section the correlation structure of the invading population is determined by the eigenvector $\vec{n}_0 = ([S]_0,[I]_0,[SS]_0,[SI]_0,[SS]_0)$ with eigenvector zero at $\beta = \beta_c$. This satisfies $M \cdot \vec{n}_0 = 0$ and it follows immediately from the resulting linear equation that, at invasion, $Q(S|I) = [SI]_0/[I]_0 = \sigma/\beta$. Substituting this into the equation det $M = 0$ gives a quadratic equation of the form $a\beta^2 + b\beta + c = 0$ where a, b and c are functions of φ and the other parameters but not β. This has an appropriate solution of the form $\beta_c = \beta_c(\varphi)$ for $0 \leq \varphi < \varphi_c$ where φ_c is given by the equation $a = 0$. This can readily be solved and gives

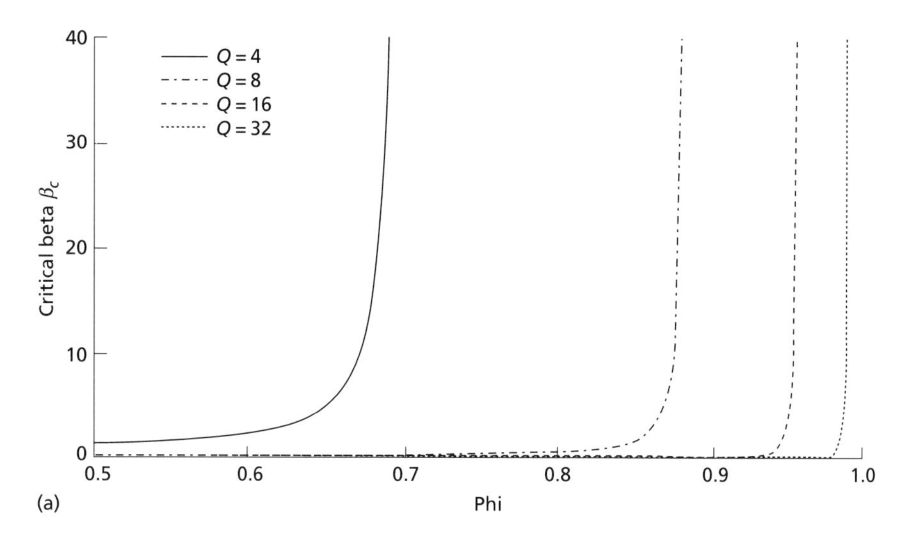

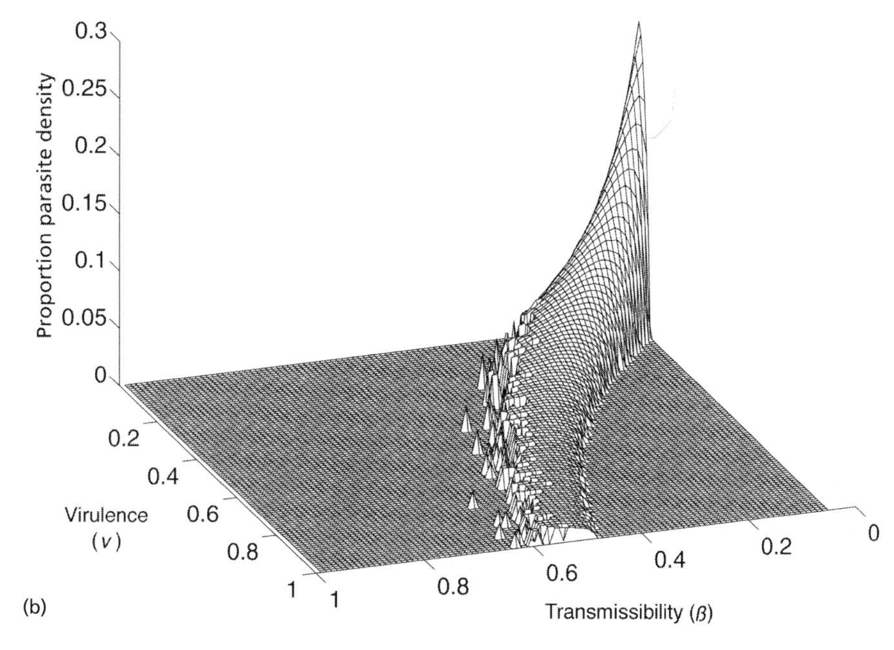

Fig. 4.2 (a) A plot of β_c against various values of φ showing the divergence of β_c to ∞. (b) The density of pathogens in stochastic simulations of the host–parasite system of p. 117 plotted against both V and τ averaged over a number of runs and including both those where the pathogen survived and where it died out.

$$\varphi_c = 1 - \frac{\sigma - \mu Q}{\sigma(Q-1)}$$

As one would expect, $\varphi_c \to 1$ as $Q \to \infty$. The dependence of β_c on φ for a variety of Q values is shown in Fig. 4.2. Notice what a massive effect this is for small Q values. Thus the situation is radically different from that for the standard SIR equation and the Bernoulli pair approximation SIR equation.

One can easily image that such a network structure will arise in situations where there are significant social structures involving families, schools or other groups. Applying these ideas to a realistic structure of this form is the next step. The importance of what I have explained here is two-fold. Firstly, that such structure can have very significant effects and secondly that it is often quite amenable to precise mathematical analysis. The calculations for this model are available as a Maple worksheet from my web site as explained in the Introduction.

Clumped networks and oscillations

In contrast to the classical mean-field SIR equation, in certain parameter ranges the clumped pair approximation displays limit cycle behaviour. In fact, for the parameter values of Fig. 4.1, as φ is increased past 0.21 the infected equilibrium solution undergoes a Hopf bifurcation in which it hands over its stability to a limit cycle which grows out of it (Morris 1997). This raises the question of whether or not this is a real phenomenon, at least to the extent that it is reflected in stochastic simulations. This is discussed in Morris (1997) where it is shown that the corresponding lattice system (with $N = 10000$) does indeed oscillate with an amplitude close to that of the pair approximation though with a slightly greater period.

Evolution to criticality in the host–parasite system of Rand *et al.*

In the paper by Rand *et al.* (1995) the authors studied a simple generic spatial, individual-based host–parasite system in which a number of interesting effects were observed. In particular, they studied evolution of virulence and transmissibility in this system. Among the biologically interesting phenomena found were the following.

1 Compared with the corresponding mean-field models, selective pressure is substantially reduced. Evolution is much slower and there is more stability to invasion by mutant hosts and pathogens. This suggests that artificial removal of pathogens by processes designed, for example, to promote host health can lead to faster evolution and can reduce evolutionary stability.

2 Critical transmissibility. Unlike the mean-field models, there exists an upper critical transmissibility β_c above which the pathogen dies out.

3 Self-evolved criticality. If the transmissibility β is allowed to mutate, it evolves to the critical value β_c. Thus the system evolves so as to put itself at the boundary of where it can exist.
There is also a lower critical transmissibility $\beta_{c,0}$ whose existence is easy to explain. The pathogen persists when $\beta_{c,0} < \beta < \beta_c$. The critical transmissibilities are functions of the other parameters of the model such as the virulence V. I now analyse this model using a pair approximation and show that the above phenomena are accessible to direct mathematical analysis. Pair approximation techniques were first applied to this system in Keeling (1995) although he did not derive the equations or calculate the evolution.

The system studied is a synchronously updated probabilistic cellular automaton and therefore the corresponding pair approximation is a map. It is more natural biologically to consider the asynchronously updated case. Luckily, the analyses of either case applies directly to the other and the results for the two systems are the same. Therefore, since the asynchronous case fits in with this exposition, I only treat that here.

In addition to the susceptible (S) and infected (I) states, sites are also allowed to be empty (ϕ). Infected hosts die at a rate V and uninfected hosts give birth into a neighbouring empty site at rate b. The events are:

1 *birth*: at edge e, $S\phi \xrightarrow{b} SS$ at rate b.
2 *death*: at site x, $I \xrightarrow{V} \phi$ at rate V constant, V is called *virulence*.
3 *infection*: at edge e, $SI \xrightarrow{\beta} II$ at rate β constant, β is called *transmissibility*.
4 *recovery*: at site x, $I \xrightarrow{v} S$ at rate v constant.

Mean-field system

I do not consider here the fully homogeneously mixed system, but the more appropriate Q-neighbourhood mean-field system where it is assumed that each individual has a neighbourhood of size Q and the neighbourhood ratios $Q(i|j)$ are the same for each neighbourhood and hence equal $\rho_i Q$. The non-trivial equilibrium $[I] = b\sigma(\alpha - 1)/(b\sigma + \beta V)\alpha$ exists provided $\beta > \beta_{c,0} = \sigma/Q$. Here $\sigma = V + v$ and $\alpha = Q\beta/\sigma$. Moreover, the selective pressure is $\mathrm{d}s = \beta^{-1}\sigma \mathrm{d}\beta - \mathrm{d}V - \mathrm{d}v$ (see Rand *et al.* 1995). This means that if the system is invaded by parasites whose parameters β, V and v differ from those of the resident by $\mathrm{d}\beta$, $\mathrm{d}V$ and $\mathrm{d}v$ then the exponential rate of growth of the invader (the invasion exponent) is given by $\mathrm{d}s$ up to terms which are quadratic in $\mathrm{d}\beta$, $\mathrm{d}V$ and $\mathrm{d}v$ and therefore very small. Thus the mean-field system should evolve so as to maximize β and minimize $\sigma = V + v$.

Pair approximation

In the work by Keeling (1995), it was discovered that in simulation of the probabilistic cellulor automaton, the following triple correlations which occur in the master equation did not satisfy the Bernoulli trials approximation but instead was well approximated by the following:

$$[ISI]=1.4\kappa\frac{[SI]^2}{S}$$

The use of such approximations was discussed on p. 110. For all the other triples that occur in the master equation the Bernoulli trials approximation was acceptable. With this and assuming constant neighbourhood size and $v = 0$ one arrives at the following pair approximation:

$$\frac{d}{dt}[S]=-\beta[SI]+b[\phi S]\quad\frac{d}{dt}[I]=\beta[SI]-V[I]$$

$$\frac{d}{dt}[SS]=b[\phi S](\kappa Q(S|\phi)+1-\beta\kappa Q(S|S))$$

$$\frac{d}{dt}[SI]=b\kappa Q(S|\phi)[\phi I]+(\beta(\kappa Q(S|S)-1.4\kappa Q(I|S)-1)-V)[SI]$$

$$\frac{d}{dt}[II]=\beta(1.4\kappa Q(I|S)+1)[SI]-V[II]$$

By direct solution one finds that there is an equilibrium with positive parasite density provided $\beta > \beta_{c,0}$ where

$$\beta_{c,0}=\frac{V(b(Q-2)+V)}{(Q-2)(b(Q-1)+V)}$$

This is the lower critical transmissibility. The equilibrium number of infecteds can be solved for but the expression is too unwieldy to include here. Its qualitative form is similar to the example shown in Fig. 4.3. Note how the parasite numbers remain bounded away from zero in this figure as β approaches the upper critical transmissibility β_c (which for this example is 0.5140671 . . .).

The upper critical transmissibility β_c

In this model what determines the upper critical transmissibility β_c is an inverted Hopf bifurcation. When β is slightly less than β_c the above solution is linearly stable. However, close to it is an unstable limit cycle. As β passes through β_c this limit cycle coincides with the equilibrium solution destroying its stability. The equilibrium solution continues to exist but is now unstable. As a result the orbit of any initial condition close to it spirals away and comes arbitrarily close to $[I] = 0$. An example of this happening just above criticality is shown in Fig. 4.3(b). I believe that these oscillations are seen in the simulations of the cellular automaton.

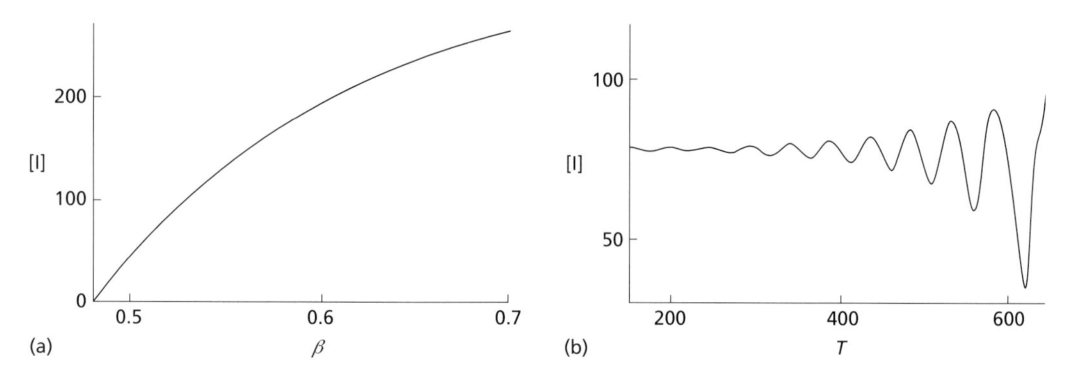

Fig. 4.3 (a) A plot of the equilibrium value of [I] against β for the host–parasite model. (b) A plot showing the way in which the solution oscillates with β is slightly bigger than the upper critical value. This solution has been started very close to the equilibrium value which became unstable at the Hopf bifurcation.

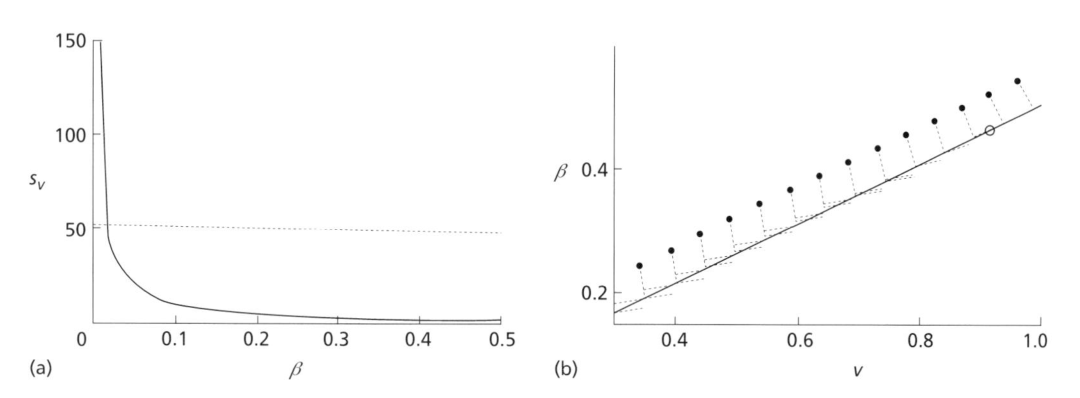

Fig. 4.4 (a) The values s_β (solid) and s_V (dotted) of the selective pressure in the β and V directions respectively. The selective pressure 1-form is then $s = s_\beta d\beta + s_V dV$. (b) This plot shows the upper-β boundary in the (β, V)-parameter space given by the upper critical transmissibility and a sample of the selective pressure kernels along it. The perpendicular arcs with circular ends show the direction of positive pressure. The ESS is where the kernels become tangent to the boundary curve.

Evolution to criticality

Figure 4.4(b) shows the upper boundary of the domain of existence of the parasites for the parameters used in Rand *et al* (1995). Now I want to address the question of the evolution of the parasite transmissibility and virulence by drawing the selective pressure ds for this system onto this diagram. To do this one calculates at each point (β,V) the straight line through this point which is tangent at (β,V) to the curve of those points which have a zero invasion exponent with respect to the resident population with parameters (β,V).

This can be calculated using the formalism for selective pressure given in Rand *et al.* (1994). These kernels are also shown in Fig. 4.4(b). I have calculated the selective pressure in the β and V directions and show the answers in Fig. 4.4(a).

It thus remains to remains to add to the diagram which of the two directions corresponding to the perpendiculars to these kernels corresponds to the direction of positive selective pressure. To do this it is only necessary to do a rough calculation using the data in Fig. 4.4(a) and the answer is shown in Fig. 4.4(b). Straight away we see the following two important facts. Firstly, that for the birth and recovery rates considered there is a single point (V_{ESS}, β_{ESS}) on the curve at which the kernel is tangent. This is marked with a circle in Fig. 4.4. Secondly, that any population starting in the interior of existence domain will evolve to the boundary given by the curve of upper critical transmissibilities and then to (V_{ESS}, β_{ESS}). This is because evolution having driven the state to the critical curve will evolve along the curve in the direction which is positive with respect to these kernels. On this curve, if $V < V_{ESS}$ then there is selective pressure to increase V and if $V > V_{ESS}$ to decrease it. Thus, for this system, the virulence and transmissibility evolve to definite intermediate values with the virulence close to 1. This is very different from the behaviour of the mean-field system. In previous models to obtain evolution to intermediate values one has been forced to postulate the existence of constraints between V and β.

Host–parasite system of Satō *et al.*

This model was considered in Satō *et al.* (1994). The states are the same as in the previous model but the model differs in that infected hosts do not recover but can die and it is assumed that the death rate is independent of infection. Moreover, only susceptibles can give birth. Thus the infection affects the birth rate but not the death rate. The events are therefore:

1 *death with replacement by susceptible*: at site x, S or $I \xrightarrow{d} S$ at rate $d = 1$ constant;

2 *birth*: at edge e, $S\phi \xrightarrow{b} SS$ at rate $b = m_S/Q$ constant;

3 *infection*: at edge e, $SI \xrightarrow{\beta} II$ at rate $\beta = m_I/Q$ constant.

In simulations of this model Satō *et al.* find that $Q(I|\phi\phi)$ is substantially less than $\kappa Q(I|\phi)$. They thus propose to use an approximation of the form $Q(I|\phi\phi) \approx \varepsilon Q(I|\phi)$ which they call an improved pair approximation. The value of ε to use must be chosen *ad hoc* and they estimate it as follows.

Consider the case where $[I] = 0$, i.e. there is no infection. For persistence of the S population they need $\beta > \beta_c = Q/\varepsilon(Q-1)$. But β_c should also be the critical value λ_c for ergodicity of the true contact process. For a two-dimensional square lattice ($Q = 4$), $\lambda_c \approx 0.4119$ so as an estimate one can take $\varepsilon \approx 0.8093$. In this case Satō *et al.* find that the parameter space of transmissibility β and relative fecundity m_I/m_S is divided into three regions so that if β^{-1} is fixed at a value less than approximately 0.61 then for small m_I/m_S the disease dies out. For slightly larger m_I/m_S the disease is endemic and for even larger values all individuals become infected and so the pathogen drives the extinction of the host.

Measles

So far all the pair approximations considered have had either a fixed point or limit cycle as their attractors. Now I discuss a case where one can obtain chaotic attractors and where the residual noise coming from the corrections can play a significant dynamical role. I briefly consider the dynamics of measles epidemics following Keeling *et al.* (1997). Not only is an understanding of measles dynamics very important from a public health perspective (it is still major killer outside the developed world), but it has also become a testbed for epidemiological ideas because as well as possessing complex dynamics, it has a very simple natural history and, in comparison to other ecological systems, there are lots of good data (see references in Keeling & Grenfell 1997).

One of the main stumbling blocks to a more complete understanding of the disease is the inability of models to match the existence of relatively violent seasonally driven epidemics with persistence at the critical population size of around 300 000–500 000 (Bartlett 1960; Black 1966). I consider what light pair approximations can throw on this and the related issue of persistence.

The basic mean-field model for this system is the SEIR equations which model the proportions of susceptible, exposed, infectious and resistant individuals in the population (Anderson & May 1992). Other, more complicated and more realistic systems such as the RAS model (Schenzle 1984) are based on these. Therefore, I start by considering the pair approximation corresponding to the SEIR equation. This involves an extra state: exposed E which corresponds to an individual who has been infected but is not yet infective to others. The events are as follows:

1 *death = replacement by susceptible*: at site x, $S, E, I, R \xrightarrow{m} S$ at rate m where $m \approx (50 \text{ years})^{-1}$;

2 *infection*: at edge e, $SI \xrightarrow{\beta} EI$ at rate β described below;

3 *onset of infectiousness*: at site x, $E \xrightarrow{a} I$ at rate a where a^{-1} is the latent period of the infection (approximately 8 days);

4 *recovery*: at site x, $I \xrightarrow{g} R$ at rate g where g^{-1} is the infectious period (approximately 5 days).

The values for the rates given here are from Anderson & May (1992).

The effective contact and transmission rate β used in the SEIR equations is often modelled as seasonally varying, $\beta = 4.93(1 + 0.28\cos(2\pi t))$ day^{-1}

where t is the time in years (values are from Olsen & Schaffer 1990). This time dependence is supposed to be due to the annual cycle of new recruitment of students to schools. It therefore corresponds to a periodic modulation of the structure of the interaction network of the population. Here it is modelled directly as such. It will be assumed that for a school-child during term-time not only is the neighbourhood size Q larger but also, since the class is highly interconnected, the proportion of triangles φ is increased. Thus the transmissibility β is kept fixed and the network structure varied instead.

The following equations follow from the master equation.

$$
\begin{aligned}
&d[SS]/dt = 2m([SE]+[SI]+[SR]) - 2\tau[SSI], \\
&\qquad d[EE]/dt = -2m[EE] + 2\tau[ESI] - 2a[EE] \\
&d[SE]/dt = m([EE]+[EI]+[ER]-[SE]) + \beta([SSI]-[ESI]-a[SE]) \\
&d[SI]/dt = m([EI]+[II]+[IR]-[SI]) - \beta([ISI]+[SI]) + a[SE] - g[SI] \\
&d[SR]/dt = m([ER]+[IR]+[RR]-[SR]) - \beta[RSI] + g[SI] \\
&d[EI]/dt = -2m[EI] + \beta([ISI]+[SI]) + a([EE]-[EI]) - g[EI] \\
&d[ER]/dt = -2m[ER] + \beta[RSI] - a[ER] + g[EI], \\
&\qquad d[II]/dt = -2m[II] + 2a[EI] - 2g[II] \\
&d[IR]/dt = -2m[IR] + a[ER] + g([II]-[IR])
\end{aligned}
\tag{4.12}
$$

To obtain the appropriate pair approximation I use the clumped pair approximation for the triples given by Equation 4.12. Both Q and φ are periodic functions of t.

The inclusion of correlations has a strong impact on the dynamics. Although age structure has been ignored it is still possible to reproduce regular biennial epidemics reminiscent of the dynamics in developed countries as well as annual cycles and chaotic transients that are stabilized by the residual stochastic corrections as in Rand & Wilson (1991) and Morris 1997. The troughs are much less deep than those of the standard SEIR model. Such dynamics are shown in Fig. 4.5. The inclusion of pairwise correlations has a dampening effect on the spread of the disease because it keeps track of the relatively high level of *I-I* correlations which inhibit spread. In addition, keeping track of the network limits each infective to at most Q susceptibles to infect and, moreover, the absence of connections implied by a high value of φ for a given Q means that large reservoirs of susceptibles can persist.

One can approximate the nine-dimensional system (Equation 4.12) by a four-dimensional system by using the fact that the correlation $C_{SE} =$

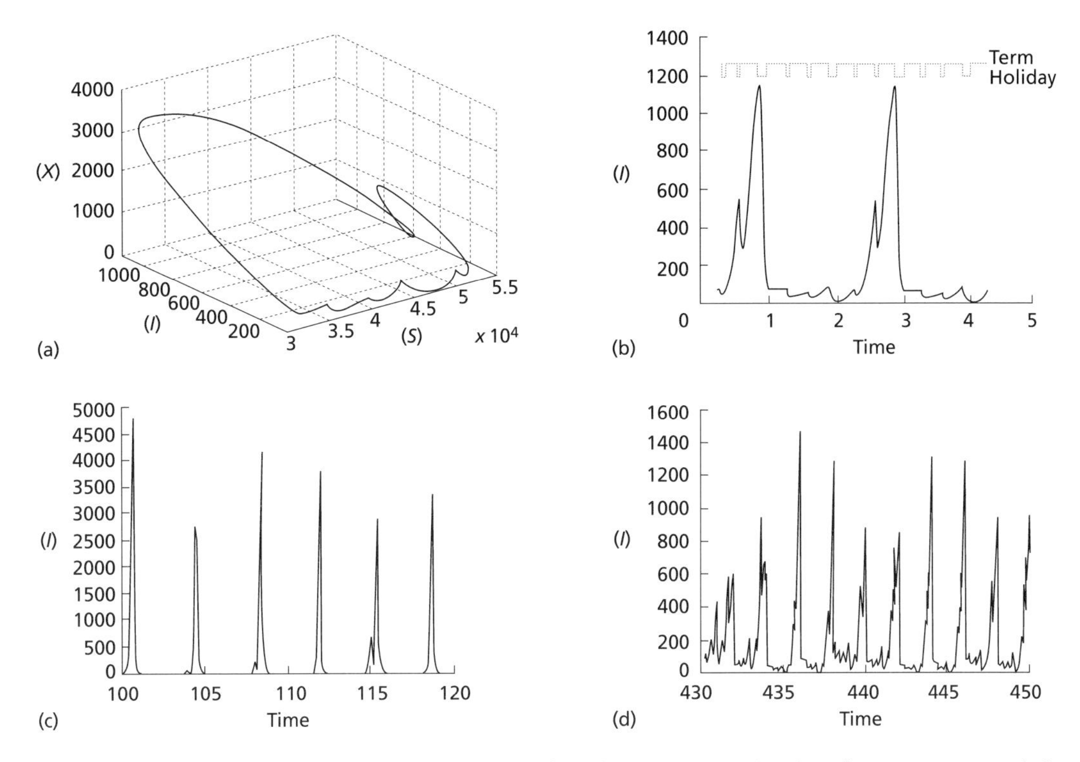

Fig. 4.5 Output from the seasonally forced SEIR pair approximation. The attractor as a period two orbit. (a,b) The attractor. (c,d) Typical transient behaviour at times of 100 and 430 years. This will be stabilized by noise as in Rand & Wilson (1991). (From Morris 1997).

$N[SE]/Q[S][E]$ is well approximated by $\kappa C_{SS} = \kappa N[SS]/Q[S]^2$ (Keeling *et al.* 1997). This is because upon becoming infectious the new infected takes over the whole neighbourhood of the exposed and if the infection does not last too long compared with the time that typical neighbours are susceptible then there is not much time for this correlation to decay. These models lack the refinements that arise when age-structure is included, but demonstrate the strong stabilizing effects that pair correlations confer.

It is generally accepted that the inclusion of age-structure into models for measles is vital to understand and predict the dynamics (Schenzle 1984; Bolker 1993). The resulting RAS (realistic age-structured) model is a considerable improvement on the original SEIR equations, predicting realistic biennial cycles. With this in mind an age-structured pair (ASP) model of Equation 4.12 was considered in Keeling *et al.* (1997).

This network for this model has four age-classes (preschool, primary, secondary and adult) and has connections within and between these classes. Within classes the connections are defined so as to add some rudimentary family structure. In particular, the single set of parameters β, Q and φ are replaced by multiple neighbourhoods each with a common transmissibility.

For example, a school-child may be a member of two subnetworks, a school subnetwork where the number of neighbours and triangular connections are large, and a family subnetwork where transmission rates are higher, but the number of neighbours far less. The network within schools may be further subdivided so that heterogeneities in the number of contacts can also be modelled.

Within the range of reasonable parameters, 2-year cycles reminiscent of the RAS model and longer, more complex cycles are possible all of which correspond well to the England and Wales data set. As in the RAS model, it is found that the majority of the dynamics is within the primary-school layer where the fluctuations in the number of cases are the largest. This is in agreement with the age-structured data from developed countries. It also appears that when the dynamics are most erratic, it is the slower transmission within the preschool layer that forms a reservoir of the disease, maintaining the epidemics through the troughs. This would imply that it is the family structure, which accounts for the majority of contacts being preschool and primary-school, that enables measles to persist.

The important improvement upon the RAS model comes about when one considers critical community size and levels of fadeout. For this one has to use a stochastic version of the equation taking into account the fluctuations that are removed from the deterministic version. The stochastic RAS model does not reproduce observed behaviour (see the references in Keeling & Grenfell 1997): for example, the critical community size exceeds 20 million.

The stochastic ASP model does much better. For example Fig. 4.6(a) shows the average number of fadeouts (defined as three or more consecutive weeks without infection) per year for the parameters used in Keeling *et al.* (1997) and compares the result to the England and Wales data and to the RAS model. For the biennial parameters the number of fadeouts is even lower with the overall behaviour being consistent.

Moreover, the ASP model predicts a critical community size of around one million which is far less than is found with the RAS or SEIR models. This can be attributed to the presence of correlations between infectious and susceptible individuals and the above-described effect of this.

Even more striking than the lower critical community size, which could be obtained from large metapopulation models or other modifications to the RAS model (Keeling & Grenfell 1997), is the agreement between the ASP model and data for small populations. This would indicate that the pair model is capturing the essential local features of epidemics, although it may be failing to reproduce all the larger scale heterogeneities.

Finally, I want to mention that the ASP model has interesting power-law scaling for the size of epidemics in small, isolated communities (see Fig. 4.6 and Keeling *et al.* 1997). Such laws have previously been discussed by

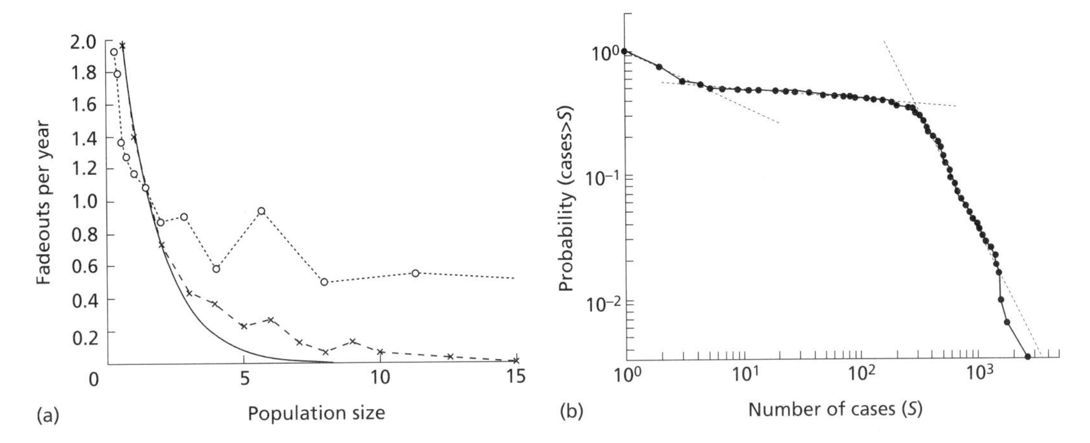

Fig. 4.6 Results from a stochastic version of the pairwise model. (a) Average number of fadeouts per year: the solid line is the best exponential fit to the prevaccination England and Wales data, the circles are the results from a stochastic version of the RAS model and the crosses are from the pairwise model. The pairwise model predicts a lower critical community size, and is an extremely good fit for small populations. (b) Power-law behaviour for the size of an epidemic in a small isolated community; population size 25 000, with on average three infectious individuals introduced every 4 years. There are three distinct sections of power-law scaling: within family, within school and between school.

Rhodes & Anderson (1996a,b) in the context of an explicit spatial model for island communities, but without the multiple scaling regimes seen here.

Spatial games

We move on to consider some simple spatial games. It is here that there has been one of the most extensive studies of how well the pair approximations describe lattice and network simulations (Morris 1997). A discussion of this will help put the above theoretical discussion into context.

I consider games in which individuals playing pure strategies play against their neighbours. There are a number of ways of formulating the basic stochastic process corresponding to different biological and social interactions and depending upon whether the dynamics operate through survival and reproduction or by learning.

Almost all treatments of evolutionary game theory (e.g. Maynard Smith 1982) assume homogeneous mixing of the population. For example, in a game with pairwise interactions it is usually assumed that the pairs interacting are drawn randomly from the whole population. Clearly, in most biological situations, this is not the case and individuals are more likely to interact with those closest to them. In this case, correlations are likely to become important because they will alter the expected payoffs. For example, in games involving cooperation, if the frequency of the cooperators is small and there is homogeneous mixing then the typical interaction is with a noncooperator and therefore the cooperators will have a low fitness. On the other hand, high correlations between cooperators will mitigate this by increasing the frequency of interactions between cooperators.

The game rules

Let E_{ij} denote the payoff to an individual playing strategy i from a contest with an individual playing j. Then, the fitness F_x of the individual at x is defined to be the average payoff to the individual against its neighbours. Examples of game processes include the following.

- *Replacement by fit strategies* where individuals playing a given strategy are replaced by individuals playing alternatives strategies in the neighbourhood at a rate proportional to the fitness of the alternatives. This could be because the individuals die and are replaced by the offspring of neighbours or because individuals change their strategies regularly according to their fitness. The events are then just pair events of the form:

  (a) *replacement*: at edge $e, ij \xrightarrow{r} jj$ at rate F_{e_j}.

- *Learning* where individuals x adopt the strategy of a more fit neighbour y at a rate proportional to the excess fitness of the neighbour:

  (b) *learning*: at edge $e, ij \xrightarrow{r} jj$ at rate $r = F_{e_j} - F_{e_i}$ if $F_{e_j} > F_{e_i}$.

- *Replacement by death* where the fitness of an individual x playing i determines its death rate and where, upon death, the individual is replaced by an offspring of one of the neighbours chosen randomly. If the death rate is given by $d_x = \exp(-\alpha F_x)$ then the events are:

  (c) *replacement through death*: at site $x, i \xrightarrow{r} j$ at rate $r = q(j|i)\, d_x i$ where $q(j|i) = Q(j|i)/Q$.

I will mainly consider the first of these processes here but the latter process will be used to study the evolution of cooperation in the Prisoner's Dilemma on p. 137. The situation where individuals play mixed strategies is also easy to model in this way but is not considered here because of lack of space.

Calculating the pair approximation

In the first part of this section my main aim is to compare pair models and simulation. Therefore I will restrict attention to games involving only two strategies which I denote by 1 and 2. Also I consider the first replacement process described above. I will take a little trouble here to calculate the equations, partly to illustrate the technique but mainly because it will be important in our discussion of the differences between regular and irregular networks.

In what follows if i is one of the strategies 1 or 2, then i' denotes the other strategy. Assuming a regular network, it is clearly enough to calculate the differential equations for $[i]$ and $[ii]$ for $i = 1$ and 2. In a ii' pair the i strategy replaces i' at rate F_{e_i}. Let F_i denote the mean fitness $Q(i|ii')E_{ii} + Q(i'|ii')E_{ii'}$. Then the master equation gives

$$\frac{\mathrm{d}}{\mathrm{d}t}[i] = \sum_{\sigma e=ii'} F_{e_i} - F_{e_{i'}} \underset{=}{Q_x \equiv} Q^{-1}(F_i - F_{i'}) \tag{4.13}$$

$$\frac{\mathrm{d}}{\mathrm{d}t}[ii] = \sum_{\sigma e=ii'} F_{e_i} Q_{e_{i'}}(i) - F_{e_{i'}} Q_{e_i}(i) \underset{=}{Q_x \equiv} Q^{-1}(F_i Q(i|i'i) - F_{i'} Q(i|ii')) \tag{4.14}$$

where the equality and approximation marked accordingly are only true when $Q_x \equiv Q$ is independent of the site x (see the discussion on p. 107). The approximation of the last line neglects the correction terms of the form $\Sigma_{\sigma e=ii'} \eta_e(k|ii') \times \eta_e(l|i'i)$ where k and l are i or i'. Because $i \neq i'$ we expect these to be small.

If the Bernoulli trials approximation is used and the correction terms ignored then the following pair approximation is obtained:

$$\frac{\mathrm{d}}{\mathrm{d}t}[ii] = 2Q^{-1}[ii']((E_{ii} - E_{i'i})\kappa Q(i|i)(\kappa Q(i|i')+1) + E_{ii'}(\kappa Q(i'|i)+1)(\kappa Q(i|i')+1) - E_{i'i'}\kappa^2 Q(i|i)Q(i'|i')) \tag{4.15}$$

This gives all equations because $[12] = NQ - [11] - [22]$ and $Q[i] = [ii] + [ii']$.

A hawk–dove game

Following Morris (1997) let us now consider the Hawk–Dove game. Hawks escalate the contest until injured or until the opponent retreats whereas doves display but then retreat at once if the opponent escalates. If the resource competed for has a value v and injury reduces fitness by an amount c then using the simplest assumptions one obtains the payoffs given in Table 4.1. I am only considering here the most interesting case where $v < c$. Recall that w_0 must be such that all the payoffs are positive. I choose the minimal such value $w_0 = (v - c)/2$. Then if all the payoffs are scaled by $2/c$ (which only changes the time-scale) the payoffs are $E_{HH} = 0$, $E_{HD} = 1 + s$, $E_{DH} = 1 - s$ and $E_{DD} = 1$ where $s = v/c$.

Regular networks

Assume now that the network is regular with $Q_x \equiv Q$. The pair approxima-

		Player *B*	
		Hawk *H*	Dove *D*
Player *A*	Hawk *H*	$w_0 + \frac{1}{2}(v - c)$	$w_0 + v$
	Dove *D*	w_0	$w_0 + v/2$

Table 4.1 The payoffs to player *A* in a contest against player *B* when *A* and *B* play the strategies shown. I denote by E_{ij} the payoff to player *A* when player *A* plays strategy *i* and player *B* plays strategy *j*.

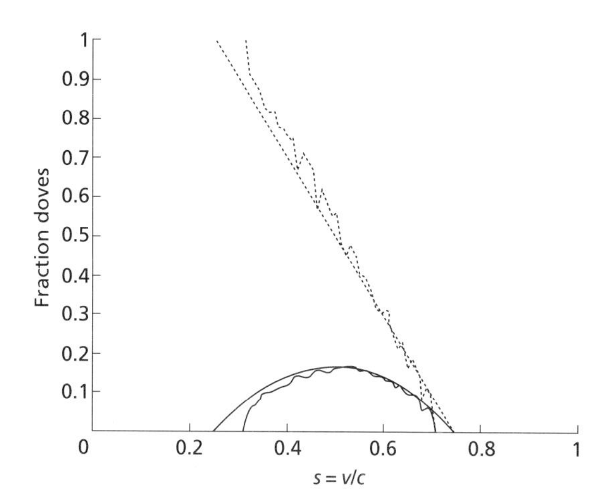

Fig. 4.7 A comparison of the data from a simulation of the game and the pair approximation. The straight line is a plot of the equilibrium value of the number of doves vs. *s* for the pair approximation (Equation 4.15). This is compared with the equilibrium values found in a simulation on a regular lattice of size $N = 2500$. Also shown is the fraction of hawk–dove pairs for both the pair approximation and the simulation. (From Morris 1997.)

tion for this system is given in Equation 4.15 and the equilibria for this are given by one of $[H] = 0, [D] = 0$ or

$$[D] = \frac{N(Q - Qs - 1)}{Q - 2},\ [H] = \frac{N(Qs - 1)}{Q - 2},\ [HH] = \frac{Q^2 Ns(Qs - 1)}{(Q - 2)(Q - 1)}$$

$$[DD] = \frac{Q^2 N(1 - s)(Q - Qs - 1)}{(Q - 2)(Q - 1)},\ [HD] = \frac{QN(Q - Qs - 1)(Qs - 1)}{(Q - 2)(Q - 1)}$$

The coexistence equilibrium only exists between $s = Q^{-1}$ and $s = 1 - Q^{-1}$. Inside this interval it is stable and as s passes out of this interval there is an exchange of stability with one of the trivial solutions $[H] = 0$ or $[D] = 0$. Thus $[H] = 0$ if $s < Q^{-1}$ and $[D] = 0$ if $s > 1 - Q^{-1}$.

For regular networks these agree very well with full spatial simulations (Morris 1997) as illustrated by Fig. 4.7. The validity of the Bernoulli trial pair approximation can be checked in some detail using the simulation. Recall from the section on deriving correlation equations (pp. 104–109) that the important thing here was our estimates for the correlation $\Gamma(i|j|k)$. In our case all of these equal either $\pm\Gamma(D|D|D)$ or $\pm\Gamma(H|H|H)$ because $\Gamma(i|j|k) = \Gamma(k|j|i)$ and, if $i \neq j$, $\Gamma(i|j|k) = -\Gamma(j|j|k)$ since $\eta(i|j) = -\eta(j|j)$ as there are only two species. A typical time series of $\Gamma(D|D|D)$ and $\Gamma(H|H|H)$ in a simulation is shown in Fig. 4.8. Note that these settle down to equilibrium values that are quite far from zero. The noisy fluctuations about the equilibrium levels are very small and scale as one would expect. I now consider the equilibrium levels.

In equilibrium

$$Q(i|i) = \frac{[ii]}{[i]} = \begin{cases} \kappa^{-1}Q(1-s) & \text{if } i = D \\ \kappa^{-1}Qs & \text{if } i = H \end{cases}$$

so assuming Bernoulli trials it follows from Equation 4.7 that

$$\Gamma(i|i|i) = Q(i|i)\left\{1 - \frac{Q(i|i)}{Q}\right\} = \begin{cases} \kappa^{-1}Q(1-s)\{1-\kappa^{-1}(1-s)\} & \text{if } i = D \\ \kappa^{-2}(Q - Qs - 1)s & \text{if } i = H \end{cases}$$

Thus if $Q = 4$, then the equilibrium levels of $\Gamma(D|D|D)$ and $\Gamma(H|H|H)$ are respectively 16/25 and 224/225 when $s = 0.4$ and 8/9 and 16/18 when $s = 0.5$ which is in general agreement with the observed values shown in Fig. 4.8. Thus in this case the Bernoulli trials approximation is good. The correction to it is well modelled by low-amplitude, zero mean noise.

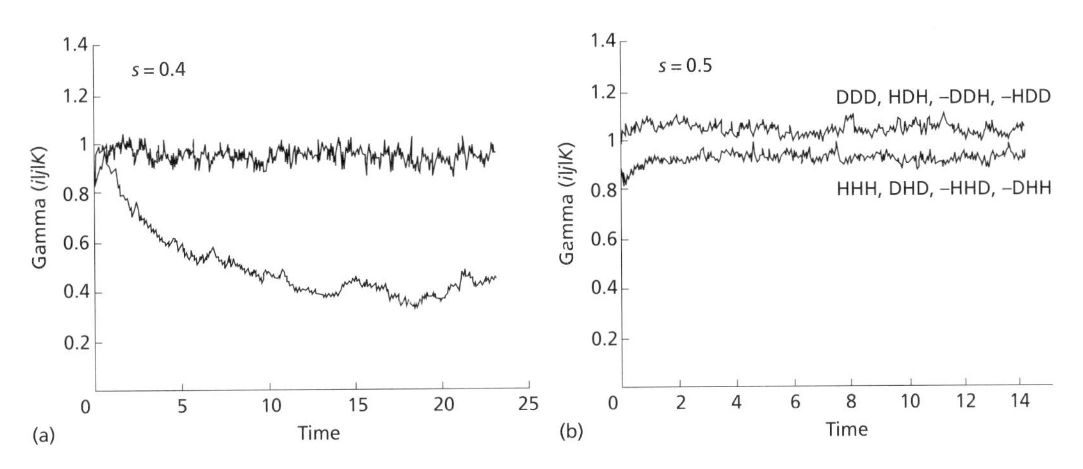

Fig. 4.8 Times series of both independent $\Gamma(i|j|k)$ for a simulation of the hawk–dove lattice system on a lattice of size $N = 2500$ with $s = 0.4$ (a) and $s = 0.5$ (b). Both time series eventually settle to small fluctuations about an equilibrium value. The labels give the value of *ijk*. (From Morris 1997.)

Irregular networks

In the case of an irregular network where Q_x varies from site to site one cannot use the same pair approximation because although Equations 4.13 and 4.14 are correct the equations for all other pair types do not follow from these unless Q is constant. More importantly, the equalities marked $Q_x \equiv Q$ are no longer valid. If this is the case then the factors $1/Q_{e_i}$ in $F_e(i)$ introduce an essential nonlinearity into the equations and because of this, for this example, the behaviour of regular networks with $Q_x \equiv Q$ differs significantly from irregular networks as shown in Fig. 4.9. This compares a simulation of the irregular network system with the equilibrium levels for the regular network pair approximation for varying s. The coexistence equilibrium for the irregular system now has a significantly smaller range of existence and the equilibrium level has a strong dependence upon s. Thus the nonlinearities introduced by the irregularity have some real effects. Luckily in many systems, such as the epidemiological ones considered previously, irregularity does not introduce such nonlinearities and both regular and irregular systems can be dealt with together. In some applications with nonlinearities one can assume that the Q_x are Poisson distributed with mean Q and make some progress in deriving equations but that is beyond the scope of this chapter.

The evolution of altruism and cooperation

In this section I discuss the role of spatial correlation in enabling altruism

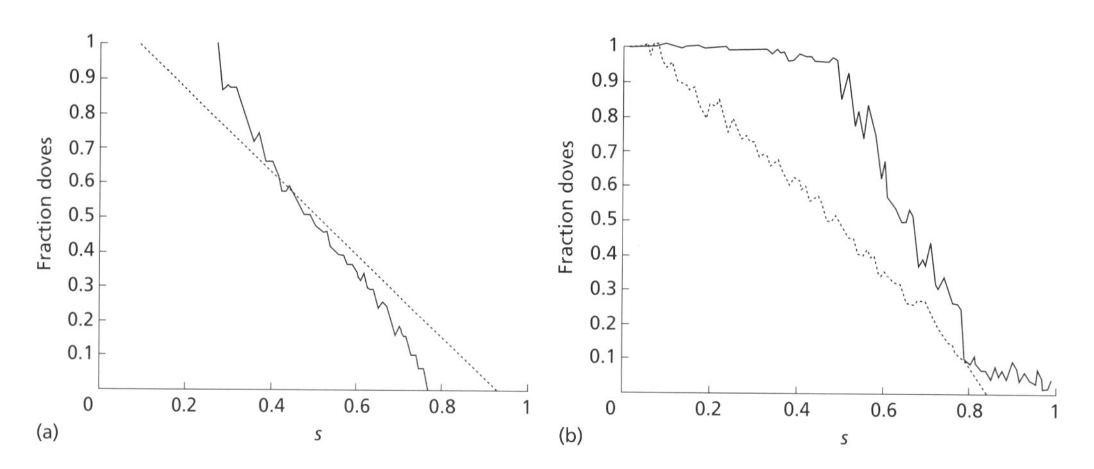

Fig. 4.9 Equilibrium composition of the population plotted against s for (a) a fixed random network and (b) a random dynamic network where births and deaths create and destroy sites in the network. Both have $Q = 12.6$. The jagged curve close to the straight line given by the pair approximation is the corresponding graph for a regular lattice (b) with $Q = 12.6$. The simulations were run for 100 000 events on a lattice with $N = 2500$. (From Morris 1997.)

and cooperation. In particular, we will see that when correlations are taken into account it is possible for altruism and cooperation to invade nonaltruistic or noncooperating populations.

Darwin in *On the Origin of Species* already realized that the evolution of both altruism and cooperation either within or between species was an obvious problem for his theory. To explain this he argued that

> this difficulty, though appearing insuperable, disappears when it is remembered that selection may be applied to the family, as well as the individual....

The problem is that while altruists get the benefit from other altruists, they must also pay the costs of altruism whereas nonaltruists get the benefit but do not have to pay the costs. Early on, the concept of intergroup selection was widely adopted to explain social behaviour in all kinds of animal groups. However, by the mid-1960s the opponents of this view, the individual selectionists, had won the day by arguing that, firstly, there are few populations that have the kind of group structure required of intergroup selection and, secondly, that even if such selection could exist, it was bound to be weaker than individual selection because groups displace each other within populations more slowly than individuals displace each other in groups.

This left open the question of how to explain altruism and cooperation. Leaving aside the possibility that they do not exist there are three current approaches to a solution associated with three different types of altruism.

1 *Kin altruism.* Kin selection may favour altruistic acts between related individuals because there is a reasonable chance that they carry a copy of each other's genes (Hamilton 1964; Maynard Smith 1964). The main result (Hamilton's rule) is that such altruism will evolve provided that relatedness times benefit averaged over the population is greater than the cost of altruism.

2 *Reciprocal altruism.* Under reciprocal altruism altruistic behaviour evolves by reciprocity among nonrelated individuals (Trivers 1971, 1985; Axelrod & Hamilton 1981). This could account for cooperation distributed widely throughout a population. It is only expected to arise when individuals have a long association with each other and can discriminate between those that reciprocate and those that do not. The dominant model for this is the Prisoner's Dilemma game although there are other approaches (Connor 1995).

3 *Unconditional altruism.* In this case individuals behave so as to contribute a benefit to all other individuals in the population that interact with them independently of whether they act in a similar way or not. In general, the altruist will have to bear a cost. An example might be a bacterium that acts altruistically by providing an enzyme that breaks down some substrate that can then be used for food by all its neighbours.

To work reciprocal altruism requires: (i) repeated interactions between each pair of interacting individuals; (ii) each participant must be able to retaliate against defection by the other; and (iii) either individual recognition must be possible or the number of partners with whom an individual interacts must be small and preferably only one.

Even when these severe conditions are satisfied one has to face a basic problem with the theory which is that in mean-field approaches cooperation can never invade a noncooperating resident population. It is quite reasonable that ensembles of cooperating strategies should be stable once they have become established at a high frequency in the population. However, because it requires a reciprocating partner to work, reciprocal altruism confers little fitness when reciprocators are rare. A number of authors have considered spatial approaches to this and related problems (Nowak & May 1992; Hutson & Vickers 1995; Ferriere & Michod 1996 and references therein). I claim that correlations provide a solution to this problem and on p. 138 I consider a model to illustrate this due to Nakamaru *et al.* (1997). This shows that provided the memory of past interactions is not too weak, cooperation can invade.

The primary explanation for the evolution of unconditional altruism involves trait groups. This essentially relies on the assumption that at some stage in their life-cycle individuals are associated with small subgroups of the population and it is there that selection acts. It is assumed that the fitness of an individual in such a group depends upon the group's make-up in such a way that the mean fitness of a given type in the population (say an altruist) is not the average of their fitnesses in each of the trait groups. Then because these groups are small there will be fluctuations in their composition and some will be dominated by altruists which can use the nonlinear fitness to maintain their number.

The problem with this explanation is that apparently very few populations have the structure and nonlinear fitness of the type required to make it work. I will discuss an approach with, I believe, much more realistic population structures. Whereas the trait group approach destroys correlations by selecting the groups randomly from the population in each generation and depends solely upon the fluctuations of the trait groups, the approach I consider depends upon the build-up of spatial correlations. The altruists form mutually beneficial clusters which the correlation equations can track and quantify. Such an approach was first discussed by Matsuda *et al.* (1992) and the approach to invasion by altruists that I discuss is due to van Baalen & Rand (1997).

Another problem is that although an altruistic individual will benefit from altruistic neighbours, when dispersal is limited it will compete with them as well. On the basis of simulations of cellular automaton models, Wilson *et al.* (1992) found that altruism is favoured only in what they considered to be a very limited and therefore unrealistic subset of the parameter domain. Taylor (1992a) has shown that if the 'spatial scale of competition' is equal to the 'spatial scale of dispersal' the benefit of altruism and the cost of local competition cancel out exactly and he claims that this is the case for viscous populations, effectively ruling out the evolution of altruism. In the model considered here this is not the case.

Unconditional altruism in viscous populations

A viscous population is one which is characterized by dispersal that is both reduced and local in nature so that individuals tend to remain in each other's neighbourhoods, with the potential to affect each other's demographic parameters. The basic process that we consider has states empty (ϕ), nonaltruist (N) and altruist (A) and, for $i = N$ or A, the events are:

1 *i-death* at site $x, i \xrightarrow{d_i} \phi$ at rate d_i;
2 *i-birth*, at edge $e, i\phi \xrightarrow{b_i} ii$ at rate b_i;
3 *i-migration*, at edge $e\ i\phi \xrightarrow{m_i} \phi i$ at rate m_i.

In these definitions the rates d_i, b_i and m_i can depend upon i and the number of neighbouring altruists and nonaltruists. I shall assume here that the birth and death rates depend linearly on the number of altruists neighbouring the individual and involves the cost of altruism so that, for an i-individual at x, $b_i = b_0 + b_1 Q_x(A) - c_i$ and $d_i = d_0 - d_1 Q_x(A) - k_i$ where $c_N = k_N = 0$ and c_A and k_A represent the cost of altruism. Also, one can envisage natural modifications that further advance altruism. I will, for example, consider an assumption whereby the migration rate increased with the number of neighbouring nonaltruist, i.e. $m_A = m_0 + m_1 Q_x(N)$.

The master equation gives the following equations in which I have neglected any mention of the correction terms. In these i is N or A and i' is the other.

$$\frac{\mathrm{d}}{\mathrm{d}t}[i] = (b_i Q(\phi \mid i) - d_i)[i]$$

$$\frac{\mathrm{d}}{\mathrm{d}t}[\phi i] = \big(a_i(Q(\phi \mid \phi i) - Q(i \mid \phi i) + 1) - a_{i'} Q(i' \mid \phi i) - \delta_{i\phi} - b_i\big)[\phi i] + \delta_{i'i}[ii'] + \delta_{ii}[ii]$$

$$\frac{\mathrm{d}}{\mathrm{d}t}[ii'] = (a_i + a_{i'})[i'\phi i] - (\delta_{ii'} + \delta_{i'i})[ii'] \quad \text{and} \tag{4.16}$$

$$\frac{\mathrm{d}}{\mathrm{d}t}[ii] = 2(a_i Q(i \mid \phi i) - a_i + b_i)[\phi i] - 2\delta_{ii}[ii]$$

In these equations, a_i and δ_{ij} are respectively the rates of arrival of i-individuals into an empty site per $i\phi$ pair and departure of them (to ϕ's) from ij pairs per ij pair.

We consider now whether a very small population of altruists can invade a resident population of nonaltruists. If we solve for the equilibrium of the above equation when altruists are absent we obtain $\rho_N = \Lambda/\Delta, \rho_{\phi N} = d_0\Lambda/b_0\Delta$ and $\rho_{\phi\phi} = p_0 d_0^2/b_0\Delta$ where $p_0 = b_0 + m_0\kappa$, $\Lambda = (d_0 - Qb_0)p_0 + b_0^2$ and $\Delta = b_0(Qp_0 - b_0 - d_0/Q)$.

Let $\vec{n} = ([\phi A], [NA], [AA])$. From the Bernoulli approximation to Equation 4.16 this satisfies that differential equation $\dot{\vec{n}} = M \cdot \vec{n}$ where

$$M = \begin{bmatrix} a_A(\overline{Q} - 2\overline{Q}(A|\phi A)) - (\alpha_N + \alpha_A)\overline{Q}(N|\phi) - \delta_A - \overline{b}_A & \delta_N & \delta_A \\ (\alpha_N + \alpha_A)\overline{Q}(N|\phi) & -\delta_A - \delta_N & 0 \\ 2\alpha_A\overline{Q}(A|\phi A) + 2\overline{b}_A & 0 & -2\delta_A \end{bmatrix}$$

where $\overline{Q} = \kappa Q$, $\overline{Q}(i|j) = \kappa Q(i|j)$, $\overline{Q}(A|\phi A) = Q(A|\phi A) - 1$ and α_i and δ_i are the Bernoulli approximations to a_i and δ_{ij} above, i.e. $\alpha_i = \overline{b}_i + m_i^\alpha$, $\delta_i = \overline{d}_i + m_i^\delta \overline{Q}(\phi|i)$, $\overline{b}_i = b_0 + b_1\overline{Q}(A|i)$ and $\overline{d}_i = d_0 + d_1\overline{Q}(A|i)$. I give the expressions for m_i^α and m_i^δ below. Let α_0 and δ_0 denote respectively the values of α_N and δ_N in the resident population, i.e. when altruists are absent.

Note that M depends upon both $Q(N|\phi)$ and $Q(A|\phi)$. These terms arose as approximations to $Q(N|\phi A)$ and $Q(A|\phi A)$ and therefore they must be regarded as being the appropriate values for the invading population alone and not for the whole population. For the whole population $Q(A|\phi)$ will be approximately zero at invasion, but may be positive within the invading population which will early on have its own local structure. The value of $Q(N|\phi)$ can be determined by noting that as $b_A, b_N \to b_0$ and $d_A, d_N \to d_0$, $Q(\phi|A) \to Q(\phi|N)$ so that $\delta_A, \delta_N \to \delta_0$ and therefore $\det M \to 0$. Solving $\det M = 0$ when $\alpha_A = \alpha_N = \alpha_0$, $\delta_A = \delta_N = \delta_0$ and $b_A = b_N = b_0$ gives $Q(N|\phi) = Q - Q(A|\phi) - \delta_0/\kappa\alpha_0$.

If we neglect the stochastic terms at invasion $\vec{n}(t) \approx \vec{n}_0 \exp \lambda t$ where λ is the dominant eigenvalue of M and $\vec{n}_0$ is the corresponding eigenvector. A straightforward calculation gives that for $\vec{n}_0$ one can take $(\delta_0, \alpha_0(\overline{Q} - Q(A|\phi A)) - \delta_0, \alpha_0\overline{Q}(A|\phi A) + b_0)$. Since $\vec{n} = [A](Q(\phi|\phi), Q(N|A), Q(A|A))$ we deduce that, at invasion, if $B_0 = \alpha_0 Q + m_0 = \alpha_0\overline{Q} - b_0$ and $C_0 = \alpha_0(\overline{Q} - \overline{Q}(A|\phi A)) - \delta_0$,

$$Q(\phi|A) = Q\frac{\delta_0}{B_0}, \; Q(N|A) = Q\frac{C_0}{B_0} \text{ and } Q(A|A) = Q\frac{\alpha_0\overline{Q}(A|\phi A) + b_0}{B_0} \tag{4.17}$$

The invasion exponent is the eigenvalue λ which, up to terms which are quadratic in $d\alpha_A$ and $d\delta_A$ (and therefore very small), is

$$\lambda = \frac{\delta_0 + \alpha_0(\overline{Q} - Q(A|\phi A))}{\alpha_0\overline{Q} + \delta_0 + b_0}\Lambda \approx \left(1 - \frac{\alpha_0\overline{Q}(A|\phi A) + b_0}{\alpha_0\overline{Q}}\right)\Lambda \tag{4.18}$$

where $\Lambda = \alpha_0^{-1}(\delta_0 d\alpha_A - \alpha_0 d\delta_A)$ and the approximation neglects terms that are proportional to $1/Q^2$. Thus we deduce that invasion will occur provided

$d\alpha_A/d\delta_A > \alpha_0\delta_0$. Consequently, evolution acts so as to increase the ratio α_A/δ_A. This has the interpretation given above as a ratio of arrivals to departures.

In deriving this result we neglected the stochastic correction terms. When one takes these into account the effect is to change the statement that condition $\lambda > 0$ implies successful invasion to the statement that it implies that there is a positive probability that the invasion will be successful. If the challenge by altruists is repeated often enough they will eventually invade.

For a moment let us assume that the migration rate $m_A = m_N = m_0$ is assumed constant. Then $m_i^\alpha = m_i^\delta = m_0$. From Equation 4.18 it is clear that if only either the birth rate or the death rate is allowed to mutate (so that either $b_A = b_0 + b_1 Q_x(A) - c$ or $d_A = d_0 + d_1 Q_x(A) - k$) then the conditions for invasion are respectively

$$b_1\overline{Q}(A|A) > c \quad \text{and} \quad d_1\overline{Q}(A|A) > k$$

Note the similarity of these to Hamilton's condition for kin selection. In our condition, the altruist–altruist correlation $\overline{Q}(A|A)$ replaces the relatedness in Hamilton's condition. Similarly, one easily obtains an invasion condition when they are allowed to mutate simultaneously.

This raises the question of how large $\overline{Q}(A|A)$ is. It is given by Equation 4.17. A detailed calculation shows that $\overline{Q}(A|\phi)$ is small so that $Q(A|A)^{-1} \approx \kappa(\alpha_0 b_0) + Q^{-1} \geq 1$ implying $Q(A|A) \leq 1$. If the migration rate is very small it is approximately 1.

The invasion condition can also be used to understand the evolution of other traits. For example, I suggested above that certain migration could aid altruism. Consider, for example, a situation where as well as the birth rate or death rate being allowed to mutate as above, a behavioural mutation is also possible that encourages higher migration when a mutant is surrounded by many nonaltruists. Assume for example that $m_A = m_0 + m_1 Q_x(N)$. Then the migration term in δ_{AA} changes to $(m_0 + m_1 \kappa Q(N|AA))Q(N|AA)$ and similarly for δ_{AN}. The factor of κ arises because one should approximate $\Sigma_{\sigma e = AA} Q_{e_A}(N) Q_{e_A}(\phi)$ by $\kappa Q(N|AA) Q(N|AA)[AA]$. Thus, the approximation to use for both in δ_A is $m_A^\delta = (m_0 + m_1 \kappa \overline{Q}(N|A))\overline{Q}(\phi|A)$. The migration term in α_A is $(m_0[\phi A] + m_1[NA\phi])/[A\phi]$ which should be approximated by $m_A^\alpha = m_0 + m_1 \kappa Q(\phi|A)$. Since we are assuming that the resident nonaltruists have a constant migration rate $m_N^\alpha = m_N^\delta = m_0$.

Consider the term Λ. By Equation 4.18, this determines whether or not $\lambda > 0$. The term $d\alpha_A$ can be written as a sum $d\alpha_A^1 + d\alpha_A^m$ where the second term corresponds to the contribution due to the change in migration and the first from births and deaths. Similarly for $d\delta_A$. But, by the expressions for m_A^α and m_A^δ, $d\alpha_A^m = m_1 \kappa Q(\phi|A)$ and $d\delta_A^m = m_1 \kappa \overline{Q}(N|A)\overline{Q}(\phi|A)$. Substituting in the value of $Q(\phi|A)$ given by Equation 4.17 we obtain that the extra contribution to λ due to this migration is given by

$$\lambda^m = m_1 \kappa \delta_0 b_0 \overline{Q}\,(N|A) / (\alpha_0 \overline{Q} + b_0) \approx m_1 \delta_0 \left(1 - \left(b_0^{-1}\,\delta_0 + 1\right) Q^{-1} - \kappa b_0^{-1}\, m_0\right)$$

where the approximation is correct up to terms which are either proportional to $Q(A|\phi)$ or quadratic in Q^{-1} and m_0. Thus we see that $d\mu^m$ is always positive and is not small. Thus the adoption of this behaviour simultaneously with the change in the birth and death rates can significantly enhance the invasion of altruists.

The overall important point is that the effect of spatial correlation coming from the population dynamics can be powerful enough to allow altruists to overcome the cost of their behaviour and invade noncooperating populations.

Cooperation in viscous populations

Now we follow the discussion of Nakamaru *et al.* (1997) to consider the effect of correlations in the Prisoner's Dilemma game. The payoffs for this game are described in Table 4.2. In this two players interact an indefinite number of times. After each interaction there is a probability w of a further one. A defection against a cooperator gets the greatest payoff T and cooperation against a defector the least. On the other hand, joint cooperation pays more than joint defection. If the precise number of games is finite and known then it pays to adopt the self-explanatory strategy *Always Defect* which we denote by *AD*. On the other hand, Axelrod has shown that the strategy *Tit For Tat* (denoted *TFT*) is very successful in simulated tournaments. A player playing this strategy cooperates on the first game and then plays whatever the opponent played on the previous game.

In a sufficiently large randomly mixing population the payoff to *TFT* and *AD* individuals are respectively

$$W_{TFT} = W_0 + \frac{R}{1-w}\rho_T + \left(S + \frac{wP}{1-w}\right)\rho_D$$

$$W_{AD} = W_0 + \left(T + \frac{wP}{1-w}\right)\rho_T + \frac{P}{1-w}\rho_D$$

Table 4.2 The payoffs satisfy $T > R > P > S$. It is also assumed that $2R > T + S$; this ensures that the payoff is greater to each of two players who cooperate than to a pair that alternately cooperate and defect. We denote by E_{ij} the payoff to player A when player A plays strategy i and player B plays strategy j.

		Player *B*	
		Cooperate *C*	Defect *D*
Player *A*	Cooperate *C*	*R*	*S*
	Defect *D*	*T*	*P*

where ρ_T and ρ_D denote respectively the proportions of the population playing *TFT* and *AD*. Therefore, $W_{TFT} > W_{AD}$ precisely when $\rho_T > \rho_0 = (1 - w)(P - S)/((S + T - 2P)w + R - S_T + P)$ and thus $\rho_T \to 0$ if the initial density of individuals playing *TFT* is less than ρ_0.

Consequently, in such populations, for a given value of w, a small population of individuals playing *TFT* can never invade a resident population playing *AD* if its size is below some threshold. The question we address is whether *TFT* can invade in a spatial population when we take account of correlations. In the underlying stochastic process used by Nakamaru *et al.* (1997) the only states are T and D corresponding to *TFT* and *AD* and the events are as follows:

1 *replacement by TFT*, at site x $D \xrightarrow{r} T$, at rate $r = M_D(Q_x(T))$

2 *replacement by AD*, at site x $T \xrightarrow{r} D$, at rate $r = M_T(Q_x(T))$

where $M_D(n) = q_x(T)\exp(-\alpha P_n(D))$ and $M_T(n) = q_x(T)\exp(-\alpha P_n(T))$ where $P_n(D) = nE_{DT} + (Q - n)E_{DD}$ and $P_n(T) = nE_{TT} + (Q - n)E_{TD}$ are the payoffs to respectively a D and a T when it has n T-neighbours.

A simple calculation using the above formalism and the Bernoulli trials substitution gives the following equation in which we neglect all the correction terms:

$$\dot{\rho}_T = -\rho_T v_{TD}(v_{TT} + v_{TD})^{Q-1} + \rho_D v_{DT}(v_{DT} + v_{DD})^{Q-1} \tag{4.19}$$

$$\dot{\rho}_{TT} = -2\kappa\rho_T v_{TT} v_{TD}(v_{TT} + v_{TD})^{Q-2} + 2\rho_D v_{DT}\{v_{DT} + v_{DD}\}(v_{DT} + v_{DD})^{Q-2} \tag{4.20}$$

where $v_{ij} = Q^{-1}Q(i|j)\exp(-E_{ij})$. The other equations are determined by the relations $\rho_D + \rho_T = 1$, $\rho_{DD} + \rho_{DT} + \rho_{TT} = 1$, $\rho_T = \rho_{DT} + \rho_{TT}$ and $\rho_D = \rho_{DT} + \rho_{DD}$.

This system is degenerate in the sense that all points on the line $\rho_{TT} = \rho_T$ are equilibria. The analysis of Nakamaru *et al.* (1997) showed that for the given values of the payoffs and $Q = 8$, if the initial condition is interior then (i) if $0 \le w < 0.49$ all trajectories converge to $\rho_T = 0$, (ii) if $w > 0.77$ then $\rho_T \to 1$ and $Q(T|T) \to Q$ and (iii) for intermediate values of w the system was bistable with convergence to one of the two attractors given in (i) and (ii). Thus, according to the model, *TFT* can invade a cooperating model provided w is large enough. Nakamaru *et al.* compared the quantitative prediction with simulations carried out on a lattice using the Moore neighbourhood and found that, while the mean-field approximation did badly, the pair approximation and simulations were consistent.

Concluding synthesis

Correlations often play a crucial role in determining the dynamics of individual-based models of populations. In disease dynamics correlations between infecteds can slow down disease spread, depress R_0 increase critical community sizes and enhance persistence. In host–parasite systems the correlated structure of both host and parasite populations can greatly alter evolutionary velocities and lead to new evolutionary stable states. Correlations in animal and plant communities can enhance ecological interactions and significantly alter dynamics. For example, in both unconditional altruism and the Prisoners' Dilemma game high correlation between altruists or cooperators can permit them to invade selfish populations.

Phenomena due to such correlations cannot be studied using mean-field models as these assume that correlations are absent. On the other hand, the full stochastic models are currently beyond analysis. I have tried to show that pair approximation of correlation equations provide a fruitful framework for the study of such systems. Using these one is able to analyse a wide range of quite complex ecological situations including those mentioned immediately above. I have explained a formalism for deriving these equations and also carefully considered the various approximation and closure schemes available. I show how to calculate the master equation and then discuss how one can close this system to get a stochastic differential equation by approximating higher order correlations by quantities involving only lower order correlations. The appropriate approximation scheme depends upon the system to be modelled.

This approach is particularly fruitful in studying invasion which is a fundamental process in ecology, evolution and epidemiology and which I have stressed. It gives new insights and enables us to tackle a range of interesting problems involving spatial structure that were previously inaccessible. It enables a deeper, more sophisticated and more realistic treatment of invasion by carefully considering the different stages of this complex process and calculating the emergent correlation structure of the invading population and the effect of this on the rate of invasion. An important aspect of invasion is that at invasion the invaders are present in small numbers and therefore stochastic effects can be very strong.

Finally, I have tried to stress calculation. Eventually, a mathematical theory needs to be able to allow calculation of quantities such as the strength of an effect or the velocity of a change. As I have tried to show, pair approximations enable you to do this in many complex ecologies.

Acknowledgements

I am particularly indebted to Minus van Baalen, Matthew Keeling and Andrew Morris. All three have contributed greatly to my understanding of these problems. I also would like to gratefully acknowledge the financial support of the EPSRC and NERC.

References

Anderson, R.M. & May, R.M. (1992) *Infectious Diseases of Humans*. Oxford University Press, Oxford.

ben-Avraham, D., Burschka, M.A. & Doering, C.R. (1990) Statics and dynamics of a diffusion-limited reaction: anomalous kinetics, nonequilibrium self-ordering, and a dynamic transition. *Journals of Statistical Physics*, **60**, 695–728.

Axelrod, R. & Hamilton, W.D. (1981) The evolution of cooperation. *Science*, **211**, 1390–1396.

van Baalen, M. & Rand, D.A. (1997) The unit of selection in viscous populations and the evolution of altruism. *Journal of Theoretical Biology* (in press).

Bartlett, M.S. (1960) The critical community size for measles in the U.S. *Journal of the Royal Statistical Society*, **A123**, 37–44.

Bezuidenhout, C. & Grimmett, G. (1990) The critical contact process dies out. *Annals of Probability*, **18**, 1462–1482.

Black, F.L. (1966) Measles endemicity in insular populations: critical community size and its evolutionary implications. *Journal of Theoretical Biology*, **11**, 207–211.

Boerlijst, M.C. & Hogeweg, P. (1991) Spiral wave structure in pre-biotic evolution–hypercycles stable against parasites. *Physica*, **D48**, 17–28.

Bolker, B.M. (1993) Chaos and complexity in measles models: a comparitative numerical study. *IMA Journal of Mathematics Applied in Medicine and Biology*, **10**, 83–95.

Bolker, B.M. & Grenfell, B.T. (1993) Chaos and biological complexity in measles dynamics. *Proceedings of the Royal Society of London*, **B251**, 75–81.

Bolker, B. & Pacala, S.W. (1997) Using moment equations to understand stochastically driven spatial pattern formation in ecological systems. *Theoretical Population Biology*, **52**, 179–197.

Connor, R.C. (1995) Altruism amoung non-relatives: alternatives to the prisoners' dilemma. *Trends in Ecology and Evolution*, **10,** 84–86.

Durrett, R. (1988) *Lecture Notes on Particle Systems and Percolation*. Wadsworth, Pacific Grove, CA.

Ferriere, R. & Michod, R.E. (1996) Invading wave of cooperation in a spatial iterated prisoners' dilemma *Proceedings of the Royal Society of London, Proceedings of the Royal Society of London*, **B259**, 77–83.

Hamilton, W.D. (1964) The genetical evolution of social behaviour. *Proceedings of the Royal Society of London*, **7**, 1–52.

Harada, Y. & Iwasa, Y. (1994) Lattice population dynamics for plants with dispersing seeds and vegetative propagation. *Research in Population Ecology*, **36**, 237–249.

Harada, Y., Ezoe, H., Iwasa, Y., Matsuda, H. & Satō, K. (1995) Population persistence and spatially limited social interaction. *Theoretical Population Biology*, **48**, 65–91.

Hutson, V.C.L. & Vickers G.T. (1995) The spatial struggle of tit-for-tat and defect. *Philosophical Transactions of the Royal Society of London*, **B348**, 393–404.

Keeling, M.J. (1995) *The ecology and evolution of spatial hostparasite systems*. PhD thesis, Warwick University.

Keeling, M.J. & Grenfell, B.T. (1997) Disease extinction and community size: modeling the persistence of measles. *Science*, **275**, 65–67.

Keeling, M.J. & Rand, D.A. (1996) Spatial correlations and local fluctuations in host–parasite ecologies. In: *From Finite to Infinite Dimensional Systems* (ed. P. Glendinning). Kluwer, Amsterdam.

Keeling, M.J., Rand, D.A. & Morris A.J. (1997) Correlation models for childhood epidemics. *Proceedings of the Royal Society of London*, **B264**, 1149–1156.

Matsuda, H., Ogita, N., Sasaki, A. & Satō, K. (1992) Statistical mechanics of population–the lattice Lotka–Volterra model. *Progress in Theoretical Physics*, **88**, 1035–1049.

Maynard Smith, J. (1964) Group selection and kin selection. *Nature*, **201**, 1145–1147.

Maynard Smith, J. (1982) *Evolution and the theory of games*. Cambridge University Press, Cambridge.

Morris, A.J. (1997) *Representing spatial interactions in simple ecological models*. PhD thesis, Warwick University.

Nakamaru, M., Matsuda, H. & Iwasa, Y. (1997) The evolution of cooperation in a lattice-structured population. *Theoretical Population Biology*, **184**, 65–81.

Nowak, M. & May, R.M. (1992) Evolutionary games and spatial chaos. *Nature*, **359**, 826–829.

Olsen, L.F. & Schaffer, W.M. (1990) Chaos versus noisy periodicity: alternative hypotheses for childhood epidemics. *Science*, **249**, 499–504.

Rand, D.A. & Wilson, H. (1991) Chaotic stochasticity: a ubiquitous source of unpredictability in epidemics. *Proceedings of the Royal Society of London*, **B246**, 179–184.

Rand, D.A., Wilson H. & McGlade, J.M. (1994) Dynamics and evolution: evolutionarily stable attractors, invasion exponents and phenotype dynamics. *Philosophical Transactions of the Royal Society of London*, **343**, 261–283.

Rand, D.A., Keeling, M. & Wilson, H. (1995) Invasion, stability and evolution to criticality in spatially extended host–pathogen systems. *Proceedings of the Royal Society of London*, **B259**, 55–63.

Rand, D.A. & Keeling, M. (1995) A spatial mechanism for the evolution and maintenance of sexual reproduction. *Oikos*, **74**, 414–424.

Rhodes, C.J. & Anderson, R.M. (1996a) Power laws governing epidemics in isolated populations. *Nature*, **381**, 600–602.

Rhodes, C.J. & Anderson, R.M. (1996b) A scaling analysis of measles epidemics in a small population. *Philosophical Transactions of the Royal Society of London*, **B351**, 251–268.

Satō, K. & Konno, N. (1995) Successional dynamic models on the 2-dimensional lattice space. *Journal of the Physical Society of Japan*, **64**, 1866–1869.

Satō, K., Matsuda, H. & Sasaki, A. (1994) Pathogen invasion and host extinction in lattice structured populations. *Journal of Mathematical Biology*, **32**, 251–268.

Schenzle, D. (1984) An age-structured model of pre- and post-vaccination measles transmission. *IMA Journal of Mathematics, Applied in Biology and Medicine*, **1**, 169–191.

Taylor, P.D. (1992a) Altruism in viscous populations—an inclusive fitness model. *Evolutionary Ecology*, **6**, 352–356.

Taylor, P.D. (1992b) Inclusive fitness in a homogeneous environment. *Proceedings of the Royal Society of London*, **259**, 299–302.

Tretyakov, A., Provata, A. & Nicolis, G. (1994) Nonlinear chemical dynamics in low-dimensional lattices and fractal sets. *Journal of Physical Chemistry*, **99**, 2770–2776.

Trivers, R.L. (1971) The evolution of reciprocal altruism. *Quarterly Review of Biology*, **46**, 35–57.

Trivers, R.L. (1985) *Social Evolution*. Benjamin-Cummings, Menlo Park, CA.

Wilson, D.S., Pollock, G.B. & Dugatkin, L.A. (1992) Can altruism evolve in a purely viscous population? *Evolutionary Ecology*, **6**, 331–341.

5: Theoretical aspects of community assembly

R. Law

Introduction

Community ecology is concerned with the properties of sets of species at given locations in time and space. Are the sets random samples of species found in the general area? If not, in what way are they nonrandom? How do the sets change over the course of time? Are there rules underlying the assembly process? What, if any, properties characterize the sets that emerge from the process of assembly? These are the sorts of questions that motivate community ecology.

Although we have a theoretical framework for community dynamics (May 1974; Pimm 1982; Yodzis 1989), the theory does not deal explicitly with the arrival and disappearance of species as communities of interacting species are assembled. This chapter gives a different theoretical basis for community ecology, to try to get closer to the central issues of turnover of species. The key is to switch from dynamics describing the flux of individuals to dynamics describing the flux of species. The idea is to change the state variable from a set $\boldsymbol{x}$ of population densities to one comprising a set S of species as shown in Fig. 5.1.

To achieve this switch, the notion of a species pool is needed, as communities have to be treated as open systems susceptible to invasions from outside: invaders may be absorbed without loss of any species, or they may lead to extinctions. Much of the chapter is concerned with a qualitative theory to establish how a successful invader affects the species composition of a community. With this in place, we can investigate the properties that emerge as communities are assembled, and the type of states to which the assembly process eventually leads.

In a sense this chapter is an unfashionable plea for simplicity. Ecologists are in the unenviable position of having on the one hand real-world systems of mind-numbing complexity and on the other hand dynamical models which, even at their simplest, behave in ways that can outstrip this natural richness. One of the reasons why I have written this chapter is to argue that, even if we leave out a lot of this complexity, we can still address some of the fundamental issues in community ecology.

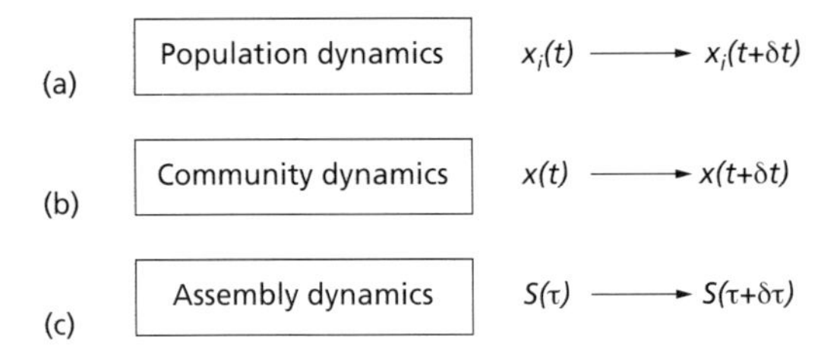

Fig. 5.1 Three levels at which to describe ecological dynamics. (a) Updates the density x_i of a single species *i* due to birth and death events in species *i* over a time period δt. (b) Updates a set *x* of densities of interacting species due to birth and death events in each of them over a time period δt. (c) Updates the species composition *S* due to arrival and extinction events of interacting species over a time period $\delta\tau$.

Regional species pools

A community is rarely, if ever, isolated from other communities. New species arrive from outside; some become established, and may drive one or more of the resident species to extinction. The structure that emerges at a local level depends as much on which species are available in the geographical region as on the processes operating locally (Ricklefs 1987; Ricklefs & Schluter 1993). This is evident, for instance, in the species richness of communities, more species usually being found locally when more species occur regionally (Cornell 1985; Ricklefs 1987; Lawton *et al.* 1993). Figure 5.2 illustrates this in guilds of cynipid gall wasps living on seven species of oak in California. Clearly the oak species with larger regional pools of gall wasps also tend to have more taxa of wasps locally (Cornell 1985). There is no sign here of a hard upper limit to species richness caused by some property internal to the community such as niche space, and we cannot expect to understand the patterns within communities simply in terms of the processes operating within them. It helps therefore to envisage a regional pool or set of species associated with a community. Such a theoretical construct has been at the back of ecologists' minds at least since the introduction of the equilibrium theory of island biogeography (MacArthur & Wilson 1967), indeed its origins go back much further (Gleason 1926). The set can be thought of in several different ways. I will define it and some related sets as follows.

S_m is the *species pool*. This is a set of *m* species: (i) having a geographical range that encompasses the location of the community; and (ii) not excludable on the grounds of the physical environment at this location (this enables us to put on one side, for instance, aquatic organisms in terrestrial environments) (Kelt *et al.* 1995).

S^+ is a subset of the species pool comprising the species which are resident (have positive densities) in the community.

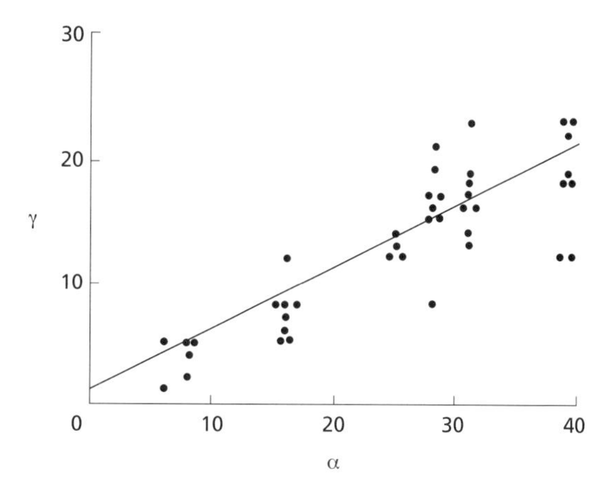

Fig. 5.2 Species richness of communities of gall wasps living on seven species of oak (*Quercus*). The graph plots the local species richness (α diversity) against the regional richness (γ diversity) associated with each species of oak (Cornell 1985).

S^0 is a subset of the species pool comprising the species which are absent (have zero densities) in the community, i.e. $S^0 = S_m \backslash S^+$.

Estimates of regional pools have rarely been made at a given site, although it would be feasible to do so for certain kinds of organisms with patience and appropriate trapping techniques. Species in the regional pool inevitably differ greatly in abundance, mobility and proximity to a given site; when making such estimates, it therefore helps to have some idea of the probability of arrival per unit time associated with each species, as this is bound to influence how the community develops. One reason why such information is important is that rather little is known about S^0. By way of contrast, much more is known about S^+ since making lists of the species present is relatively straightforward. This means that at present we are rarely in a position to say whether a species is absent from a community because it is being excluded or simply because it has yet to reach it in large enough numbers to become established. An interesting exception to this are the four species of cuckoo dove documented by Diamond (1975) on the islands around New Guinea because in certain cases the outcome of attempted introductions is known.

The regional pool is no more than the union of the sets of species present in communities in some neighbourhood; to separate local and regional processes is, in a sense, artificial (Cornell & Lawton 1992; Warren 1996). It ought to be possible to develop a theoretical framework which is spatially explicit, comprising a landscape made up of a mosaic of local communities, with movement of individuals across the mosaic (Karieva & Wennergren 1995). The regional pool of species is then made internal to the system. However, we retain the separation for the sake of simplicity in this chapter.

State variables in community ecology

Several recent reviews have argued that all is not well within community ecology (Shrader-Frechette & McCoy 1993; Weiner 1995; Grimm & Wissel 1997; see also Chapter 11): there is a lack of agreement about the meaning of some basic terms, a major gap between theory and empirical work, and the theoretical foundations of the subject in particular seem insecure. One source of confusion is that community ecologists have not really achieved a consensus as to what they should be measuring, or to put it in the language of dynamical systems, what the state variables of community ecology are. This is a fundamental problem that needs to be resolved to develop a strong foundation to the subject.

If you were to ask ecologists which state variable is central to population ecology, there would be general agreement that some measure of abundance such as population size or density lies at the heart of the subject. This provides a focus for the work of empiricists and theorists alike.

However, matters are not so clear cut in community ecology. Here a major part of empirical research is concerned with qualitative issues to do with the presence or absence of species, such as the sets of species found living together, and the rules by which these sets change. Yet much of the theory about community dynamics does not tackle directly the turnover of species in communities; it is an extension of population dynamics, in which the abundance of a single species is replaced by a set of abundances of two or more species. Since interacting species affect each other's population dynamics (e.g. predators kill their prey), such an extension seems natural. It leads to a coupled system of equations describing how the abundance of the species change together (May 1974; Pimm 1982; Yodzis 1989), where the state variable is the set of abundances. But this provides both too much and too little information to illuminate our understanding of turnover of species in natural communities. Too much, because many of the basic issues are qualitative rather than quantitative; and too little, because the dynamics of high-dimensional nonlinear dynamical systems are usually too complex to interpret. We are, if you like, paying the price of moving up a level in the organization of life while retaining the state variable appropriate to the lower level.

Another strand of theory comes from Markovian models of community assembly, based on the probability with which one kind of community changes to another (Horn 1975; Usher 1992). Here the choice of community state is essentially subjective; for instance it may be based on the presence or absence of certain indicator species. The state variable is a vector for the probability of each state, and this is updated on the basis of transition probabilities between states and an assumption that the process is a Markov chain. This approach comes rather closer to the interests of many empiricists because it is concerned with changes in species composition, but it leaves open the question as to how the states should be defined.

I am going to suggest a way out of this impasse that takes the states as being those subsets of species from the regional pool that have the property of *persistence*. The reason for suggesting this is that these are the only sets of species which could have more than a transient existence. We can then concentrate on the transitions from one set of persistent species to another, through the entry and loss of species from communities. I will refer to such turnover of species as *assembly dynamics* (Fig. 5.1), to distinguish it from the turnover of individuals within species, dealt with by the more familiar study of population dynamics of interacting species (often called community dynamics). Ultimately community theory will need to consider processes on transients of community dynamics as well, but this will be harder to do, and it is better to begin with more tractable goals.

At a practical level, to determine what sets of species from an m-species pool can persist, one may have to check up to $2^m - 1$ subsets of species (ignoring the empty set). This might seem prohibitive for all but the smallest species pools, but the problem is not quite as serious as it might seem, because ecologists are almost always looking at some subset defined on functional or taxonomic grounds. Table 5.1 illustrates the search for combinations of species with the ability to coexist in a pool of six protists forming a simple food web (Weatherby *et al.* 1998). The species were chosen in the absence of a priori knowledge about coexistence, although we knew they could be maintained under the culture conditions of the experiment when fed on bacteria or other protists from the pool. The experiment started with replicated microcosms of all $2^6 - 1 = 63$ combinations in which the species could occur. Within 100 days, most had collapsed to one of the eight sets shown in Table 5.1, and there was little indication of any further change. These eight appear to comprise the sets with the property of persistence, and are the states one would expect to find over prolonged periods of time.

Table 5.1 Sets of species, from a regional pool of six protists, shown to be able to persist in laboratory microcosms (Weatherby *et al.* 1998).

0 species	1 species	2 species
{}	{*B*}	{*B*, *P*}
	{*P*}	{*B*, *C*}
	{*C*}	{*P*, *T*}
	{*T*}	

The species are: *B*: *Blepharisma japonicum*, *C*: *Colpidium striatum*, *P*: *Paramecium caudatum*, *T*: *Tetrahymena pyriformis.* Two other species that were not present in any set at the end are *Amoeba proteus* and *Euplotes patella.* The empty set {} is included because all species went to extinction from certain initial sets.

Criteria for coexistence

What is less straightforward is to develop a qualitative theory for coexistence of species on which models for assembly dynamics can be built. This section describes some of the steps that have been made in this direction.

Whether a set of species can coexist is primarily a matter of how their population dynamics are coupled. We will consider this in the context of some set S of n species with population densities denoted by $\boldsymbol{x} = \{x_i \mid i \in S\}$. The deterministic dynamics can be described by a system of autonomous ordinary differential equations of the general form

$$\dot{x}_i = x_i \cdot f_i(\boldsymbol{x}) \quad \text{for } i \in S \tag{5.1}$$

where $\dot{x}_i$ is the rate of change of density of species i with respect to time, and $f_i(\boldsymbol{x})$ is its per capita rate of increase. The framework in which the properties of the system are most readily understood is an n-dimensional non-negative phase space. The state of the system is given by a point in the space, and a solution of Equation 5.1 describes an orbit or trajectory through the space. The reason for writing Equation 5.1 in this form is to make it explicit that the boundary of the phase space is invariant; this means that species i cannot appear spontaneously from nowhere, i.e. there is no immigration.

The criterion for coexistence of n species most widely used in theoretical ecology is the existence of an equilibrium point $\hat{\boldsymbol{x}}$, at which all n species have positive densities, with the property of *asymptotic stability* in the sense of Lyapunov (see Simmons 1972; Pimm 1982; Yodzis 1989; Darzin 1992), sometimes referred to as local or neighbourhood stability in ecology. This has the advantage of tractability. There is a straightforward test for asymptotic stability based on the eigenvalues of the Jacobian matrix J, evaluated at the equilibrium point,

$$J = [a_{ij}] \text{ with } a_{ij} = \frac{\partial}{\partial x_j}(x_i \cdot f_i(\boldsymbol{x}))\bigg|_{\boldsymbol{x}=\hat{\boldsymbol{x}}} \tag{5.2}$$

in that the real parts of the eigenvalues of J should all be strictly negative.

Asymptotic stability does however have certain drawbacks as a criterion for coexistence. First, we are replacing Equation 5.1 with a linear system which applies only in a very small region around the equilibrium point. What happens to the orbits if the fixed point is unstable is not known; in particular, we do not know whether the orbits tend to the boundary causing extinction of one or more species. Second, issues of coexistence are not settled in the neighbourhood of an interior equilibrium point, but close to the boundary of the phase space, where at least one species has a density close to zero. Coexistence depends on the tendency for the species to increase in density when rare, and the relationship this has with asymptotic stability of $\hat{\boldsymbol{x}}$ is simply not known.

Global asymptotic stability of $\hat{\boldsymbol{x}}$ would seem to be a more powerful alternative, and this has been used on a number of occasions to tackle problems in community ecology (e.g. Goh 1977; Case & Casten 1979). If $\hat{\boldsymbol{x}}$ can be shown to have the property of global asymptotic stability, then no orbit can tend from the interior to the boundary, and the species will certainly coexist. To establish global asymptotic stability, one needs to be able to write down a Lyapunov function, $L(\boldsymbol{x})$, applicable to the whole interior of the phase space. Such a function has the properties: (a) $L(\hat{\boldsymbol{x}}) = 0$; (b) $L(\boldsymbol{x})$ is positive definite; and (c) for an orbit $\boldsymbol{x}(t)$, $\dot{L}(\boldsymbol{x})$ is negative definite.

One drawback to the use of global asymptotic stability as a criterion for coexistence, and indeed to any method which assumes that the asymptotic state is an equilibrium point, is that there are other sorts of attractors, such as periodic and chaotic attractors, on which all the species may coexist even if the equilibrium point is unstable. To put it another way, proving instability of an equilibrium point is not equivalent to showing that these species cannot coexist. It seems particularly important to allow for the possibility of nonequilibrium attractors at the present time when there is so much uncertainty about the relative importance of equilibrium and nonequilibrium asymptotic states in ecology (Hastings *et al.* 1993).

A way out of these difficulties is to use a criterion for coexistence that does not depend on what kind of attractor is involved. Ideas along these lines were discussed by theorists a long time ago (Lewontin 1969; Maynard Smith 1969; Holling 1973), and it was suggested that a criterion of dynamic boundedness might be appropriate (Lewontin 1969). To understand what this means, we need to draw a distinction between the interior of the phase space where all n species have positive densities, and the boundary where at least one species has zero density. The system is dynamically bounded if, in the neighbourhood of the boundary, all orbits are moving away from the boundary. These ideas were made formal and precise as a notion of permanence. This was done first in the context of hypercycles (Schuster *et al.* 1979), but mathematicians have applied the notion subsequently to a number of ecological contexts (reviewed by Hofbauer & Sigmund 1988; Hutson & Schmitt 1992), although it has yet to enter into the thinking of ecologists in a substantive way.

Permanence can be defined as follows. Consider a 'skin' of thickness $\delta > 0$ around the boundary of the phase space. Equations 5.1 are said to be permanent if all orbits not initially in the boundary eventually remain at least at a distance $d > \delta$ from the boundary; to be realistic we also require that no population density should tend to infinity (Hofbauer & Sigmund 1988, pp. 97, 160). Another way of thinking of this is that the boundary of the phase space is a repellor to all orbits that start in the interior of the space (infinity being thought of as part of the boundary). Permanence provides a global criterion for coexistence which places no restriction on the kind of attractor. It is concerned only with the tendency of orbits to move away from the boundary of the phase space, and is a natural formalization of the notion of dynamic boundedness (Lewontin 1969). Other notions of boundedness exist in the literature, including one of stochastic boundedness (Chesson & Ellner 1989), but these do not lead to a test that can be applied to a system of n species.

In fact, it is not obvious that a test for permanence could be constructed for an n-species system with dynamics as in Equation 5.1 in the way that eigenvalues of the Jacobian matrix can be used to test for asymptotic stability of equilibrium points. We are, after all, asking that *all* orbits close to the boundary of the n-species system should be moving away from the boundary. However, by making one major simplification, a sufficient condition for permanence of n species can be constructed; the assumption is that the dynamics are of Lotka–Volterra type, i.e. putting

$$f_i(\boldsymbol{x}) = r_i + \sum_{j \in S} \alpha_{ij} \cdot x_j \tag{5.3}$$

into Equation 5.1 where r_i and α_{ij} are parameters that define the configuration of the community. The reason why this helps lies in an averaging property of the orbits of Lotka–Volterra systems which applies as long as the densities remain strictly positive and finite, namely that:

$$\lim_{T \to \infty} \frac{1}{T} \int_0^T g_i(\boldsymbol{x}(t))\,\mathrm{d}t = g_i(\hat{\boldsymbol{x}}) \tag{5.4}$$

where $g_i(\boldsymbol{x})$ is a function of $\boldsymbol{x}$, and $\hat{\boldsymbol{x}}$ is defined by $f_i(\hat{\boldsymbol{x}}) = 0$ for all i. The simplest example of such a function is $g_i(\boldsymbol{x}) = x_i$; applying Equation 5.4 in this case says that the time average of the densities tends to $\hat{\boldsymbol{x}}$. Figure 5.3 illustrates this for a three-species system with a cyclic attractor. The time average of the densities has been added to the phase space; it can be seen that the average converges to a point, this point being the equilibrium point at which all three species are present. The task of testing for permanence of an n-species system is now greatly simplified because we need only check whether orbits are moving away from the boundary close to a relatively small number of boundary equilibria.

The test for permanence, like that for global stability, is based on proving the existence of a Lyapunov function, but the important properties now relate to what happens close to the boundary of the phase space. The function which is a candidate for a Lyapunov function is:

$$P(\boldsymbol{x}) = \prod_{i \in S} x_i^{h_i} \quad \text{for some choice of } h_i > 0 \tag{5.5}$$

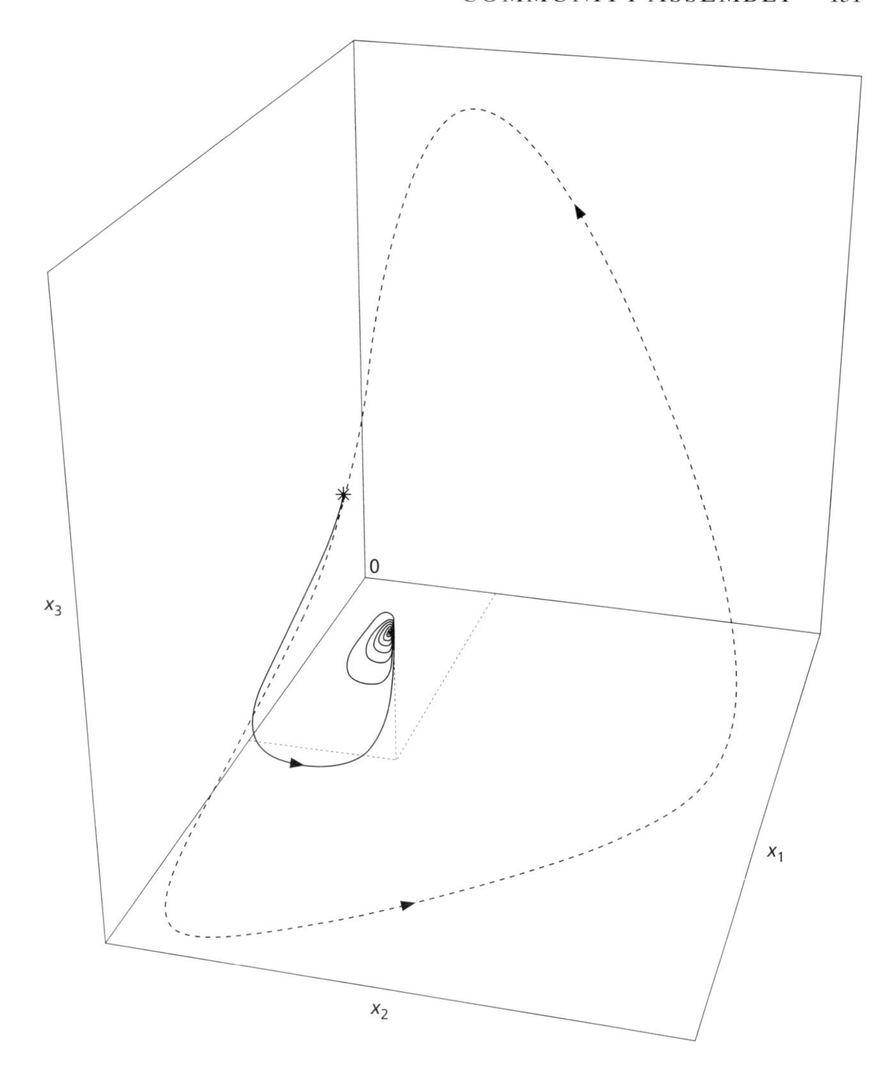

Fig. 5.3 The time averaging property of Lotka–Volterra orbits. The discontinuous line describes the periodic attractor of this system of two self-supporting and one consumer species, with densities x_1, x_2 and x_3 respectively. The continuous line is the time average of the densities, starting from the point labelled *; the time average tends to the interior equilibrium point. (Law & Morton 1996.)

This may be thought of as a surface with height zero in the boundary and positive elsewhere. For orbits close to the boundary to be moving away from it, the derivative with respect to time must be positive, i.e.

$$\dot{P}(\boldsymbol{x}) = P(\boldsymbol{x}) \cdot \varphi(\boldsymbol{x}) > 0, \quad \text{where} \quad \varphi(\boldsymbol{x}) = \sum_{i \in S} h_i \cdot f_i(\boldsymbol{x}) \tag{5.6}$$

We can now exploit the result in Equation 5.4, and concentrate on the time average of orbits in the boundary, testing whether orbits starting close to boundary equilibrium points $\hat{\boldsymbol{x}}^{(1)}$, $\hat{\boldsymbol{x}}^{(2)}$, . . . move further away from the boundary. This greatly simplifies matters because the number of boundary equilibrium points is relatively small; a $P(\boldsymbol{x})$ satisfying $\varphi(\hat{\boldsymbol{x}}^{(1)}) > 0$, $\varphi(\hat{\boldsymbol{x}}^{(2)}) > 0$, . . ., is known as an *average* Lyapunov function. The test for permanence can be written as a linear program (Jansen 1987):

$$
\begin{aligned}
&\text{Minimize} && z \\
&\text{subject to} && \sum_{i \in S} f_i(\hat{\boldsymbol{x}}^{(1)}) \cdot h_i + z \geq 0 \\
& && \sum_{i \in S} f_i(\hat{\boldsymbol{x}}^{(2)}) \cdot h_i + z \geq 0 \\
& && \vdots \\
&\text{with} && \sum h_i \leq 1 \\
&\text{and} && h_i > 0 \quad \text{for all } i \in S
\end{aligned}
\tag{5.7}
$$

If $z < 0$ at the solution of the linear program, $P(\boldsymbol{x})$ is an average Lyapunov function, because all the $\varphi(\boldsymbol{x})$ functions evaluated at the boundary equilibria are positive. In a sense what the linear program is doing is testing for the existence of a set of h_is that make $P(\boldsymbol{x})$ an average Lyapunov function. If $P(\boldsymbol{x})$ is an average Lyapunov function, the set S is permanent. (Actually, matters are not quite as simple as this, because we must also be sure that there are no orbits on which densities can tend to infinity; but this will hold for any realistic system.)

Permanence has its own limitations, and is still not an ideal criterion for coexistence. First, it says only that the 'skin' around the boundary should satisfy $\delta > 0$; δ could be very small, in which case orbits could get close enough to the boundary in a stochastic world for extinction to be possible (see Chapter 2). Second, it needs only one orbit to tend to the boundary for a sytem not to be permanent; local interior attractors are compatible with a nonpermanent system (Case 1995). Third, when $n > 3$, a test for permanence is only possible with Equation 5.3 in place. It would be highly desirable to relax the requirement for a Lotka–Volterra system, but at present this would leave us in uncharted mathematical territory. Just how serious a restriction it is is debatable, since many other systems can be transformed to Lotka–Volterra form (Peschel & Mende 1986). Fourth, the test for permanence has the limitation that it is sufficient (rather than necessary and sufficient) for permanence when there are more than three species present. The reason for using permanence is that there appears to be no better alternative at the moment.

As a footnote to this discussion of coexistence, it ought to be pointed out that the use of stability concepts in ecology is, frankly, a mess. A recent count by Grimm & Wissel (1997) found 163 definitions of 70 different stability concepts. Yet underlying this confusion they could only find three fundamentally different properties: (a) the tendency to remain unchanged; (b) the tendency to return to the reference state (or dynamic) after a temporary disturbance; and (c) persistence through time of an ecological system. Permanence clearly falls into the third category; since it has a precise mathematical meaning, it would be helpful to retain this in ecology.

Methods for sequential assembly of communities

The sections above argue that the states in which communities are found may be taken in the long term as subsets of the regional pool with the property of persistence. In experimental and natural systems, persistence is demonstrated operationally by the capacity of species to live together over prolonged periods of time. The longer the period the better, so that the system gets close to an asymptotic state; one may be misled if the system is on a transient orbit. In theoretical work, permanence provides a criterion for persistence (with the caveats noted above).

With this in mind, we turn to the transitions between states. These transitions hold the key to assembly dynamics, as they indicate what replacements of species are allowable as new species arrive. Determining the replacements is relatively tractable in experimental systems. A resident community is seeded with new species; in principle it is straightforward to establish what community emerges at each step, simply by waiting until the species composition stops changing. By doing this repeatedly, a sequence of communities is obtained, indicating a path of community assembly. An iterative procedure along these lines has been used to follow community development in a number of experimental studies on multispecies microcosms (Robinson & Dickerson 1987; Robinson & Edgemon 1988; Drake 1991; Drake *et al.* 1993), although these studies may have allowed introductions too frequently for the community to reach its asymptotic species composition at each step (Grover & Lawton 1994). The results of Robinson & Edgemon (1988) in fact show how great an effect on the community a high rate of introduction of new species can have.

It ought also to be possible to use the same general approach to investigate assembly dynamics in more abstract numerical analyses. This has been attempted on a number of occasions (Post & Pimm 1983; Rummel & Roughgarden 1985; Mithen & Lawton 1986; Drake 1990; Law & Morton 1996). The basic requirements for doing this are: (a) a condition for successful establishment of a new species; and (b) an algorithm to give the species composition of the new community. We will consider these below but, before doing so, note two assumptions usually made (sometimes implicitly) to facilitate the analysis.

The first assumption is a separation of the time-scale on which community dynamics takes place, from the longer time-scale of assembly dynamics. In other words, arrival and establishment of new species are relatively rare phenomena. This assumption is introduced to make theoretical work possible, because the behaviour of new species along all possible transient paths of the resident community is too complicated to deal with. A time-scale separation ensures that the community is close to an attractor when new species arrive, and the behaviour of these new species can be determined simply on the basis of what they do close to this attractor. In keeping with these comments, I will refer to the time-scale on which new species arrive as τ, there being n_τ species present at time τ, denoted by the set S_τ^+, and $m - n_\tau$ absent, denoted by S_τ^0.

The second assumption is that species are independent in their probabilities of arrival. Together with the first assumption, this means in effect that one species arrives at a time. This is not essential, but it does simplify the analysis, because there are only the effects of $m - n_\tau$ single-species arrivals to consider, as opposed to $2^{m-n_\tau} - 1$ arrivals of all combinations of the missing species.

Conditions for establishment

Given the time-scale separation above, the densities $\boldsymbol{x}_\tau$ of the resident species are taken as being close to an attractor A_τ at the time of arrival of the next species. A species j is now chosen in accordance with the probability distribution of arrivals; clearly it is only going to contribute a new species if it comes from the set S_τ^0. The fate of this and the resident species is determined in a phase space of dimension $n_\tau + 1$, comprising the resident community augmented by the new arrival, with densities denoted by the set $\{\boldsymbol{x}_\tau, x_j\}$. It will help below to have some name for this larger system of $n_\tau + 1$ species; I will call it the *augmented* system.

For species j to establish itself, it has to be able to increase in the $n_\tau + 1$ space close to what was the attractor, A_τ, of the resident species (which now lies in the boundary of the augmented system). It is not enough for species j to do this close to what was a single point on A_τ, because it may increase in some places and decrease in others; we need the average of its per capita rate of increase along A_τ to be positive, i.e.

$$\lim_{T\to\infty} \frac{1}{T} \int_0^T f_j(\boldsymbol{x}_\tau, x_j)\, \mathrm{d}t > 0 \quad \text{where} \quad \boldsymbol{x}_\tau \in A_\tau,\ x_j = 0 \tag{5.8}$$

Another expression is that the Lyapunov exponent should be positive:

$$\lim_{T\to\infty}\frac{1}{T}\log\frac{|\Delta x(T)|}{|\Delta x(0)|}>0 \tag{5.9}$$

The Lyapunov exponent is obtained from two trajectories, one being the orbit $\boldsymbol{x}'(t)$ obtained after adding a small density of species j at $t = 0$, and the other $\boldsymbol{x}(t)$ being a reference orbit without this addition; $|\Delta x(t)| = |\boldsymbol{x}'(t) - \boldsymbol{x}(t)|$ is a measure of the distance between them at time t (Baker & Gollub 1990; Metz *et al.* 1992). The inequalities shown in Equations 5.8 and 5.9 are equivalent (Dieckmann & Law 1996). We can go further in the case of a Lotka–Volterra system, and exploit the result on the time averages of orbits in Equation 5.4, to obtain the condition

$$f_j(\hat{\boldsymbol{x}}_\tau, 0) > 0 \tag{5.10}$$

irrespective of the form of the attractor A_τ. Inequality in Equation 5.10 has been used in several studies of the assembly of Lotka–Volterra communities; the earlier studies assumed that the attractor of the resident species was the equilibrium point (Post & Pimm 1983; Drake 1990), but we now know that this assumption is not needed (Law & Morton 1996).

It ought to be noted that the conditions above miss the demographic stochasticity associated with the initial growth of new species. One of the safest ecological generalizations is that a new species will rarely become established. As an empirical rule, Williamson & Fitter (1996a) noted that only about 10% of imported species are found in the wild, and of these only about 10% establish self-sustaining populations. There are good grounds to expect stochastic effects of small population size to play a part in this; individuals are most likely to arrive in small numbers, and there is a substantial probability that the population will die out, even if the birth rate exceeds the death rate (see Chapter 2). Models of assembly dynamics will eventually need to allow for demographic stochasticity; without it, the rate at which assembly takes place is artificially inflated.

Algorithms for finding the new community

What happens if the new species j becomes established? To what set of species does the system tend? It may be that species j is absorbed without any extinctions. Or it may be that the orbits head for some part of the boundary of the phase space of the augmented system, leading to extinction of one or more species. The problem is to work out what the composition of the community eventually becomes, i.e. what the set $S^+_{\tau+1}$ is. Broadly, there are two ways to tackle this. We may follow the orbit after adding j to see what $S^+_{\tau+1}$ it leads to; alternatively, we can try to jump directly to $S^+_{\tau+1}$. In other words, we may follow the *transient*, or try to determine the *asymptotic state* directly.

The method using transients is similar to that of experimental studies. We start with the resident community on its attractor, add a small amount of species j to it, and integrate the augmented system until there is no further change in species composition. The species which remain comprise the set $S^+_{\tau+1}$. Such an approach has been used on several occasions for continuous-time systems (Taylor 1988; Case 1990) and related discrete-time ones (Rummel & Roughgarden 1985). There is much to recommend this approach, not least because it can be applied to any dynamical system of the form of Equation 5.1; no assumption such as the Lotka–Volterra dynamic in Equation 5.3 is needed. However, it does have the drawback of being computationally demanding, particularly if there are large numbers of species.

The slowness of numerical integration has led to a search for algorithms that give the asymptotic state without having to go through the transient. One such algorithm was developed by Post & Pimm (1983), Mithen & Lawton (1986) and Drake (1990) for assembly of communities with Lotka–Volterra dynamics. This starts by eliminating species with negative densities at the equilibrium point of the augmented system. When a set of species has been found with positive equilibrium densities, the equilibrium point is checked for asymptotic stability; if it is asymptotically stable, these species are taken to be the set $S^+_{\tau+1}$. This method is heuristic and cannot be justified on mathematical grounds. A check on its results against those from numerical integration has shown that it often gives the new community incorrectly (Morton *et al.* 1996). Luh and Pimm (1993) tried out a method which ignored dynamics altogether, transitions between communities being decided by various randomization schemes.

Another algorithm, making use of permanence, can be applied to communities with Lotka–Volterra dynamics (Law & Morton 1996). This checks the augmented system for permanence; if it is permanent, species j coexists with the residents, so $S^+_{\tau+1}$ is simply the augmented community. If the augmented system is not permanent, there must be orbits that tend from the interior of the $n_\tau + 1$ space to the boundary. The problem is to find which part of the boundary this is. It has to be a part in which the species with nonzero densities coexist, i.e. are permanent. It has also to be part of the boundary which attracts orbits from the interior; the per capita rates of increase of the other species should all be negative on the average. Such a part of the boundary can be called an *attracting subsystem*. As long as there is only one attracting subsystem of the augmented system, this becomes $S^+_{\tau+1}$. This does not of course exhaust all the possibilities; there could, after all, be more than one attracting subsystem, or there could be none. But it is an interesting property of the numerical results to date that such cases are rather unusual (Law & Morton 1996); where they do arise, it is always possible to revert to numerical integration.

Table 5.2 An augmented system taken apart into its subsystems to find an attracting subsystem. The resident community is $S_\tau = \{1, 3\}$; Species 2 invades, but augmented system is not permanent; the attracting subsystem is {2}, so $S^+_{\tau+1} = \{2\}$ (Law & Morton, 1996).

Subsystem	Equil.	Perm.	f_1	f_2	f_3
{}	Yes	Yes	0.86	1.35	−1.01
{1}	Yes	Yes	0	0.49	0.52
{2}	Yes	Yes	−0.1	0	−0.65
{3}	No	—	—	—	—
{1, 2}	No	—	—	—	—
{1, 3}	Yes	Yes	0	0.77	0
{2, 3}	No	—	—	—	—
{1, 2, 3}	No	—	—	—	—

1 The f_i values are the per capita rates of increase at the equilibrium of the corresponding subsystem.
2 Dynamics are of Lotka–Volterra type and defined by the per capita rates of increase of the species, as follows:

$$f_1 = 0.86 - 0.97x_1 - 0.77x_2 - 0.35x_3$$

$$f_2 = 1.35 - 0.97x_1 - 1.08x_2 - 0.02x_3$$

$$f_3 = -1.01 + 1.73x_1 + 0.29x_2$$

Table 5.2 provides an illustration of the algorithm based on permanence, comprising three species with Lotka–Volterra dynamics. Species 1 and 2 are self-supporting, and Species 3 is a consumer. The resident community is $\{1,3\}$ at the start of the iteration. Species 2 now arrives; its per capita rate of increase is positive at the equilibrium point of {1, 3} and, according to the inequality in Equation 5.10, it becomes established. However, the augmented system $\{1,2,3\}$ is not permanent, and in fact there is no equilibrium point in the interior of the augmented system. To find the subsystem to which the orbits tend, we take apart the augmented system as shown in Table 5.2, and search for an attracting subsystem. This turns out to be $\{2\}$ because: (a) $\{2\}$ is permanent; and (b) Species 1 and 3 both have negative per capita rates of increase on the attractor of $\{2\}$. (The attractor of $\{2\}$ is in fact the equilibrium point; but remember from Equation 5.8 that we could use the per capita rate of increase of Species 1 and 3 evaluated at this point even if the attractor of $\{2\}$ was not an equilibrium point.) Figure 5.4 shows the transient dynamics obtained by numerical integration; the system moves from the set $\{1,3\}$ to $\{2\}$ as predicted by the permanence algorithm.

Assembly dynamics

In the section above I have put in place methods to determine the outcome when new species are added to communities. This goes some way to providing a formal framework for modelling assembly dynamics. With it we can address some longstanding questions about the succession of ecological communities.

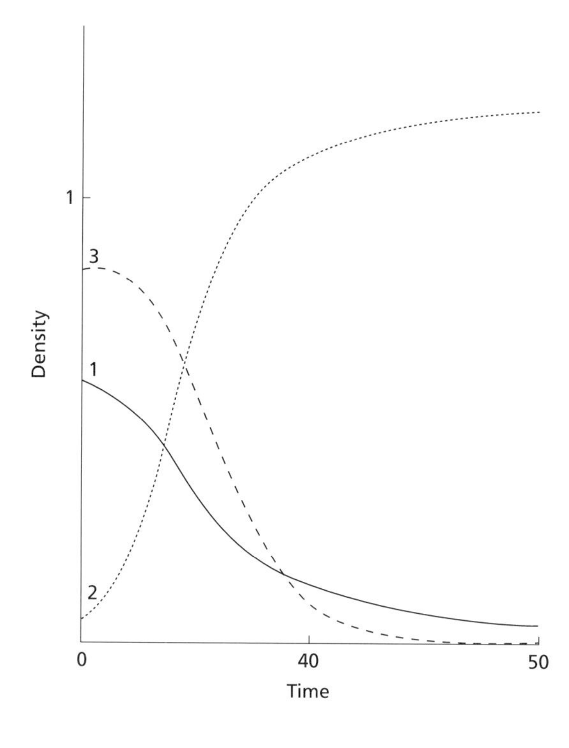

Fig. 5.4 Numerical integration of a 3-species system comprising two self-supporting and one consumer species with densities x_1, x_2 and x_3 respectively. The system starts with Species 1 and 3 into which a small density of Species 2 is introduced, and shows the replacement of {1, 3} by {2}. (From Law & Morton 1996.)

Are there general properties of communities which change during succession? Can we go beyond the description of many special cases of succession and find principles that are common to them all? This issue has fascinated ecologists since Clements' (1916) vision of succession as a highly ordered sequence of vegetational change. For instance, Margalef (1963) argued that succession is characterized by a decline in the energy flow per unit biomass and by an increase in the information content, and Odum (1969) suggested that greater buffering from environmental perturbations (homeostasis) plays a key role. But such suggestions have not received empirical support, and research subsequently retreated from speculations about general emergent properties and concentrated on studies at the smaller scale of population dynamics (Glenn-Lewin *et al.* 1992). The only recent argument I know of is Pimm's (1991, p. 251 *et seq.*) suggestion that there might exist some function, analogous to the adaptive landscape of evolutionary biology, on which communities move uphill during the course of succession. However, if such a function exists, we have no idea as to what it might be.

Models of assembly dynamics might help us to understand what if anything guides the path of succession. To avoid raising unjustified hopes, let me make it clear that no answers are going to be offered here. Rather I am going to point out certain properties of community assembly that would need to be present in giving an answer.

Table 5.3 Three assembly sequences from a single species pool. The resident community at the start of the iteration and the new arrival are shown. New species are picked at random from a pool of 15 species (numbers 1 to 5 are basal species and 6 to 15 consumers), the next community being determined by the permanence algorithm (p. 155).

	Sequence 1		Sequence 2		Sequence 3	
Iteration	Resident community	New species	Resident community	New species	Resident community	New species
1	{}	5	{}	3	{}	3
2	{5}	7	{3}	15	{3}	13
3	{5, 7}	8	{3, 15}	4	{3, 13}	4
4	{5, 8}	10	{3, 15, 4}	13	{13, 4}	3
5	{5, 8}	1	{4, 13}	12	{13, 4}	14
6	{5, 8, 1}	9	{4, 12}	5	{13, 4}	14
7	{5, 8, 1}	12	{12, 5}	6	{13, 4}	7
8	{5, 8, 1, 12}	13	{12, 5}	14	{13, 4}	11
9	{5, 8, 1, 12}	9	{5, 14}	1	{13, 4}	2
10	{5, 8, 1, 12}		{5, 14, 1}		{13, 4, 2}	

Succession is indeterminate

This is an obvious point, but it is important to dispel any notion of succession as an ordered, deterministic sequence of communities. First, as long recognized, the chance arrival of different species causes differences in the local resident communities (Gleason 1926). Second, these local communities place different restrictions on the species that can enter next. The combined effect of this is to generate variation among the local communities. Table 5.3 illustrates this with three assembly sequences constructed by drawing species at random from a single species pool. After a small number of iterations, the communities have very little in common.

Such indeterminacy is well documented in empirical studies of succession (Miles 1987). As an illustration, consider Miles' (1985) outline of the transitions among the major types of vegetation in the mountains of north-west Scotland (Fig. 5.5). Although not resolved to the level of individual species, this indicates that many alternative paths of succession are possible, 26 transitions being included out of a possible total of 56. Simpler alternatives could of course be envisaged in theory; for instance, it might be the case that only one or two species from the regional pool could establish themselves at any time, in which case the number of alternative paths would become very much smaller. This would come closer to the kind of picture Clements (1916) had in mind, but such systems are not well documented empirically.

Succession then is best envisaged as a stochastic process. It helps to have a mental image of an ensemble comprising many assembly sequences

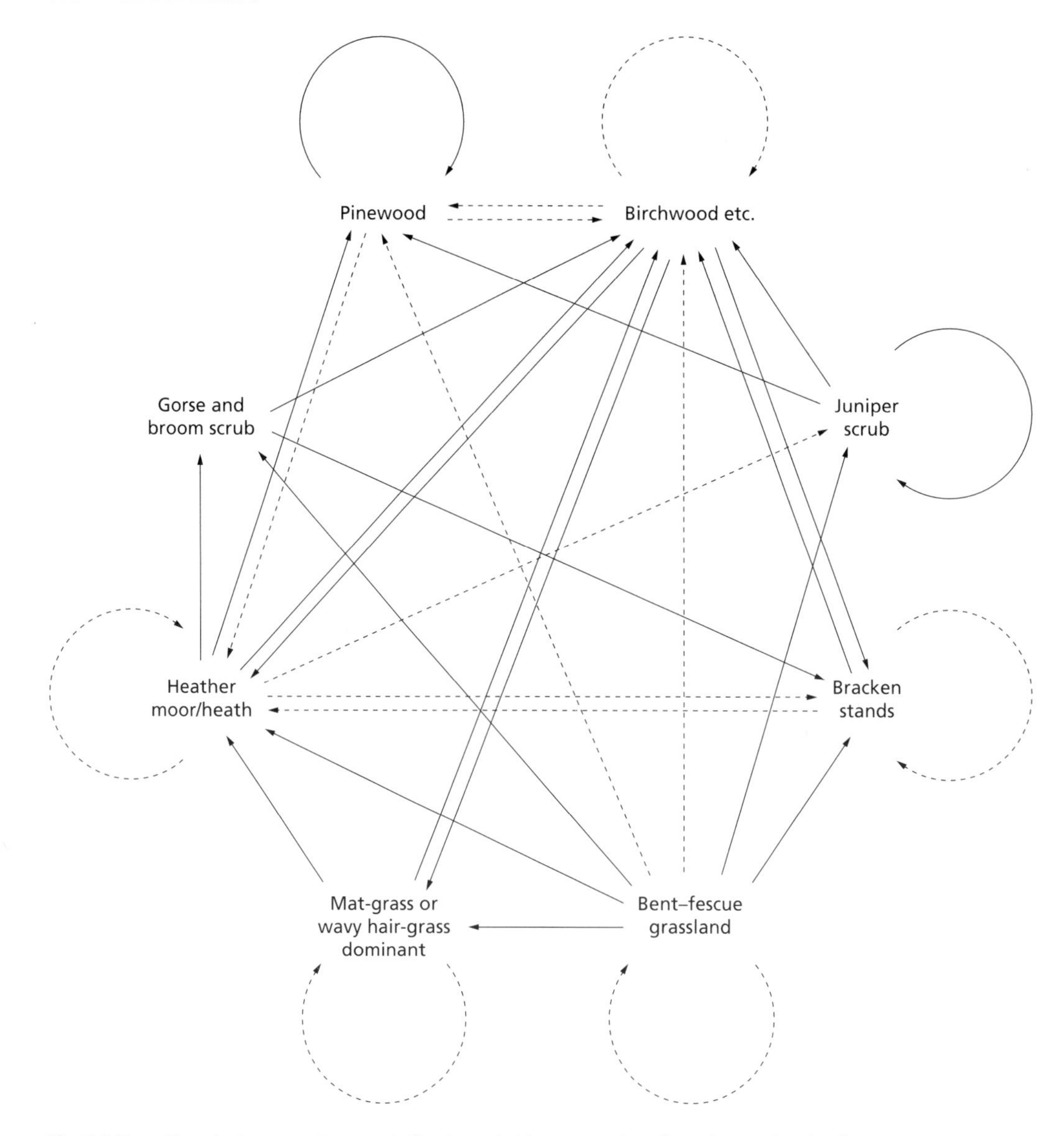

Fig. 5.5 Transitions between major vegetation types in the mountains of north-west Scotland under low grazing and without burning. The more common transitions are shown as continuous arrows, the less common as discontinuous arrows. (From Miles 1985.)

developing from a single species pool. If we are seeking general properties about succession, they are most likely to be found as averages across the ensemble rather than as features of individual sequences. The results on invasion resistance discussed below are of this kind. However, before dealing with them, it is worth mentioning the following simple property, which often applies to the replacement of one community by another.

Successful invaders are present in the community following invasion

When communities with dynamics given by Equations 5.1 and 5.3 are assembled using the permanence algorithm, species which succeed in invading remain present once the asymptotic state of the augmented system is reached. You can see this in Table 5.3. To take an example, Species, 8, which invades the resident community {5, 7} in the first assembly sequence, is present in the resulting community {5, 8}. Species 10 on the other hand never gets started, as it does not satisfy the inequality in Equation 5.10 for invasion (see Table 5.4 for the sign of every f_i). Although there are certain caveats about this result to keep in mind (Case 1995; Law & Morton 1996), it is typically the case that successful invaders remain present in the community which subsequently emerges. Notice that it is often the case that one or more species becomes extinct when a new arrival becomes established; the important point is that the new species is not itself among the ones that disappear. It may be lost later on, but this requires invasion by further species, as in the case of Species 15 driven out by Species 13 in the second sequence of Table 5.3.

This property is surprising. A new species might, after all, destroy the resource base on which it depends for its existence. We could dismiss the property as a special feature of Lotka–Volterra systems assembled by the permanence algorithm, as it does not necessarily apply outside this framework. On the other hand, it is remarkable how rare it is to find counterexamples in natural communities. Proper documentation has yet to be carried out, but there appear to be only a few exceptions all of which are on islands. These exceptions include rabbits that build large populations on islands and occasionally destroy the vegetation on which they depend, a process which Williamson & Fitter (1996b) refer to as 'boom and bust'; Flux (1994) cites an example of this on one of the islands in Hawaii. The fact that these examples come from islands suggests that we may be seeing an additional effect of small system size; if the transient passes close to the boundary of the phase space in which $x_j = 0$, demographic stochasticity could lead to loss of the new species j, even if this part of the boundary was not an attractor in the deterministic system.

Invasion resistance increases

Invasion resistance is a joint property of the set S_τ^+ of resident species and the set S_τ^0 of those absent in iteration τ, and is obtained by classifying each species j in S_τ^0 according to the sign of its per capita rate of increase at the equilibrium point $\hat{\boldsymbol{x}}_\tau$ of the resident community (see inequality in Equation 5.10):

Table 5.4 Invasion resistance of the first assembly sequence of Table 5.3.

			Sign of per capita rate of increase at equlibrium point of the resident community:														
Iteration	Resident community	Invasion resistance	f_1	f_2	f_3	f_4	f_5	f_6	f_7	f_8	f_9	f_{10}	f_{11}	f_{12}	f_{13}	f_{14}	f_{15}
1	{}	0.67	+	+	+	+	+	−	−	−	−	−	−	−	−	−	−
2	{5}	0.07	+	+	+	+	0	+	+	+	+	−	+	+	+	+	+
3	{5, 7}	0.15	+	+	+	+	0	−	0	+	+	+	−	+	+	+	+
4	{5, 8}	0.85	+	+	−	−	0	−	−	0	−	−	−	−	−	−	−
5	{5, 8}	0.85	+	+	−	−	0	−	−	0	−	−	−	−	−	−	−
6	{5, 8, 1}	0.75	0	+	−	−	0	+	−	0	−	−	−	+	−	−	−
7	{5, 8, 1}	0.75	0	+	−	−	0	+	−	0	−	−	−	+	−	−	−
8	{5, 8, 1, 12}	1.00	0	−	−	−	0	−	−	0	−	−	−	0	−	−	−

$$f_j(\hat{\boldsymbol{x}}_\tau, 0)\begin{cases} >0 \Rightarrow \text{species } j \text{ invades} \\ <0 \Rightarrow \text{species } j \text{ does not invade} \end{cases} \tag{5.11}$$

The proportion of species in S_τ^0 which do not invade measures the invasion resistance of the resident community. Table 5.4 shows the kind of thing that can happen during a single assembly sequence; this starts with invasion resistance at 0.67 and ends at 1.00. (The final state at which there is complete invasion resistance we will look at in more detail in the next section.) Notice that it is in no way preordained that invasion resistance should increase monotonically. In fact it starts by going down, as no consumer species can invade until at least one basal species is resident, and it also falls when Species 1 invades later in the assembly sequence.

If there is any trend in invasion resistance, it is to be found in the average properties over an ensemble of assembly sequences from a species pool. Figure 5.6 shows one such pattern based on the average of 100 assembly sequences; this has the property that, after an initial drop, there is a gradual rise to an asymptotic value of 1. During the period in which invasion resistance is increasing, more and more assembly sequences are reaching a completely invasion-resistant state, so the asymptotic value of 1 is inevitable. What is perhaps more interesting is whether those communities which have still to reach this state show such a trend. You can see from Fig. 5.6 that, after removing assembly sequences which have reached a completely invasion-resistant state, the same trend is still present at the earlier stages. Later on the remaining assembly sequences get to a state where approximately one species can still invade; there is then a shuffling of community composition, until the set of resident species happens to 'hit' an invasion-resistant state, after which there is no further change.

Increasing invasion resistance is one of the clearest results to emerge so

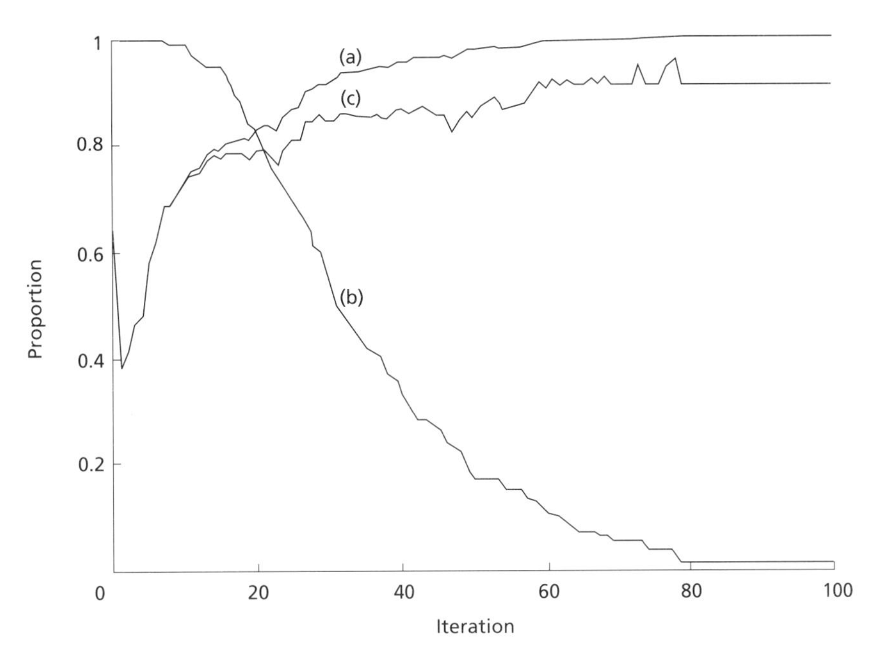

Fig. 5.6 A pattern of invasion resistance during community assembly. This is based on 100 assembly sequences from a species pool of five basal and ten consumer species: (a) the average invasion resistance; (b) the proportion of assembly sequences which have yet to reach a completely invasion-resistant state; (c) the average invasion resistance obtained after removing assembly sequences that have reached an invasion-resistant state.

far from modelling community assembly (Post & Pimm 1983; Drake 1990; Law & Morton 1996). It argues for a role of history in the process of succession in the following sense. Suppose you have two communities from the same species pool, one taken early and the other late in an assembly sequence. The two communities may have the same number of species but, on the average, the older one should be better able to resist invasion by further species from the regional pool. Evidently there is some directionality to the adjustments in community structure taking place as species enter and replace one another. Perhaps Pimm's (1991, p. 251 *et seq.*) suggestion that there might be a function of communities which increases during assembly is not as improbable as it might seem at first sight. However, we will need to seek such a function at the level of the ensemble of assembly sequences, rather than at the level of individual sequences.

Unfortunately, we do not have data from replicated systems of successional communities on which to test these ideas. Perhaps this is not surprising, because it is not enough to investigate a single assembly sequence; we have to look at statistical properties across a large number of such assembly sequences from a single pool. Hard though this sounds, it ought not to be neglected. At the end of the recent major international research investment into biological invasions organized by SCOPE, ecologists were little the wiser as to what makes a community invasible (Drake *et al.* 1989; Gilpin 1990). Models of community assembly suggest there may be general answers to this core question, as long as we accept that they have to be of a statistical kind.

Endstates of assembly sequences

The previous section has shown that the number of paths which assembly sequences can follow during succession is potentially very large. What eventually happens to these paths? Do they end up reaching a myriad of different states? Do they finally settle down to one or two states? Can we distinguish between states of qualitatively different types? These are questions first and foremost about the species pool, the set S_m. We are in effect looking for subsets of S_m that entrap assembly sequences, preventing further change in community composition (Drake 1990). I will refer to these trapping sets as *permanent endstates*, the qualifier 'permanent' being added to make it clear that they are found by the permanence algorithm for community assembly.

Kinds of endstate

To date, two kinds of permanent endstate have been shown to exist (Law & Morton 1996; Morton & Law 1997). The first is a resident community that:

1 can be reached by the permanence-based assembly; and
2 is uninvadable by any other species from the regional pool.

The set {1, 5, 8, 12} in Tables 5.3 and 5.4 is an example; it was shown to be attainable by sequential invasions in Table 5.3, and Table 5.4 shows that it is uninvadable by any other species from the set S_m. We may call this a *permanent endpoint*, as it comprises exactly one community; it represents an achievable and unchanging configuration of species from the species pool. Endpoints have an obvious parallel with the ecological notion of a climax community, as states beyond which no further change in community composition is to be expected in the absence of changes in the physical conditions.

The second kind of endstate is more subtle and complicated. It comprises a set which is itself the union of three or more subsets; these subsets are communities which replace one another in a cyclic or more complicated sequence. The set may be called a *permanent endcycle* if the communities which make up the subsets have the following properties:

1 at least one can be reached by the permanence-based assembly;
2 each is permanent in its own phase space;
3 each is invadable by at least one of the other species in the endcycle;
4 each is uninvadable by every species that is not in the endcycle.

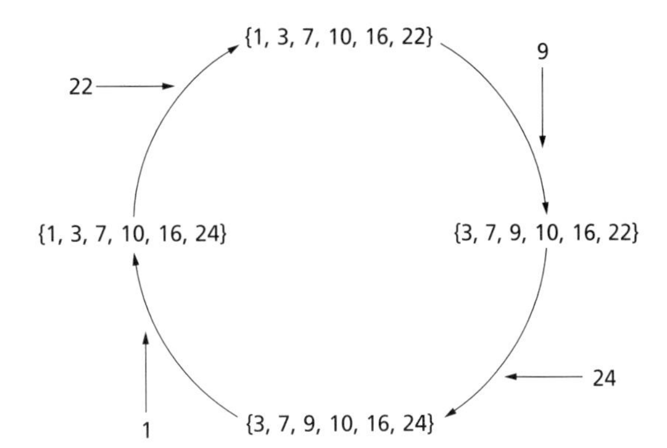

Fig. 5.7 A permanent endcycle of a species pool comprising 24 species. Basal species: 1, 3, 7, 9, 10; consumers: 16, 22, 24. (From Morton & Law 1997.)

Figure 5.7 is an example of a permanent endcycle, from a species pool of 12 basal and 12 consumer species with Lotka–Volterra dynamics (Morton & Law 1997). Each six-species community can be invaded by only one species, and the sequence of invasions gives rise to a loop. The union of these communities, the set of eight species {1, 3, 7, 9, 10, 16, 22, 24}, constitutes a permanent endcycle of this species pool.

Permanent endcycles have some mathematical interest through their relationship with a class of solutions to Equation 5.1 quite distinct from the equilibrium points, periodic and chaotic orbits mentioned above. These are *heteroclinic cycles*, cyclic sequences of saddle-like boundary equilibria. Heteroclinic cycles come about naturally in Equation 5.1 due to the invariance of the boundary (Hofbauer 1994); they were first noted by May & Leonard (1975) in systems of three competing species, and have subsequently been found in a number of other settings (e.g. Kirlinger 1989; Law & Morton 1993). The heteroclinic cycle in Fig. 5.7 is relatively straightforward, as it involves only four parts of the boundary. Writing the corresponding boundary equilibria as $\hat{\boldsymbol{x}}^{(k)}$ ($k = 1, \ldots, 4$), we have the property that, at $\hat{\boldsymbol{x}}^{(k)}$, there is exactly one species j which belongs to the endcycle and satisfies $f_j(\hat{\boldsymbol{x}}^{(k)}) > 0$ (Table 5.5). It is this simple arrangement of the per capita rates of increase that is responsible for the cyclic replacement of one species by another seen in Fig. 5.7. However, heteroclinic cycles can be a great deal more complicated than this example suggests. If more parts of the boundary are involved, and if there is more than one species with $f_j(\hat{\boldsymbol{x}}^{(k)}) > 0$ at the corresponding boundary equilibria, a complex network of interconnected communities can result. Another point to note is that a heteroclinic cycle may or may not be an attractor. The criteria defining a permanent endcycle above do not deal with this; there may be an attractor comprising the whole endcycle, some part of it, or none of it; see Hofbauer (1994) for the criteria for stability of heteroclinic cycles. The ecological implications of permanent endcycles are thus potentially important, because

Table 5.5 Information on the endcycle of Fig. 5.6.

Boundary index (k)	Resident community	Sign of per capita rate of increase:			
		$f_1(\hat{x}^{(k)})$	$f_{22}(\hat{x}^{(k)})$	$f_9(\hat{x}^{(k)})$	$f_{24}(\hat{x}^{(k)})$
(1)	{3, 7, 9, 10, 16, 24}	+	–	0	0
(2)	{1, 3, 7, 10, 16, 24}	0	+	–	0
(3)	{1, 3, 7, 10, 16, 22}	0	0	+	–
(4)	{3, 7, 9, 10, 16, 22}	–	0	0	+

their presence implies a form of stability that may take many generations to observe.

Cyclic endstates are related to mosaic cycles sometimes observed in ecological communities, where one species replaces another in a cyclic manner (see Chapters 1 & 11; Remmert 1991). Such phenomena have been reported on a number of occasions, in heathlands (Watt 1947, 1955; Gimingham 1988), forest ecosystems (Remmert 1991), and on rocky shores (Hartnoll & Hawkins 1985). But the scales at which mosaic cycles have been observed are small, and are often attributable to death of individual plants or animals. It may be best to envisage such cycles as operating within communities, and to model them in a spatially explicit framework (Wissel 1992; Rand 1994; Hendry & McGlade 1995; see also Chapters 1, 3 & 11). A cyclic endstate in contrast sits at one point (community) on the cycle until the chance arrival of another species shifts the system to another point on the cycle. It goes around the endcycle in a series of stops and starts.

Number of endstates

How many endstates we find obviously depends on S_m, and is not easy to answer in a general way. If all species are independent, there is only one endstate. On the other hand, if interactions are strong, different assembly sequences could get trapped at different endstates, depending on the order in which species have arrived (Drake 1990). The bistable case of two competing species, where interspecific competition is stronger than intraspecific competition, is a simple case in point; whichever species arrives first creates a community uninvadable by the other, giving a 'priority effect' in favour of the species which arrives first (Grover & Lawton 1994). This same effect can be seen even more clearly in the spatial models developed to examine invasion sequences.

Some models of community assembly have suggested that the number of endstates could be large, assembly sequences often ending up at different states (Drake 1990). For the kind of species pools used, this seems to have been an artefact of the algorithm for community assembly (Morton *et al.* 1996), although we cannot rule out the possibility of other kinds of species pools that do have this property. Pools constructed from simple empirical rules about size-based predator–prey interactions (Cohen *et al.* 1993) do sometimes contain more than one endstate, but the number is small (Law & Morton 1996; Morton & Law 1997).

This small number of endstates is in keeping with informal observations on real communities. We are not confronted with an inexplicable mosaic of different communities dominated by local priority effects. There are usually striking resemblances among disjunct communities which live in similar environments and share a common species pool; a well-documented case of this is the plant species composition of calcareous grasslands in England (Mitchley & Grubb 1986). This is not to deny that alternative endstates can be found in nature; for example some coral communities in Jamaica were altered by a hurricane in 1980 to a state in which staghorn corals subsequently remained rare or absent (Knowlton *et al.* 1990). However, cases of this kind are noteworthy because they seem to be relatively unusual.

Concluding comments

This chapter has been as much about what to leave out of a theory of community ecology as what to include. If we carry over too much from population dynamics when moving up to the level of the community, we run the risk of generating dynamical systems too complex to provide insights into many important community processes. We need to put in place a theoretical framework which has the turnover of species at its core, rather than the turnover of individuals within species. At the same time, this framework does need to be consistent with processes operating at the smaller scale of population dynamics. This is one reason for taking as states the sets of species with the property of persistence. Population dynamics then provide the critical qualitative information needed on presence and absence of species.

This is no more than a first step towards a formal framework for studying assembly dynamics. In view of the inherent uncertainty about paths of community assembly, it seems most appropriate to think eventually in terms of a stochastic process, with the state variable $P(C_i,\tau)$ being the probability that the system is in state C_i at time τ. The probability per unit time of moving from state C_i to C_j, $w(C_j \mid C_i)$, can then be seen as a Markov process depending on:

1 the arrival probability per unit time of each species in the regional pool; and

2 the set of resident species which results from each arrival (determined by the population dynamics).

We may then investigate major features of the stochastic process, perhaps in terms of its summary statistics. It may not be easy to do this, as there is no obvious local coupling of one state to another; species composition can undergo major changes following invasion events.

Analysis of the stochastic process would still leave many questions unanswered. What, for instance, is the effect of removing the separation of time-scales between community dynamics and assembly dynamics so that assembly can take place on the transients of community dynamics? What are the consequences of allowing simultaneous arrival of species? What happens in a world where the external environment changes through time and space? What would be the effect of making the system size small enough for stochastic effects to dominate population dynamics? What happens when the process is embedded in an explicit spatial domain? And then there is a nontrivial matter of evolution generated by the interactions within communities. These are important issues, and ones which will be made sharper by a formal framework for the dynamics of community assembly.

Acknowledgements

I thank colleagues who have helped my understanding of issues in this chapter and have played a major part in developing many of the ideas it contains, in particular, J.C. Blackford, U. Dieckmann, V. Grimm, J. Hofbauer, V. Hutson, R.D. Morton, and M.H. Williamson. The background to the chapter was developed while in receipt of grants GR3/8205, GR3/08890 and GST/02981 from the Natural Environment Research Council, and the writing was supported by the Wissenschaftskolleg zu Berlin.

References

Baker, G.L. & Gollub, J.P. (1990) *Chaotic Dynamics: an Introduction*. Cambridge University Press, Cambridge.

Case, T.J. (1990) Invasion resistance arises in strongly interacting species-rich model competition communities. *Proceedings of the National Academy of Sciences USA*, **87**, 9610–9614.

Case, T.J. (1995) Surprising behavior from a familiar model and implications for competition theory. *American Naturalist*, **146**, 961–966.

Case, T.J. & Casten, R.G. (1979) Global stability and multiple domains of attraction in ecological systems. *American Naturalist*, **113**, 705–714.

Chesson, P.L. & Ellner, S. (1989) Invasibility and stochastic boundedness. *Journal of Mathematical Biology*, **27**, 117–138.

Clements, F.E. (1916) *Plant Succession: an Analysis of the Development of Vegetation*. Carnegie Institute Washington Publication 242, Washington DC.

Cohen, J.E., Pimm, S.L., Yodzis, P. & Saldaña, J. (1993) Body sizes of animal predators and animal prey in food webs. *Journal of Animal Ecology*, **62**, 67–78.

Cornell, H.V. (1985) Species assemblages of cynipid gall wasps are not saturated. *American Naturalist*, **126**, 565–569.

Cornell, H.V. & Lawton, J.H. (1992) Species interactions, local and regional processes, and

limits to the richness of ecological communities: a theoretical perspective. *Journal of Animal Ecology*, **61**, 1–12.
Darzin, P.G. (1992) *Nonlinear systems*. Cambridge University Press, Cambridge.
Diamond, J.M. (1975) Assembly of species communities. In: *Ecology and Evolution of Communities* (eds M.L. Cody & J.M. Diamond), pp. 342–444. Belknap Press, Cambridge, MA.
Dieckmann, U. & Law, R. (1996) The dynamical theory of coevolution: a derivation from stochastic ecological processes. *Journal of Mathematical Biology*, **34**, 579–612.
Drake, J.A. (1990) The mechanics of community assembly and succession. *Journal of Theoretical Biology*, **147**, 213–233.
Drake, J.A. (1991) Community-assembly mechanics and the structure of an experimental species ensemble. *American Naturalist*, **137**, 1–26.
Drake, J.A. *et al.* (1989) *Biological Invasions: a Global Perspective*. Scientific Committee on Problems of the Environment. Wiley, New York.
Drake, J.A. *et al.* (1993) The construction and assembly of an ecological landscape. *Journal of Animal Ecology*, **62**, 117–130.
Flux, J.E. (1994) World distribution. In: *The European Rabbit: the History and Biology of a Successful Invader* (eds H.V. Thompson & C.M. King), pp. 8–21. Oxford University Press, Oxford.
Gilpin, M. (1990) Ecological prediction. *Science*, **248**, 88–89.
Gimingham, C.H. (1988) A reappraisal of cyclical processes in *Calluna* heath. *Vegetatio*, **77**, 61–64.
Gleason, H.A. (1926) The individualistic concept of the plant association. *Bulletin of the Torrey Botanical Club*, **53**, 7–26.
Glenn-Lewin, D.C., Peet, R.K. & Veblen, T.T. (eds) (1992) *Plant Succession: Theory and Prediction*. Chapman & Hall, London.
Goh, B.S. (1977) Global stability in many-species systems. *American Naturalist*, **111**, 135–142.
Grimm, V. & Wissel, C. (1997) Babel, or the ecological stability discussions: an inventory and analysis of terminology and a guide for avoiding confusion. *Oecologia*, **109**, 323–334.
Grover, J.P. & Lawton, J.H. (1994) Experimental studies on community convergence and alternative stable states: comments on a paper by Drake *et al. Journal of Animal Ecology*, **63**, 484–487.
Hartnoll, R.G. & Hawkins, S.J. (1985) Patchiness and fluctuations on moderately exposed rocky shores. *Ophelia*, **24**, 53–63.
Hastings, A., Hom, C.L., Ellner, S., Turchin, P. & Godfray, H.C.J. (1993) Chaos in ecology: is mother nature a strange attractor? *Annual Review of Ecology and Systematics*, **24**, 1–33.
Hendry, R. & McGlade, J.M. (1995) The role of memory in ecological systems. *Proceedings of the Royal Society of London*, **B259**, 153–159.
Hofbauer, J. (1994) Heteroclinic cycles in ecological differential equations. *Tatra Mountains Mathematical Publications*, **4**, 1–14.
Hofbauer, J. & Sigmund, K. (1988) *The Theory of Evolution and Dynamical Systems*. Cambridge University Press, Cambridge.
Holling, C.S. (1973) Resilience and stability of ecological systems. *Annual Review of Ecology and Systematics*, **4**, 1–23.
Horn, H.S. (1975) Markovian properties of forest succession. In: *Ecology and Evolution of Communities* (eds M.L. Cody & J.M. Diamond), pp. 196–211. Belknap Press, Cambridge, MA.
Hutson, V. & Schmitt, K. (1992) Permanence and the dynamics of biological systems. *Mathematical Biosciences*, **111**, 1–71.
Jansen, W. (1987) A permanence theorem for replicator and Lotka–Volterra systems. *Mathematical Biosciences*, **82**, 411–422.

Karieva, P. & Wennergren, U. (1995) Connecting landscape patterns to ecosystem and population processes. *Nature*, **373**, 299–302.

Kelt, D.A., Taper, M.L., Meserve, P.L. (1995) Assessing the impact of competition on community assembly: a case study using small mammals. *Ecology*, **76**, 1283–1296.

Kirlinger, G. (1989) Two predators feeding on two prey species: a result on permanence. *Mathematical Biosciences*, **96**, 1–32.

Knowlton, N., Lang, J.C. & Keller, B.D. (1990) *Case study of natural population collapse: post-hurricane predation on Jamaican staghorn corals*. Smithsonian Contributions to the Marine Sciences, 31. Smithsonian Institution Press, Washington DC.

Law, R. & Morton, R.D. (1993) Alternative permanent states of ecological communities. *Ecology*, **74**, 1347–1361.

Law, R. & Morton, R.D. (1996) Permanence and the assembly of ecological communities. *Ecology*, **77**, 762–775.

Lawton, J.H., Lewinsohn, T.M. & Compton, S.G. (1993) Patterns of diversity for the insect herbivores on bracken. In: *Species Diversity in Ecological Communities* (eds R.E. Ricklefs & D. Schluter), pp. 178–184. University of Chicago Press, Chicago, IL.

Lewontin, R.C. (1969) The meaning of stability. In: *Diversity and Stability in Ecological Systems* (eds G.M. Woodwell & H.H. Smith), pp. 13–24. Symposium 22, Brookhaven National Laboratory, Upton, NY.

Luh, H.-K. & Pimm, S.L. (1993) The assembly of ecological communities: a minimalist approach. *Journal of Animal Ecology*, **62**, 749–765.

MacArthur, R.H. & Wilson, E.O. (1967) *The Theory of Island Biogeography*. Princeton University Press, Princeton.

Margalef, R. (1963) On certain unifying principles in ecology. *American Naturalist*, **97**, 357–374.

May, R.M. (1974) *Stability and Complexity in Model Ecosystems*, 2nd edn. Princeton University Press, Princeton.

May, R.M. & Leonard, W.J. (1975) Nonlinear aspects of competition between three species. *SIAM Journal of Applied Mathematics*, **29**, 243–253.

Maynard Smith, J. (1969) The status of neo-Darwinism. In: *Towards a Theoretical Biology* (ed. C.H. Waddington), pp. 82–89. Edinburgh University Press, Edinburgh.

Metz, J.A.J., Nisbet, R.M. & Geritz, S.A.H. (1992) How should we define 'fitness' for general ecological scenarios? *Trends in Evolution and Ecology*, **7**, 198–202.

Miles, J. (1985) The pedogenic effects of different species and vegetation types and the implications of succession. *Journal of Soil Science*, **36**, 571–584.

Miles, J. (1987) Vegetation succession: past and present perceptions. In: *Colonization, Succession and Stability* (eds A.J. Gray, M.J. Crawley & P.J. Edwards), pp. 1–29. Blackwell Scientific Publications, Oxford.

Mitchley, J. & Grubb, P.J. (1986) Control of relative abundance of perennials in chalk grassland in southern England I. Constancy of rank order and results of pot- and field-experiments on the role of interference. *Journal of Ecology*, **74**, 1139–1166.

Mithen, S.J. & Lawton, J.H. (1986) Food-web models that generate constant predator–prey ratios. *Oecologia (Berlin)*, **69**, 542–550.

Morton, R.D. & Law, R. (1997) Regional species pools and the assembly of local ecological communities. *Journal of Theoretical Biology*, **187**, 321–331.

Morton, R.D., Law, R., Pimm, S.L. & Drake, J.A. (1996) On models for assembling ecological communities. *Oikos*, **75**, 493–499.

Odum, E.P. (1969) The strategy of ecosystem development. *Science*, **164**, 262–270.

Peschel, M. & Mende, W. (1986) *The Predator–Prey Model. Do We Live in a Volterra World?* Akademie-Verlag, Berlin.

Pimm, S.L. (1982) *Food Webs*. Chapman & Hall, London.

Pimm, S.L. (1991) *The Balance of Nature?* University of Chicago Press, Chicago.

Post, W.M. & Pimm, S.L. (1983) Community assembly and food web stability. *Mathematical Biosciences*, **64**, 169–192.

Rand, D.A. (1994) Measuring and characterizing spatial patterns, dynamics and chaos in spatially extended dynamical systems and ecologies. *Philosophical Transactions of the Royal Society of London*, **A348**, 497–514.

Remmert, H. (1991) *The Mosaic-cycle Concept of Ecosystems*. Springer-Verlag, Berlin.

Ricklefs, R.E. (1987) Community diversity: relative roles of local and regional processes. *Science*, **235**, 167–171.

Ricklefs, R.E. & Schluter, D. (eds) (1993) *Species Diversity in Ecological Communities*. University of Chicago Press, Chicago.

Robinson, J.V. & Dickerson, J.E. (1987) Does invasion sequence affect community structure? *Ecology*, **68**, 587–595.

Robinson, J.V. & Edgemon, M.A. (1988) An experimental evaluation of the effect of invasion history on community structure. *Ecology*, **69**, 1410–1417.

Rummel, J.D. & Roughgarden, J. (1985) A theory of faunal buildup for competition communities. *Evolution*, **39**, 1009–1033.

Schuster, P., Sigmund, K. & Wolff, R. (1979) Dynamical systems under constant organization. III. Cooperative and competitive behavior of hypercycles. *Journal of Differential Equations*, **32**, 357–368.

Shrader-Frechette, K.S. & McCoy, E.D. (1993) *Method in Ecology*. Cambridge University Press, Cambridge.

Simmons, G.F. (1972) *Differential Equations*. Tata McGraw-Hill, New Delhi.

Taylor, P.J. (1988) The construction and turnover of complex community models having generalized Lotka–Volterra dynamics. *Journal of Theoretical Biology*, **135**, 569–588.

Usher, M.B. (1992) Statistical models of succession. In: *Plant Succession: Theory and Application* (eds D.C. Glenn-Lewin, R.K. Peet & T.T. Veblen), pp. 215–248. Chapman & Hall, London.

Warren, P.H. (1996) The effects of between-habitat dispersal rate on protist communities and metacommunities in microcosms at two spatial scales. *Oecologia*, **105**, 132–140.

Watt, A.S. (1947) Pattern and process in the plant community. *Journal of Ecology*, **35**, 1–22.

Watt, A.S. (1955) Bracken versus heather: a study in plant sociology. *Journal of Ecology*, **43**, 490–506.

Weatherby, A., Warren, P.H. & Law, R. (1998) Coexistence and collapse: an experimental investigation of the persistent communities of a protist species pool. *Journal of Animal Ecology*, **67**, 554–566.

Weiner, J. (1995) On the practice of ecology. *Journal of Ecology*, **83**, 153–158.

Williamson, M. & Fitter, A.H. (1996a) The varying success of invaders. *Ecology*, **77**, 1661–1666.

Williamson, M. & Fitter, A.H. (1996b) The characters of successful invaders. *Biological Conservation*, **78**, 163–170.

Wissel, C. (1992) Modelling the mosaic cycle of a middle European beech forest. *Ecological Modelling*, **63**, 29–43.

Yodzis, P. (1989) *Introduction to Theoretical Ecology*. Harper & Row, New York.

6: The dynamics of the flows of matter and energy

S.L. Pimm

Introduction

In the last paragraph of *The Origin of Species*, Darwin's famous description of a 'tangled bank' begs an obvious question. How do we understand nature given the consequences of species being 'dependent upon each other in so complex a manner?' Ecologists have employed at least two intertwined approaches since the first edition of *Theoretical Ecology* (May 1976). The first describes the *food webs* that depict that tangled bank—which species interact or the 'ecologically flexible scaffolding around which communities are assembled and organized' (Paine 1966; see also Chapter 5). The second asks what the flows of energy and matter over this scaffolding mean to ecosystem dynamics. It is on this second subject that I shall concentrate, but some comments about the first are in order.

Food web patterns

There must be bounds on the 'manner' of the bank's tangle. Food webs are likely to be 'patterned'—meaning that they may be statistically unusual in various ways. (Assign predator and prey at random and the webs so produced have far more loops of the kind A eats B eats C eats A than one ever sees in nature.) The first statistical analyses of empirical food web patterns appeared in the late 1970s (Cohen 1977; Pimm & Lawton 1978, 1980; Pimm 1980) with books following at regular intervals (Cohen 1978; Pimm 1982, 1991; DeAngelis *et al.* 1983; Cohen *et al.* 1990; Polis & Winemiller 1996). There are also reviews: Lawton (1989), Schoener (1989), Pimm *et al.* (1991), Pollis (1991), and Hall & Raffaelli (1993). Inspection of these sources shows that the list of putative patterns is long, the existence of some of them is controversial, different mechanisms for them have been proposed, and the mechanisms debated.

The existence of food patterns motivates our understanding of their consequences to the flows of matter and energy through ecosystems and their dynamics. (These flows and dynamics also constrain the food patterns, of course. Some food web patterns are energetically impossible.) This chapter comprises three examples that investigate the relationship between the patterns of energy and matter flows and ecosystem dynamics. Each example originates from a hitherto unappreciated source of questions about food web dynamics: Gilbert and Sullivan's opera, *The Mikado*, (1885).

1 What happens when we subject an ecosystem to a 'short, sharp shock'? How quickly will the system recover? I will show that the answer, in part, depends on how many trophic levels the food web contains.

2 Why is it that 'A is happy and B is not'? If we alter the biomass of a species (perhaps by harvesting it) or increase the production of a plant species (perhaps by increasing an essential nutrient) what will be the consequences? In the latter case, the herbivores may consume the increased plant production, the increased herbivore production might be consumed by its predators, and so on up the food chain. More food for B may make A happy, and not B. I will show that which species are changed, and which are not, depends on the structure of the food web.

3 Do some species have redundant roles in an ecosystem, so that their loss will not seriously impair the system's dynamics? Is it correct that 'it really doesn't matter whom you put upon the list, there would none of them be missed'? The alternative hypothesis is that more diverse ecosystems could have a greater chance of containing species that survive or that can even thrive during a disturbance that kills off other species. I will show that highly connected and simple food webs do differ in their responses to disturbances.

All three questions are about dynamical stability and the patterns of energy and matter flows over food webs. The first involves *resilience*—the rate at which systems recover from transient shocks. Questions **2** and **3** are about *resistance*—the degree to which systems are unchanged when we alter their components.

We can define resistance and resilience for many kinds of ecological variables. These include population sizes, the biomass of a species or a group of species, or the total mass of phosphorus one or more species contain. Consider plant biomass—the total weight of all the plants. Biomass is resistant to change, if following some disturbance it declines only slightly. Figure 6.1 shows two sets of systems that experience different declines in biomass when subjected to the same disturbance. Resistance is a ratio (biomass after disturbance/normal biomass) and so it is dimensionless.

Resilience measures how quickly biomass recovers following the disturbance. Figure 6.1 anticipates a commonsense notion (I shall justify it later) that the rate of recovery will also depend on how much the biomass has been altered. The greater the alteration, the faster is the change. This means that the dimensions of resilience should be the change per unit time divided by the change itself. Resilience has dimensions of 1/time. The

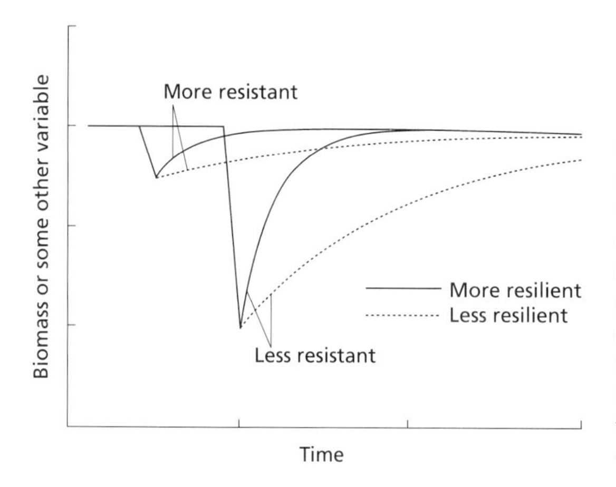

Fig. 6.1 Two systems showing differing degrees of resistance: the degree to which biomass or some other variable is reduced by a disturbance. For a given resistance, the figure shows two different rates of recovery, or resilience. Resilience is scaled according to a given displacement from equilibrium values.

figure shows that biomass can recover quickly or slowly from each of two different declines.

Short, sharp shocks

How quickly will an ecosystem recover when we subject it to transient shock? Consider a lake. Now let's shock its phytoplankton with a pulse of nutrients. The intuition is that the phytoplankton would first increase, then decrease as they use up the nutrients. Alone, they could probably recover normal levels quickly. If the lake also contained phytoplankton, zooplankton and fish, the increased phytoplankton would cause the zooplankton to increase, and then the increased zooplankton would cause the fish to increase. Now, while the fish remain unusually abundant, their prey, the zooplankton, will be rare. Consequently, their prey, the phytoplankton, will be usually abundant.

Simply, no component of the system can return to equilibrium until all the others do so. To study only one component is to hear the plop of the stone in the pond, but not see the ripples spread. The recovery time is likely to depend on the length of the food chain. The longer the chain, the further those ripples have to travel.

The tinker-toy models

How should we model resilience and the intuition of the previous paragraphs? Let's keep the model building and analysis simple to understand the principles. First, consider an equation for the phytoplankton limited by nutrients:

$$dX_1/dt = X_1(b_1 - a_{11}X_1) \tag{6.1}$$

The growth rate of the phytoplankton, dX_1/dt, depends on the size of its population, X_1, its intrinsic growth rate, b_1, and a limitation imposed by the shortage of nutrients, a_{11}. (This is the familiar '*r* and *K*' population model in a different guise. It is one that allows us to add other trophic levels more readily.) The population size approaches '*K*' its equilibrium value, b_1/a_{11} from any value of X_1. The question is how fast does it approach that value?

Now let's add another trophic level, the zooplankton:

$$\begin{aligned} dX_1/dt &= X_1(b_1 - a_{11}X_1 - a_{12}X_2) \\ dX_2/dt &= X_2(-b_2 + a_{21}X_1) \end{aligned} \tag{6.2}$$

The phytoplankton now suffers predation from the zooplankton. The zooplankton die off if there are insufficient phytoplankton to support them, that is, when $(-b_2 + a_{21}X_1) < 0$, or $X_1 < b_2/a_{21}$).

We can keep on adding levels; the three trophic level model is:

$$\begin{aligned} dX_1/dt &= X_1(b_1 - a_{11}X_1 - a_{12}X_2) \\ dX_2/dt &= X_2(-b_2 + a_{21}X_1 - a_{23}X_3) \\ dX_3/dt &= X_3(-b_3 + a_{32}X_2) \end{aligned} \tag{6.3}$$

These models are in the Lotka–Volterra family. Each set contains an equation for each species in the system. Each equation contains a growth or death term in the absence of other species (b_i) and interaction terms for every interspecific interaction. These terms are of the form $a_{ij}X_iX_j$—one that derives from the assumption that species collide with each other randomly like molecules in a gas. When species j eats species i, the effect of species j on i is (negative) a_{ij} and the effect of species i on j is (positive) a_{ji}.

Setting each $dX_i/dt = 0$ and applying some simple algebra gives the new equilibrium densities for two and three trophic level models. (I'll return to these calculations in the next section.)

The more difficult problem is how fast will the densities return to equilibrium. One way to explore the equations' behaviour is to simulate them. We can replace the differential equations with calculations using small but finite time steps, Δt. The smaller the step, the closer these finite difference equations will approximate the differential equations (that is $\Delta X/\Delta t \approx dX/dt$ for small Δt). The idea is that

$$X_{t+\Delta t} = X_t + \Delta X \tag{6.4}$$

The finite difference approximation for Equation 6.1 would be:

$$\Delta X_1 = \Delta t \cdot X_1(b_1 - a_{11} \cdot X_1) \tag{6.5}$$

As an example, I have investigated the three species system:

$$dX_1/dt = X_1(1.0 - 0.01X_1 - 0.1X_2)$$

$$dX_2/dt = X_2(-1.0 + 0.02X_1 - 0.1X_3) \qquad (6.6)$$

$$dX_3/dt = X_3(-1.0 + 0.5X_2)$$

using a time step, $\Delta t = 0.1$, and initial values of the three species of 50, 10, and 3. The choice of step length for time may require care, as if it is too long the dynamics will not reflect those of the differential equation.

You can do this at home (or wherever you keep your computer). Put these first three numbers into row 1 of a spreadsheet, into columns A, B, and C respectively, thus:

50 10 3

Add three formulae into the three columns of the next row:

Row 2, column A = A1 + (0.1)*A1*(1 – (0.01*A1) – (0.1*B1))

Row 2, column B = B1 + (0.1)*B1*(–1 + (0.02*A1) – (0.1*C1))

Row 2, column C = C1 + (0.1)*C1*(–1 + (0.5*B1))

and ask the computer to calculate the new values, which are:

47.5 9.7 4.2

Spreadsheets have the convenient feature of now allowing you to 'fill down' the calculations for as many rows as one wants. The next row will contain the formulas for A3, B3, and C3, and calculate them as:

45.386 9.244 5.817

and so on for 500 rows.

Figure 6.2 shows the results along with the other two simulations for just species 1 and for species 1 and species 2. (One gets these results by simply starting out with 50, 0, 0 in the first case, and 50, 10, and 0 in the second.) In both simulations and nature, zero animals one time means zero animals the next.

These simulations raise several issues. First, how do we characterize the rate of return to equilibrium in these models, especially since the return is direct in the one species model, but oscillates in the other two? Second, since the initial starting conditions were entirely arbitrary, will the results differ if one used different initial conditions, but the same set of parameters? Finally, although the pattern seems obvious here—the more trophic levels the longer the system takes to return—is this result typical over all sensible sets of parameters?

Consider what could happen. Imagine that the further a species' density

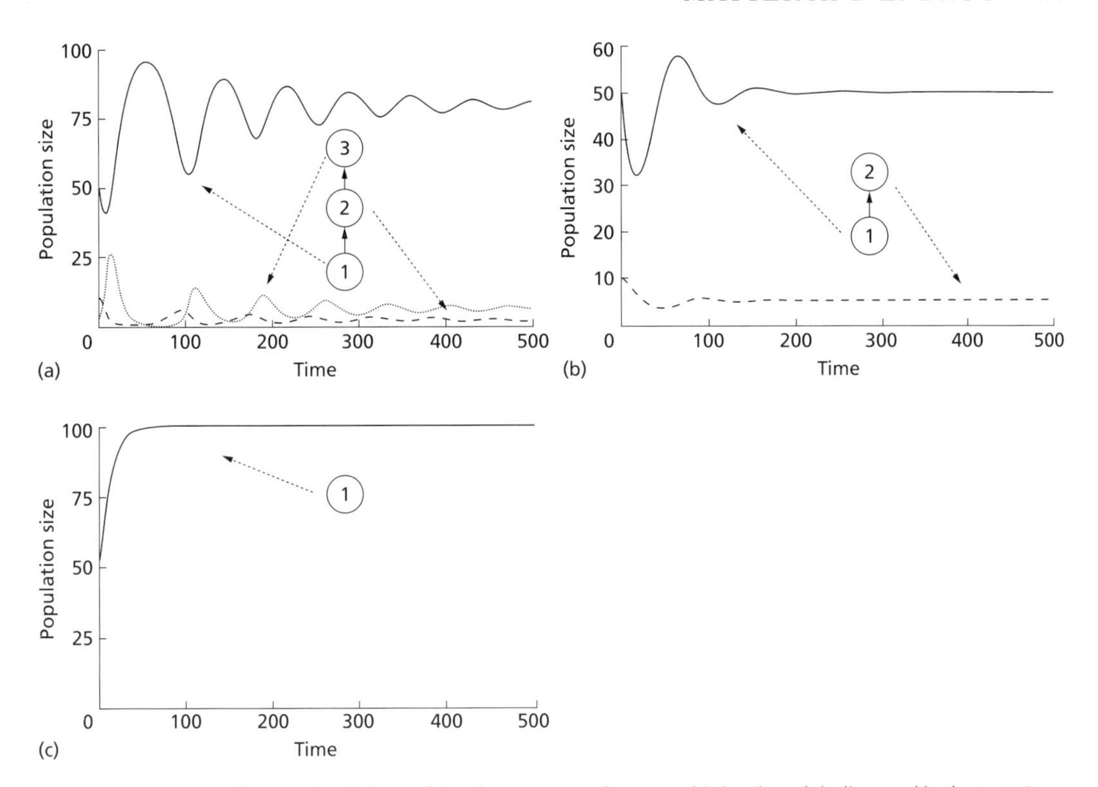

Fig. 6.2 Simulations of the three, two, and one trophic level models discussed in the text. By inspection, one sees that the rate of return to equilibrium gets longer as the number of trophic levels increases.

is away from equilibrium (call this X_1^*), the faster it will move towards it. To make things as simple as possible, assume that the rate of change (dX_1/dt) is directly proportional to the distance away from equilibrium,

$$dX_1/dt = k(X_1 - X_1^*) \tag{6.7}$$

where k is a constant. Such processes where the rate of change depends proportionally on what is changing are ubiquitous. They lead to the familiar exponential decay of the displacement from equilibrium at time t, $(X_1 - X_1^*)_t$ starting from an initial displacement of $(X_1 - X_1^*)_0$:

$$(X_1 - X_1^*)_t = (X_1 - X_1^*)_0 e^{-kt} \tag{6.8}$$

A useful metric is the *return time*. It measures how long a displacement from equilibrium will take to diminish to $1/e$ ($\approx 37\%$) of its initial value. Resilient systems have short return times. In this example, the return time is $1/k$, for at that time $(X_1 - X_1^*)_t/(X_1 - X_1^*)_0 = e^{(-k.(1/k)} = 1/e$. Notice that this return time does not depend on what was the initial condition. Plotting the logs of the deviation from equilibrium against time will yield a plot with the slope $-k$.

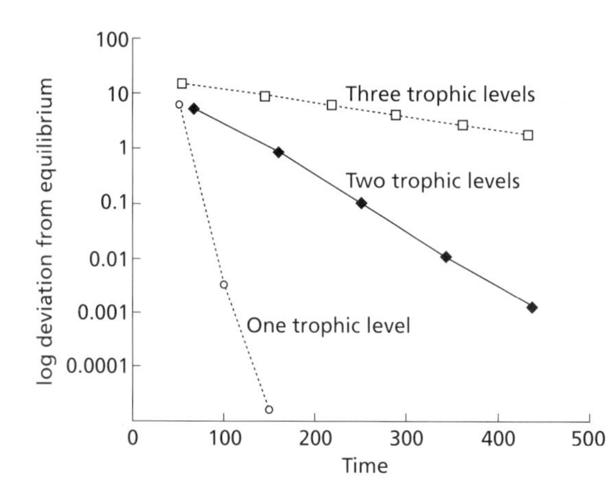

Fig. 6.3 The logarithms of the distance from equilibrium decline as a near-linear function of time for the models in Fig. 6.2. The slopes confirm that longer food chains have longer return times.

How does this simplest, 'what could happen' scenario apply to the more complex model for three food chains? Figure 6.3 plots the logarithms of the deviations from the equilibrium density of species 1 for each of the three models. In the one trophic level model, I've just used a few sample values. For the other two, I have used the maximum deviations during each oscillation. The results are extraordinarily encouraging. The logs of the deviations do decline in an almost linear fashion and they support the intuition that deviations from equilibrium decay fastest in systems with fewer trophic levels.

All this can be made analytical by a dull, yet daunting process described in Pimm (1982) and any book on differential equations. The process informs us that the return to equilibrium depends upon a set of $n \times n$ values, for an n-species model. Each value is the effect of a change in the density of one species (X_j) on the growth rate of another (dX_i/dt). For the Lotka–Volterra models, the typical value at equilibrium is $a_{ij}X_i^*$—a result we will use presently to estimate the resilience of a natural system.

The algebraic efforts show that the recipe works whether the return to equilibrium is direct or oscillatory, as my example suggests. They also show that the recipe doesn't always work. There can be transients, where the return to equilibrium is fast, then settles down to a more modest pace. In such cases, plots such as those in Fig. 6.3 will not be linear except until these transients have had time to dissipate. This problem apart, the results suggest that we can reduce each model's output to a single characterization: the return time. And with this characterization, we can see if the results hold for other sets of parameters.

We can build a computer program that selects parameters randomly over intervals that are hopefully plausible, if still entirely hypothetical. The program can build hundreds of such randomly selected models of a given food web structure. We can investigate models with extraordinary numbers of species–four, eight, even 12 species are possible, numbers a mere two orders of magnitude simpler than the simplest natural systems—and a variety of different food web patterns. For each model, the computer can reproduce the simple algebra that calculates the equilibrium densities and the more complex calculations that yield the return time of each model. Then, for models with the same number of trophic levels, we can plot the frequency distributions of these return times.

For one species per trophic level (as Equations 6.1–6.3), we find that, on average, the more levels there are, the longer are the return times on average (Pimm & Lawton 1977). Adding a few more species per level doesn't alter this conclusion (Pimm 1982).

I posed the question about a nutrient pulse in a lake. How might we take the insights from these simple, tinker-toy models and make them more realistic? Nutrients in real systems do not merely flow through trophic levels. In many systems they are recycled. Dying plants and animals may release the nutrients back into the lake helped by detritus-eating bacteria and other organisms. DeAngelis (1991) has explored the interplay of nutrient recycling and food web structure in detail with models that provide a more satisfactory accounting of where the nutrients go. Nutrient cycles may be 'tight.' A molecule of phosphorus may pass through the food web several times before washing out of the system. Or the cycles may be leaky, when the transit times will be short. When nutrients are held tightly, then so will be the shocks to the system. They, too, will take a long time to disappear. Conversely, leaky systems lose the shocks quickly.

These explicit models of nutrients allow this complication. A system with just phytoplankton may bind the nutrients into a tight cycle, whereas a system with more trophic levels may be more leaky. DeAngelis argues that long food chains could be more leaky and this can counter their tendency towards longer return times. The reverse could be true, too.

Even these model food webs are caricatures of even the simplest natural food webs and their parameters are complete fabrications. So what do real systems resemble and how large are their parameters?

Getting real

These are the questions Carpenter *et al.* (1992) address. They chose a small, (1.2 ha), steep-sided (depth 18.5 m) experimental lake in Wisconsin, USA and estimated both the flows and the stocks of phosphorus. Phosphorus is often a limiting nutrient in lakes (Schindler 1978); it makes sense to model it, rather than population biomasses. They aggregated the species in the web into six compartments: dissolved phosphorus, seston (mainly algae, but also the associated bacteria and protozoa), herbivorous zooplankton, *Chaoborus* (a predatory midge that feeds on the herbivores), planktivorous fish (that also feed on herbivores and also on the *Chaoborus*), and piscivorous fish. In 1984, planktivorous minnows were at the top of the food web. During 1985, over 50 kg of minnows were removed and replaced by a similar mass of piscivorous largemouth bass. This reconfigured the food web, adding an extra trophic level without much changing the total amount of phosphorus.

How does one obtain the parameters needed to calculate resilience from the field? One approach takes the biomass of a species to be the average annual population biomass of the species; call this X_i^*. (Carpenter *et al.* used the mass of phosphorus, but I'll use the word 'biomass' in what follows for simplicity.) We can also readily measure the feeding rate, F_{ij}, of a predator of biomass X_j on its prey of biomass X_i. The feeding rate F_{ij} should equal the rate of loss of prey X_i. From the assumptions we have made earlier, we can equate this rate with $a_{ij}X_i^*X_j$. Thus, $F_{ij}/X_j = a_{ij}X_j$, the effect of predator X_j on prey X_i's growth rate. (Recall from above that it is on such effects that the system's resilience depends.)

What is the effect of the prey on the predator's growth rate? While the prey species looses F_{ij} to the predator per year, the predator's gain is much smaller. (When a rabbit is running for its life, the fox chasing it is merely running for its supper.) The predator's gain must be reduced by the fraction of the prey's tissues that it can assimilate (the assimilation efficiency) and the fraction of the assimilated tissues that it can convert into new biomass (the production efficiency). We know these efficiencies reasonably well for many groups of species. They allow calculation of the per unit effect of prey X_i on predator X_j's growth rate.

Carpenter *et al.* essentially followed this recipe except that they also had to include other inputs to compartments. In both years a major input was from the fish feeding in the lake's small, but important littoral zone. There were flows of phosphorus back into the water column and losses, mainly to the benthic sediments. Figure 6.4 summarizes their calculations.

Adding the extra trophic level alters the distribution of the phosphorus (an issue to which we will return in the next section). Freed from predation from planktivores, *Chaoborus* increases dramatically and its prey, the herbivorous zooplankton, decrease. The major phosphorus stocks move to higher trophic levels (piscivores, *Chaoborus*) in 1986 and these species have longer lives than the zooplankton and seston.

Carpenter *et al.*'s final step was to use these numbers to calculate both the resilience and rate of nutrient recycling in each year. In 1986, the recycling was slightly tighter than in 1984. As theory predicts, the addition of the extra trophic level increased the estimated return time from 28 days in 1984, to over 200 days in 1986. Even after these heroic efforts, Carpenter *et al.* still only predicted resilience; they did not estimate it directly. Nonetheless, the results are interesting for several reasons.

There are at least two reasons for the increase in return time with trophic

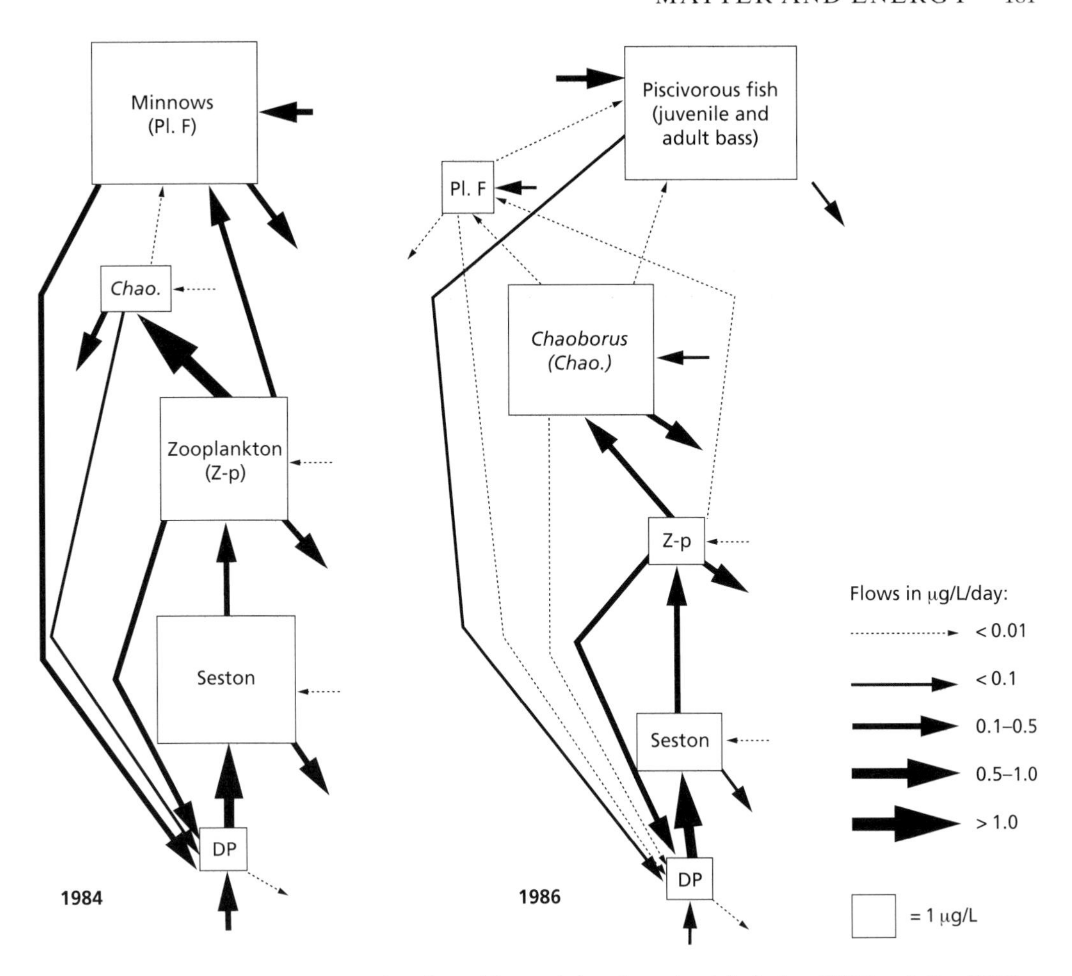

Fig. 6.4 Estimates of stocks and flows of phosphorus in a lake before (1984) and after (1986) the experimental addition of a trophic level. Not only is the distribution of phosphorus moved to higher levels, the return time is estimated to be an order of magnitude higher. A nutrient pulse would move quickly to higher levels in the 1986 system and tend to remain there for longer. The boxes and arrows are proportional to the stocks and flows of phosphorus as shown in the legend. After Pimm (1993) and original data in Carpenter *et al.* (1992).

levels. First, the number of levels itself may matter. More levels must often slow the transit of nutrients through the system, even if the converse is possible. Second, species at higher trophic levels are larger and longer-lived in aquatic ecosystems. In this lake, the experiment moved the phosphorus stocks from algae (which live days) to fish (which live years). This alone should slow the nutrient transfers.

If this second mechanism is by far the more important, then these results may not apply to terrestrial ecosystems. Again, there are two parts to the argument: the first involves the distribution of biomass with trophic levels, the second, how species live.

Terrestrial ecosystems often show a marked pyramid of biomasses. Terrestrial plants outweigh their herbivores often by factors of 10. The herbivores outweigh their predators similarly. Adding an extra trophic level would move the distribution of biomass upwards only very slightly. In contrast, Fig. 6.4 shows that the stocks of phosphorus (which reflect biomass and vice versa) can be greater at higher than lower levels in aquatic systems.

Secondly, the dynamically slower compartments need not always be at higher trophic levels—as they are in many pelagic aquatic webs. The relative life spans of species at different levels differ from system to system. Trees live longer than the insects they house, which live shorter spans than the birds and mammals that eat them. Elsewhere, long-lived birds and mammals may eat insects that live typically for a year and feed on annual plants.

In sum, adding a trophic level increased the system's resilience. However, the lake results are stacked in favour of finding an increase in return times with increasing levels. It's not yet clear whether other systems will behave differently.

A is happy, B is not

In the previous example, we considered how quickly a system would recover from some transient change. What happens if the change is permanent? Suppose we increase the flow of a limiting nutrient into an ecosystem. Perhaps the plants alone will increase in biomass, soaking up atmospheric carbon in the process. If so, should we consider sprinkling iron filings over large areas of the oceans—where iron limits the phytoplankton (Martin *et al.* 1990)—with the hope of ameliorating the increase in atmospheric CO_2? The alternative is that the herbivores might consume all the new plant growth, the carnivores all the new herbivores, and so on up the food chain. Energetically profligate birds and mammals sit atop food chains and, unlike the phosphorus retained in the previous example, might quickly respire most of the carbon back into the system. Metaphorically, giving species A more nutrients need not make it any happier, because it may be some other species that is the chief beneficiary.

More tinker toys

What do the Lotka–Volterra models predict? Consider the models for one, two, and three trophic levels (Equations 6.1, 6.2, and 6.3). Suppose that enriching the system with some limiting nutrient is equivalent to making the limitation on plant growth (a_{11}) smaller. In the model with one trophic level (Equation 6.1) this has the simple effect of directly increasing the equilibrium density of X_1 (= X_1^*), for it equals b_1/a_{11}. For the two trophic level model, setting the predator's growth rate to zero gives $X_1^* = -b_2/a_{21}$. This means that the equilibrium density of X_1 *does not change* as we add increasing amounts of a limiting nutrient. The herbivore exactly consumes the increased plant productivity and it *does* increase. In the three trophic level model, reducing a_{11} increases both the plant and the carnivore but not the herbivore.

Simply, with one species per trophic level, the top species will always increase in biomass with plant productivity increases. Thus, plants will increase if there are only plants. Herbivores should increase if there are only plants and herbivores. Plants and carnivores, but not herbivores, increase if there are three levels. The patterns alternate depending on whether the number of levels is odd or even.

At this juncture, I invoke Mark Williamson's first law of ecological modelling, which states 'that [it] is possible to provide a theoretical model of almost any circumstance' (quoted thus in Pimm 1991). Abrams (1993) and Arditi & Michalski (1996) confirm Williamson's intuition. The effect of increases in plant productivity can have almost any effect. It depends on which of several specific models one uses to describe the interspecific interactions. (The Lotka–Volterra is but one formulation.) In ratio-dependent models, for instance, the rate of consumption depends on the ratio of the densities of the consumers to the consumed (Arditi & Ginzburg 1989; Arditi *et al.* 1991). Productivity increases lead to increases in plants *and* herbivores.

Even with Lotka–Volterra models, the results also depend on whether one models simple food chains (Equations 6.1–6.3) or more complex food webs (Pimm 1991; Abrams 1993). With some nonlinear models, the effect of enrichment may be to destabilize the interactions. This may lead to cyclical changes in phytoplankton and their herbivores (Rosenzweig 1971) or even a loss of a top trophic level as productivity increases (Abrams & Roth 1994).

I will explore these ideas at two extreme scales. The first considers the argument of Oksanen *et al.* (1981) that we may see the simple Lotka–Volterra predictions writ large across the planet's ecosystems. The second is Leibold and Wilbur's (1992) small-scale experiment using artificial frog ponds.

Large-scale patterns

Oksanen *et al.* (1981) suggest that the most barren ecosystems support one functional trophic level, with only sporadic, opportunistic herbivory. Across these systems, changing plant production should increase plant biomass, for it is Equation 6.1 that describes them. More productive systems support herbivores that can regulate plant production. As in Equation 6.2, herbivore biomass, but not plant biomass, will increase along gradients of plant productivity. Equation 6.3 applies to only the most productive systems, where predators regulate the herbivores. Over large gradients in productivity there should be stepwise increases in plant, herbivore and carnivore biomass as systems change from functionally one to three trophic levels.

Compilations of terrestrial herbivore biomasses and plant productivities (McNaughton *et al.* 1989; Moen & Oksanen 1991; Cyr & Pace 1993) show that the latter range from ~10 to ~1000 g carbon $m^{-2} year^{-1}$. Aquatic values span a similar range. The consequences are the subject of debate. McNaughton *et al.* argue that herbivore biomass (HB) increases much faster than plant productivity (PP): HB scales as $PP^{1.52}$. Moen and Oksanen select data with different criteria, and recommend a model with two slopes: HB scaling as $PP^{2.2}$ for the less productive systems and HB scaling as $PP^{1.01}$ for the more productive systems.

Stepwise increases in biomass are reminiscent of the theory of Oksanen *et al.*, but there are no ranges of productivity where herbivore biomass is constant. Interestingly, an order of magnitude increase in phosphorus input had no effect on phytoplankton biomass in one study (Persson 1994). Yet further increases in phosphorus led to a sharp increase in phytoplankton biomass.

The paucity and quality of these data cloud the interpretations. Some ecosystems are possible outliers (a problem also noted by Cyr and Pace). Why some ecosystems are special in this way is not clear. Even excluding them, the range of biomasses for any given productivity often spans two orders of magnitude. Simply, for a given plant productivity, herbivore biomasses vary substantially. We shall now see that the details of food web structure provide one, if surely not the only, explanation for this variability.

Frog ponds

Leibold and Wilbur (1992) experimentally manipulated both food web structure and nutrient inputs. To mimic natural ponds they filled 1000-litre tanks with water and pine straw and introduced an assortment of algae and zooplankton. To half of the tanks they added nitrogen and phosphorus to create high nutrient levels. Additional factors in the analysis included the addition or absence of *Daphnia* and the addition or absence of *Rana* tadpoles. *Daphnia*, like the other zooplankton present, graze the phytoplankton. *Rana* graze the periphyton, the attached algae and associated organisms. The phyoplankton and the periphyton compete for nutrients.

Without *Rana*, the increase in nutrients led to an increase in zooplankton, but not the phytoplankton, which *decreased* slightly. The ungrazed periphyton also increased (Fig. 6.5a,b).

With *Rana* present, the results are different (Fig. 6.5c,d). With low nutri-

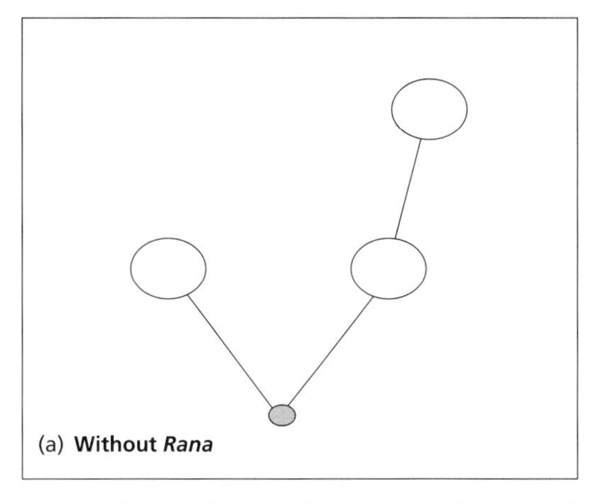

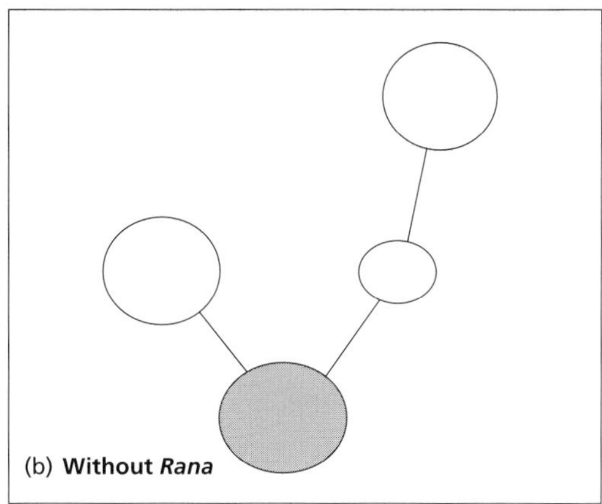

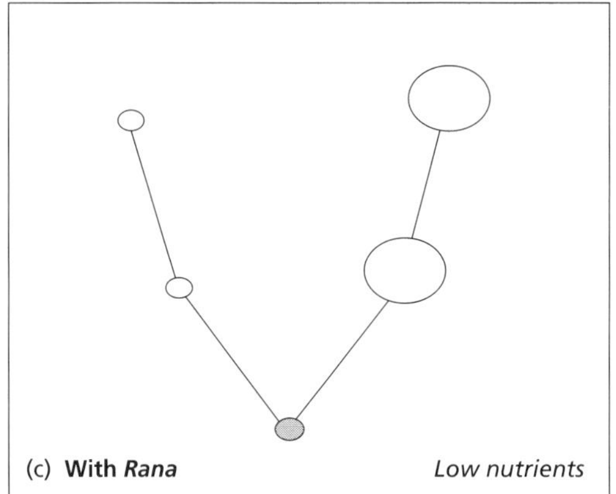

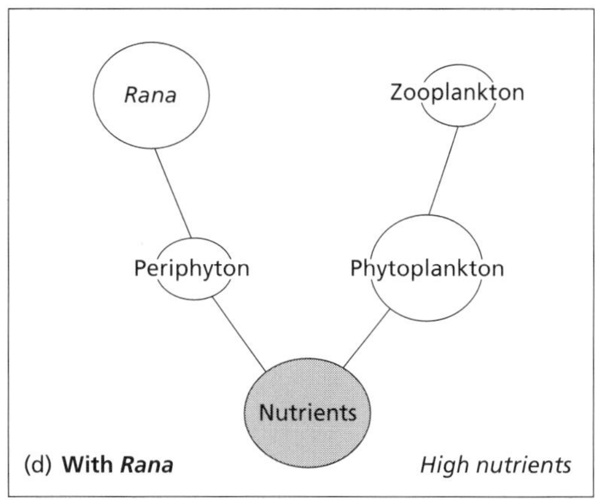

Fig. 6.5 Four food webs showing the relative biomasses of nutrients, periphyton, phytoplankton, *Rana* tadpoles, and zooplankton in experimental ponds. The size of each circle is proportional to relative biomasses within each group. Without *Rana* (a,b), increased nutrient levels (b) increase the ungrazed periphyton, the zooplankton, but not the phytoplankton it grazes. With *Rana* (c,d), increased nutrients (d) increase all groups except the zooplankton. Low nutrient levels are (a,c), high nutrient levels (b,d).

ents, *Rana* grazed the periphyton to low levels. Increased nutrients led to increased biomass of both periphyton and *Rana*. The periphyton did not reach the high biomass seen without *Rana*, while the competing phytoplankton reached its highest biomass. The zooplankton at high nutrient levels were less abundant than when *Rana* was absent. Indeed, with *Daphnia* absent, the zooplankton were only slightly more abundant in the high nutrient treatment. They were, however, substantially more abundant in the treatments in which *Daphnia* and *Rana* were present.

Without *Rana*, the results fit the Lotka–Volterra predictions for increasing nutrients. The periphyton was not grazed and it increased. The phytoplankton was grazed and it did not increase, but the zooplankton did. With *Rana* present, both plant groups increased. One herbivore, *Rana*, increased, while the herbivorous zooplankton either did or did not depending on its composition.

The conclusion is simple. The simple, one-species-per-trophic-level models work when the systems are indeed that simple. Real ecosystems are much more complex, this complexity alters the results, and so the consequences of increasing plant production are likely to be complicated.

Extensions

The preceding ideas are the points of departure for a wide literature on food web dynamics. So far, we have asked what are the effects of changing the nutrient inputs to an ecosystem. The more general question is what happens to any component of an ecosystem when we change any other.

Bottom–up vs top–down

The aquatic literature calls this topic the problem of 'bottom–up' vs 'top–down' control (McQueen *et al.* 1989). We have seen the 'bottom–up control' in some systems, where herbivore biomass can increase even faster than the increase in plant production. In other systems the changes in nutrient inputs have no effect on plant biomass because it is controlled 'top–down' by the herbivores. In such systems, herbivore removal should have a major effect on plant biomass. We have already seen the top–down consequences of adding a trophic level to a lake food chain in the previous section (Fig. 6.4).

Unexpected effects

The experimental literature discusses 'unexpected effects', the changes in densities or biomasses that appear to be in the wrong direction. For instance, one removes a predator and finds that some of its prey species decrease in abundance (rather than increase). These changes occur in a quarter to a half of all species removal experiments in different habitats and regions (Sih *et al.* 1985; Pimm 1991).

Indirect consequences

Theoretical studies anticipate these changes. The removal of a predator A may lead to the unexpected *decrease* of its prey D. Removing A causes an increase in prey species B. Species B and species D compete, perhaps for space, as in many intertidal systems, but perhaps for their shared prey species, C. B's increase thus may, in turn, deplete C on which D depends. Species A is an enemy (a predator) of an enemy (competitor B) and so a friend. At least theoretically, the indirect effects of A on B, B on C, C on D can overwhelm the direct effect of A on D. The possibility of this pattern (called 'keystone predation') and other complexities is now widely accepted (Abrams *et al.* 1996).

The question of whether indirect effects do dominate direct effects is an active area of study. Menge's (1995) examination of the results of many experiments finds that keystone predation is the most common of nine defined patterns of indirect interaction. He concludes that direct and indirect effects are of broadly similar magnitude.

Species: would any of them be missed?

I have fudged (but I am not alone). In the material presented so far I have modelled individual species, and then implied that the resulting dynamics might apply to groups of similar species or even all the species at a trophic level. Ecological modellers do this routinely—I know of no ecosystem model that doesn't lump species in this way. Yet, the effects of increasing nutrient input have already shown us that results can differ as a function of the model's complexity.

The practical question is that monocultures of trees and cereals may superficially resemble the diverse forests and prairies they replaced, but do they function in similar ways? Nature, Elton (1958) argued, works better because of its complexity.

The theory

I shall describe experiments that consider how complexity alters the effects of two kinds of disturbance on plant communities: a drought and the addition of herbivores. The specific questions are surrogates for general concerns about all trophic levels. A 'drought' is a metaphor for any disturbance that kills off a portion of species. For 'herbivore' read the addition of anything that eats anything else.

Resistance to drought

The intuition is that it will be an evil wind that blows no good. Consider two plant species that compete for space, nutrients, light, or other resources. Suppose species A severely retards the growth of species B, but that A is more susceptible to drought. A decline in the biomass of species A may effect an almost equal increase in the biomass of B. It is even possible, theoretically, that the increase in B might exceed the decrease in A. Species that are the beneficiaries of the disturbance thus compensate for the more vulnerable species. These trade-offs maintain the resistance of the system as a whole. A model that treats A + B as if it were just one species would obviously give the wrong answer; the two species combined have different dynamics.

Resistance to grazing

McNaughton's field experiments in 1977 in the Serengeti National Park of Tanzania motivated King & Pimm (1983) to develop the first theory of how complexity affects resistance. Their computer models established a set of structured food webs composed of one grazer, a variable number of plant species, and different degrees of competition between them. They chose the model parameters at random over ecologically sensible intervals (as discussed above), keeping only those models that allowed the coexistence of the grazer and all the plant species. Removing the grazer from each model, they simulated the density changes until the species densities reached an equilibrium.

The previous argument about drought, competition and resistance applies to the effects of a grazer. A grazer can only reduce the biomass of a single-species stand. If two or more competing species are present, however, the grazer may reduce one species' biomass, but a less palatable competitor may increase and offset the change. With several species present, King & Pimm (1983) found that resistance increased as the number of competitive interactions between the species increased.

We are not able to test these theories directly. In the field, it is practically impossible to work out the extent of the competition between all the species present in an experimental plot. Indirectly, we might suppose that the more species present, the more there are likely to be competitive interactions. More species would mean more competition and so greater resistance.

The uncertainty with this argument is not with the theory but with the species' natural history. Species A and B must be similar in their needs if one is to have such a strong competitive effect on the other. Yet, we must also assume that they are differentially susceptible to drought or herbivores. Lawton & Brown (1993) sensibly raised the alternative hypothesis that similar species may have broadly substitutable roles. What is bad for one species will likely be bad for its competitors. Many species may add little to a system's resistance. Only experiments can resolve the hypotheses.

The experiments

Tilman & Downing's (1994) experiments were in grasslands in Minnesota, USA. They manipulated the number of species present by varying the amount of nitrogen fertilizer applied to over 200 plots each of 16 m^2. (Nitrogen reduces the number of plant species that coexist.) They started measuring biomass in 1987, and fortuitously, in 1988 there was a severe drought. It reduced biomass differently on the different plots. The species-rich plots produced half their predrought biomass, while the species-poor plots produced only about an eighth or less (Fig. 6.6).

McNaughton (1985) extended his earlier experiments in the Serengeti grasslands. He identified areas with different natural diversities, then fenced plots of ~16 m² of them to prevent grazing by migrant herds of zebra, gazelles and wildebeest.

McNaughton used a technical index of diversity, not the number of species. The index recognizes that a set of species with relative abundances of 25, 25, 25, and 25 appear more diverse than one with relative abundances of 97, 1, 1, and 1. Those three rare species in the second example may be too rare to do anything for resistance during a disturbance and so might discount them accordingly. (Conversely, these three rare species may be just those species that are rare because of competition from the most common species and will be the ones that increase after disturbance.) We can think of the index as the (natural) logarithm of an equivalent number of equally abundant species: 4 in the first example, a little more than 1 in the second. To

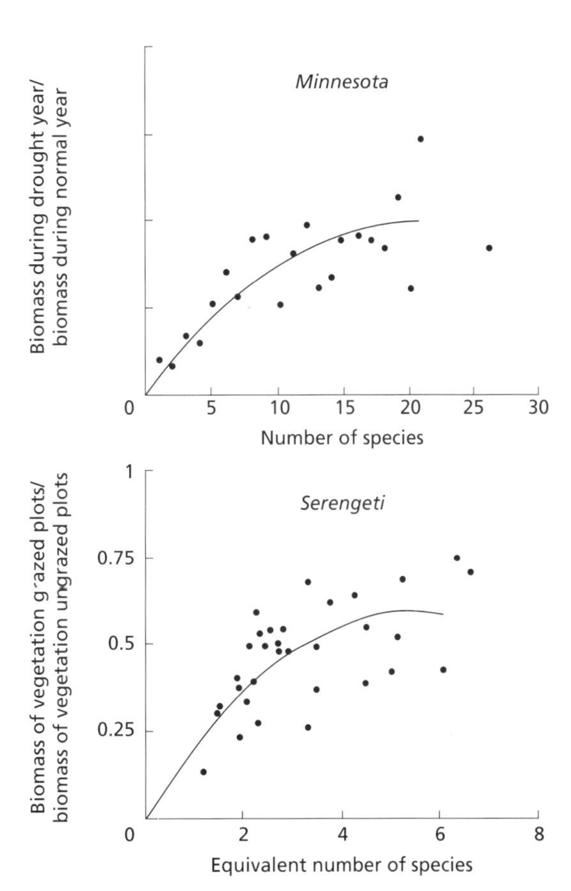

Fig. 6.6 Species-rich systems are more resistant to change than species-poor systems. In both experiments, resistance has a curvilinear relationship with species richness. This suggests that as species are lost, the contribution to resistance of those remaining increases. After Lockwood & Pimm (1994), and original data in McNaughton (1985) and Tilman & Downing (1994).

compare McNaughton's study with Tilman and Downing, I replace the index value (I) with an 'equivalent number of species' equal to e^I.

McNaughton measured the plant biomass after the grazers had passed by, then compared it to the biomass where he had excluded the grazers. The grazers reduce the biomass of diverse areas by a quarter, while the less diverse areas lost three-quarters of their biomass (Fig. 6.6).

No one doubts that as the number of species declines, systems must eventually become less resistant. The debate is about how many species we need: when exactly do species begin to matter? Consider a graph of resistance vs the number of species in the system. Lockwood & Pimm (1994) found that in both studies a quadratic regression gave a statistically better fit than the linear one. The data are significantly curved (Fig. 6.6). This means that some species do appear redundant when there are many species present—the curve flattens out at higher numbers of species. However, as the number of species is reduced, the curve is steeper, meaning that species become progressively more important in maintaining resistance as their numbers decline. Interestingly, the curve's peak in each study is clearly near the maximum diversity. There is no wide plateau of species richness where species do not matter.

The consequences of the tangled bank

My three examples demonstrate that the dynamics of the flows of matter and energy depend on food web structure, both the specific patterns and its overall complexity. Each example has practical consequences.

Notice that some systems may spend almost all of the time recovering from some previous shock. Systems with very long recovery times may never come close to their equilibria if the shocks are too frequent. Simply, the shock may be short and sharp, but the consequences need not be.

Many applied ecological problems are ones of resistance: what will be the consequences to one species (or set of species) if we harvest, pollute, or otherwise destroy some other species (or set of species)? Will the consequences be driven by simple, predictable direct effects? Or will it be the more complex, hard–to–predict, indirect effects that arise from the propagation of complex flows of matter and energy through the food web? If the latter, we must understand those flows if we are to answer the question of why A is happy and B is not.

Finally, if species richness increases resistance, disturbances to species-poor systems should be catastrophic. This prediction is being tested by perhaps the largest ecological experiment: agriculture. In many areas, annual plants grown in expansive monocultures for food production now replace the diverse, natural grasslands (Jackson 1980). In grasslands, droughts are regular and harsh, while other years are excessively wet. The native grasslands through their species richness might withstand these climatic extremes. Droughts, in contrast to their relative innocuous effect on the native prairies, inflict serious wounds on these new single-species systems.

We are losing the planet's biological diversity at an unprecedented rate (Pimm *et al.* 1995). What are all those species good for anyway? There are many compelling answers. The results from this chapter add the additional question: how many species are enough to preserve the ecosystem processes on which we depend?

References

Abrams, P.A. (1993) Effect of increased productivity on the abundances of trophic levels. *American Naturalist*, **141**, 351–371.

Abrams, P.A. & Roth, J. (1994) The responses of unstable food chains to enrichment. *Evolutionary Ecology*, **8**, 150–171.

Abrams, P., Mege, B.A., Mittlebach, G.G., Spiller, D. & Yodzis, P. (1996) The role of indirect effects in food webs. In: *Food Webs: Integration of Patterns and Dynamics* (eds G.A. Polis & K.O. Winemiller), pp. 371–395. Chapman & Hall, London.

Arditi, R. & Ginzburg, L.R. (1989) Coupling in predator-prey dynamics: ratio-dependence. *Journal of Theoretical Biology*, **139**, 311–326.

Arditi, R. & Michalski, J. (1996) Nonlinear food web models and their responses to increased basal productivity. In: *Food Webs: Integration of Patterns and Dynamics* (eds G.A. Polis & K.O. Winemiller), pp. 122–133. Chapman & Hall, London.

Arditi, R., Ginzburg, L.R. & Akçakaya, H.R. (1991) Variation in plankton densities among lakes: a case for ratio-dependent predation models. *American Naturalist*, **138**, 1287–1296.

Carpenter, S.R., Kraft, C.E., Wright, R., He, X., Soranno, P.A. & Hodgson, J.R. (1992) Resilience and resistance of a lake phosphorous cycle before and after food web manipulation. *American Naturalist*, **140**, 781–798.

Cohen, J.E. (1977) Ratio of prey to predators in community webs. *Nature*, **270**, 165–167.

Cohen, J.E. (1978) *Food Webs and Niche Space.* Princeton University Press, Princeton, NJ.

Cohen, J.E., Briand, F. & Newman, C.M. (1990) *Community Food Webs: Data and Theory.* Springer-Verlag, New York.

Cyr, H. & Pace, M.L. (1993) Magnitude and patterns of herbivory in aquatic and terrestrial ecosystems. *Nature*, **361**, 148–150.

DeAngelis, D.L. (1991) *Dynamics of Nutrient Cycling and Food Webs.* Chapman & Hall, London.

DeAngelis, D.L., Post, W.M. & Sugihara, G. (eds) (1983) Current trends in food web theory: report on a food web workshop, ORNL-5983. Oak Ridge National Laboratory, Oak Ridge, TN.

Elton, C.S. (1958) *The Ecology of Invasions by Animals and Plants.* Chapman & Hall, London.

Gilbert, W.S. & Sullivan, A. (1885) *The Mikado.* Publisher TK, London.

Hall, S.J. & Raffaeli, D. (1993) Food webs: theory and reality. In: *Advances in Ecological Research* (eds M. Begon & A.H. Fitter), pp. 187–239. Academic Press, London.

Jackson, W. (1980) *New Roots for Agriculture.* University of Nebraska Press, Lincoln, NB.

King, A.W. & Pimm, S.L. (1983) Complexity and stability: a reconciliation of theoretical and experimental results. *American Naturalist*, **122**, 229–239.

Lawton, J.H. (1989) Food webs. In: *Ecological Concepts* (ed. J.M. Cherrett), pp. 43–78. Blackwell Scientific Publications, Oxford.

Lawton, J.H. & Brown, V.K. (1993) Redundancy in ecosystems. In: *Biodiversity and Ecosystem Function* (eds E.D. Schulze & H.A. Mooney), pp. 255–270. Springer-Verlag, Berlin.
Leibold, M.A. & Wilbur, H.M. (1992) Interactions between food-web structure and nutrients on pond organisms. *Nature*, **360**, 341–343.
Lockwood, J.L. & Pimm, S.L. (1994) Species: would any of them be missed? *Current Biology*, **4**, 455–457.
McNaughton, S.J. (1977) Diversity and stability of ecological communities: a comment on the role of empiricism in ecology. *American Naturalist*, **111**, 515–525.
McNaughton, S.J. (1985) Ecology of a grazing ecosystem: the Serengeti. *Ecological Monographs*, **55**, 259–294.
McNaughton, S.J., Oesterheld, M., Frank, D.A. & Williams, K.J. (1989) Ecosystem-level patterns of primary productivity and herbivory in terrestrial habitats. *Nature*, **341**, 142–144.
McQueen, D.J., Johannes, M.R.S., Post, J.R., Stewart, T.J. & Lean, D.R.S. (1989) Bottom-up and top-down impacts on freshwater pelagic community structure. *Ecological Monographs*, **59**, 289–309.
Martin, J.H., Gordon, R.M. & Fitzwater, S.E. (1990) Iron in Antarctic waters. *Nature*, **345**, 156–158.
May, R.M. (ed.) (1976) *Theoretical Ecology: Principles and Applications.* Blackwell Scientific Publications, Oxford.
Menge, B.A. (1995) Indirect effects in marine rocky intertidal interaction webs: patterns and importance. *Ecological Monographs*, **65**, 21–74.
Moen, J. & Oksanen, L. (1991) Ecosystem trends. *Natrue*, **353**, 510.
Oksanen, L., Fretwell, S.D., Arruda, J. & Niemelä, P. (1981) Exploitation ecosystems in gradients of primary productivity. *American Naturalist*, **118**, 240–261.
Paine, R.T. (1966) Preface. In: *Food Webs: Integration of Patterns and Dynamics* (eds G.A. Polis & K.O. Winemiller), pp. ix–x. Chapman and Hall, London.
Persson, L. (1994) Natural patterns of shifts in fish communities — mechanisms and constraints on perturbation sustenance. In: *Rehabilitation of Freshwater Fisheries* (ed. I.G. Cowx), pp. 421–434. Fishing News Books, Oxford.
Pimm, S.L. (1980) The properties of food webs. *Ecology*, **61**, 219–225.
Pimm, S.L. (1982) *Food Webs.* Chapman and Hall, London.
Pimm, S.L. (1991) *The Balance of Nature? Ecological Issues in the Conservation of Species and Communities.* University of Chicago Press, Chicago, IL.
Pimm, S.L. & Lawton, J.H. (1977) The number of trophic levels in ecological communities. *Nature*, **268**, 329–331.
Pimm, S.L. & Lawton, J.H. (1978) On feeding on more than one trophic level. *Nature*, **275**, 542–544.
Pimm, S.L. & Lawton, J.H. (1980) Are food webs compartmented? *Journal of Animal Ecology*, **49**, 879–898.
Pimm, S.L., Lawton, J.H. & Cohen, J.E. (1991) Food webs patterns and their consequences. *Nature*, **350**, 669–674.
Pimm, S.L., Russell, G.J., Gittleman, J.L. & Brooks, T.M. (1995) The future of biodiversity. *Science*, **269**, 347–350.
Polis, G.A. (1991) Complex trophic interactions in deserts: a critique of food web theory. *American Naturalist*, **138**, 123–155.
Polis, G.A. & Winemiller, K.O. (eds) (1996) *Food Webs: Integration of Patterns and Dynamics.* Chapman & Hall, London.
Rosenzweig, M.L. (1971) Paradox of enrichment: destabilization of exploitation ecosystems in ecological time. *Science*, **171**, 385–387.
Schindler, D.W. (1978) Factors regulating phytoplankton production and standing crop in the world's freshwaters. *Limnology and Oceanography*, **23**, 478–486.
Schoener, T.W. (1989) Food webs from the small to the large. *Ecology*, **70**, 1559–1589.

Sih, A., Crowley, P., McPeek, M., Petranka, J. & Strohmeier, K. (1985) Predation, competition, and prey communities: a review of field experiments. *Annual Reviews of Ecology and Systematics*, **16**, 269–311.

Tilman, D. & Downing, J.A. (1994) Biodiversity and stability in grasslands. *Nature*, **367**, 363–365.

7: Population and evolutionary dynamics of consumer-resource systems

W.M. Getz

Introduction

Consumer-resource systems can be regarded as any system in which a population of organisms exploits either an energy flux, a detritus or nutrient pool, or another population. This definition then includes plants growing in sunlight, parasites growing in or on hosts, predators exploiting prey, herbivores grazing or browsing on plants, and even humans harvesting trees, fish or game. For such a vast array of systems, no single best paradigm exists for modelling the dynamics of the populations under consideration.

Here I will focus on two paradigms: a discrete-time paradigm that is suitable for modelling populations with nonoverlapping generations and a continuous-time paradigm that is suitable for modelling the flow of biomass from the resource to the consumer, and its conversion from resource mass (biomass, nutrients, or energy) into consumer mass. I will also consider the extension of the discrete-time paradigm to include overlapping generations (age structure) and the application of the models to resource management problems.

The discrete-time models fall within the class of iterative maps

$$\boldsymbol{x}_{t+1} = \psi(\boldsymbol{x}_t), \quad t = 0, 1, 2 \ldots \tag{7.1}$$

where, for the systems considered here, the elements of the vector $\boldsymbol{x}_t = (x_{1t}, x_{2t}, \ldots, x_{nt})'$ ($'$denotes the transpose of a vector) typically represent the densities of individuals in n biological populations at time $t = 0, 1, 2, 3, \ldots$ or the densities of individuals in n classes (e.g., age or size) of one population at time t, or even some combination of several populations each with several classes of individuals. The continuous-time models fall within the class of vector differential equation systems

$$\frac{d\boldsymbol{x}}{dt} = \psi(\boldsymbol{x}), \tag{7.2}$$

where, for the systems considered here, the elements of the vector $\boldsymbol{x}(t) = (x_1(t), x_2(t), \ldots, x_n(t))'$ typically represent the biomass densities of n interacting biological populations at time

$t \in (0, \infty)$.

A number of different modelling paradigms fall within the ambit of Equations 7.1 and 7.2. For example, the well-known Lotka–Volterra models (May, 1981b) are a special case of Equation 7.2, while the discrete logistic food web models of Berryman *et al.* (1995) are a special case of Equation 7.1. Here I will only consider population models conforming to the two paradigms mentioned above, although in the next section I will discuss theory used to conduct evolutionary analyses of these models in the context of general Equations 7.1 and 7.2. This theory, based on the notion of evolutionarily stable strategies (ESS), is presented in various forms elsewhere in this book, but is elaborated here for completeness and in a way that is most compatible with its application to the types of systems considered here.

ESS method of analysis

Natural selection is a dynamical process that continually takes place as long as heritable variation exists in traits that influence the fitness of individuals in the same population. A powerful method for analysing the evolution of such traits, phenotypes, or 'strategies' in ecological settings is through the application of ESS dynamics, as propounded by Vincent, Brown, Cressman, Hines, and others (Vincent & Brown 1984; Cressman *et al.* 1986; Brown & Vincent 1987; Cressman & Dash 1987; Hines 1987; Vincent & Brown 1988). In the simplest of terms, an ESS is the strategy or trait that pervades a 'resident' population and prevents all small populations playing any 'mutant' strategy or bearing any 'mutant' trait from invading (Maynard Smith 1982). A computation of the ESS values of parameters describing the dynamics of biological populations is based on an analysis of competing phenotypes, especially of their ability to invade and coexist with or exclude one another. Under a number of assumptions that are not unreasonable for theoretical investigations (e.g. weak selection, additive genetic variance; see Taylor 1989), this analysis of competing phenotypes produces results that are equivalent to those obtained from a genetic model that explicitly considers that the fate of individual alleles are associated with particular traits.

Two approaches can be taken to finding ESS solutions: the standard invasion approach in which the dynamics of a mutant invading a resident phenotype are analysed, and the augmented evolutionary dynamics approach in which an equation describing the evolution of the trait or strategy is added to the basic population dynamics model (Vincent 1990; Getz & Kaitala 1993; Vincent *et al.* 1993). First, I will present these approaches in the context of models that fall within the ambit of the discrete iterative map represented by Equation 7.1. In general, the methods outlined here could apply to a subset

$n - m$ $(n > m)$ populations or classes $(x_{m+1}, \ldots, x_n)$ that are evolving within a background of m populations or classes $(x_1, \ldots, x_m)$: i.e. we have n interacting populations or classes of one or several populations, of which only $n - m$ are evolving.

Invasion analysis: constant and periodic systems

To keep the presentation and notation simple we will consider a background population at density x_t and a resident population at density y_t that is susceptible to invasion by a mutant population at density z_t, i.e. we have a two-species interaction in which a population at density y_t is able to evolve, while a population at density x_t is unable to evolve. The ecological dynamics of this interaction is modelled by the equations:

$$\begin{aligned} x_{t+1} &= F(x_t, y_t)x_t \\ y_{t+1} &= G(x_t, y_t)y_t \end{aligned} \tag{7.3}$$

To model the evolutionary dynamics, however, we need specifically to identify the parameters associated with the per-capita growth rate function $G(\cdot)$ that represent the particular traits or strategies played by the different phenotypes under consideration. Here we focus on the simple case of a single strategy ε evolving in a population where the effects of density depend purely on the sum of the densities of the resident and mutant populations (i.e. on $y_t + z_t$). In this case, Equation 7.3 is extended to

$$\begin{aligned} x_{t+1} &= F(x_t, y_t + z_t)x_t \\ y_{t+1} &= G(x_t, y_t + z_t; \varepsilon_y)y_t \\ z_{t+1} &= G(x_t, y_t + z_t; \varepsilon_z)z_t \end{aligned} \tag{7.4}$$

where $G(x_t, y_t + z_t; \varepsilon)$ is the per-capita growth rate function for a phenotype of the evolving population playing a strategy ε. Note that a more general analysis, as proposed by Brown & Vincent (1987), would not only allow for the possibility of a vector of strategies, but also a coalition of resident phenotypes resisting invasion of a mutant phenotype. Brown and Vincent call the function $G(\cdot; \varepsilon)$ a fitness generating function, because at different population densities, represented by the variables in the first part of the argument of G, the fitness (as measured by the per-capita growth rate for that time step) is generated by the total density of the evolving population (in our case the density of one resident and one mutant phenotype) and by the strategy ε played by the particular phenotype in question.

The ability of a phenotype playing strategy ε_z to invade this system is determined by the stability properties of the equilibrium solution $(\bar{x}, \bar{y}, 0)$

of the system in Equation 7.4. From standard discrete equation stability theory (Luenberger 1979), it is known that the eigenvalues λ which satisfy the following characteristic equation obtained from the Jacobian stability matrix associated with the system in Equation 7.4 determine the local stability properties of the equilibrium solution $(\bar{x},\bar{y},0)$ to the system:

$$(G(\bar{x},\bar{y};\varepsilon_z)-\lambda)\times\det\begin{pmatrix} \frac{\partial F(\bar{x},\bar{y};\varepsilon_y)}{\partial x}\bar{x}+1-\lambda & \frac{\partial F(\bar{x},\bar{y};\varepsilon_y)}{\partial y}\bar{x} \\ \frac{\partial G(\bar{x},\bar{y};\varepsilon_y)}{\partial x}\bar{y} & \frac{\partial G(\bar{x},\bar{y};\varepsilon_y)}{\partial y}\bar{y}+1-\lambda \end{pmatrix}=0 \quad (7.5)$$

where $\det(\cdot)=0$ provides the characteristic equation for determining the stability properties of the system represented by Equation 7.3. Thus it follows from stability theory that the equilibrium $(\bar{x},\bar{y},0)$ of the extended system in Equation 7.4 is locally stable if the equilibrium $(\bar{x},\bar{y})$ of the reduced system in Equations 7.3 is stable and $|G(\bar{x},\bar{y};\varepsilon_z)| < 1$. The equilibrium $(\bar{x},\bar{y},0)$ of the extended system is unstable, however, if $|G(\bar{x},\bar{y};\varepsilon_z)| > 1$. If we consider the logarithm of $|G(\bar{x},\bar{y};\varepsilon_z)|$ these conditions imply the following result (see Vincent & Brown 1988, and Vincent *et al.* 1996 for the same statement in a different context). If a two-species interaction, modelled by Equations 7.3, is at a stable equilibrium (x^*,y^*) corresponding to the evolving species playing a strategy ε^*, then ε^* is an ESS only if it maximizes

$$I_1(\varepsilon)=\ln|G(x^*,y^*,\varepsilon)| \quad (7.6)$$

where $I_1(\varepsilon^*)=0$, i.e. ε^* is an ESS only if $I_1(\varepsilon)\leq 0$ for all possible values of ε.

For the continuous-time analogy of the system in Equations 7.4 (cf. Equation 7.2), the model takes the form

$$\frac{dx}{dt}=xf(x,y+z)$$

$$\frac{dy}{dt}=yg(x,y+z;\varepsilon_y) \quad (7.7)$$

$$\frac{dz}{dt}=zg(x,y+z;\varepsilon_z)$$

In this case, however, the stability condition relates to the sign of $g(x^*,y^*;\varepsilon_z)$ rather than the sign of the logarithm of its modulus. The same statement on whether the parameter ε^* is an ESS in the context of a function $I_1(\varepsilon)$ applies if we now define

$$I_1(\varepsilon)=g(x^*,y^*,\varepsilon) \quad (7.8)$$

In the case where the attractor is a periodic solution $(x^*(t),y^*(t))$ of period T, the condition for an ESS now applies to the integral

$$I_1(\varepsilon)\neq\int_0^T g(x^*(t),y^*(t),\varepsilon)\mathrm{d}t \tag{7.9}$$

taking on a maximum of zero.

Invasion analysis: chaotic systems

Discrete maps, such as Equations 7.3, are known to admit chaotic solutions, especially when the effects of density dependence are severe (May & Oster, 1976; Schoombie & Getz, 1998). Suppose Equation 7.3 is chaotic (i.e. it has a strange or fractal attractor rather than a point or multiple point attractor) when the resident phenotype is playing ε^* and suppose the initial values (x_0,y_0) are arbitrarily close to the attractor. Then the ensuing solution, i.e. the sequence of specific points (x_t^*,y_t^*), $t=1,2,3,\ldots$, can be used to calculate the sequence of values $G(x_t^*,y_t^*,\hat{\varepsilon})$ which represent that rate at which an arbitrarily small mutant population would invade the resident pupulation at each point t in time, where $\hat{\varepsilon}$ is the strategy played by the mutant population. The quantity

$$I_\infty(\varepsilon)=\lim_{T\to\infty}\frac{1}{T}\ln\left|\prod_{t=0}^{T-1}G(x_t^*,y_t^*;\varepsilon)\right| \tag{7.10}$$

is the logarithm of the geometric mean of the growth rates of an infinitesimally small mutant population trying to invade a resident population at each point in time. This limit may not always exist; but when it does the condition $(\ln|G(x^*,y^*,\varepsilon^*)|)\leq 0$ for $\varepsilon\neq\varepsilon^*$ now generalizes to (Ferrière & Gatto 1993; Rand *et al.* 1994).

$$I_\infty(\varepsilon)\leq 0 \tag{7.11}$$

where $I_\infty(\varepsilon)$ is defined by Expression 7.10. Conversely, the mutant playing the strategy $\hat{\varepsilon}$ will be able to invade the resident population playing the strategy ε^* whenever $I_\infty(\hat{\varepsilon})$ exists and satisfies the inequality

$$I_\infty(\hat{\varepsilon})>0 \tag{7.12}$$

The quantity $I_\infty(\hat{\varepsilon})$ is called the invasion or Lyapunov exponent of the system (e.g. see Metz *et al.* 1992), and its sign for all $\hat{\varepsilon}\neq\varepsilon^*$ determines whether or not ε^* is an ESS.

The continuous-time equivalent of the invasion exponent $I_\infty(\varepsilon)$ can also be considered to obtain necessary and/or sufficient conditions for ε to be an ESS using measure-theoretic arguments (Rand *et al.* 1994; Ferrière & Gatto 1995). These details will not be elaborated here.

Parameter evolution

A second approach to analysing the evolutionary fate of a strategy $\varepsilon_t, t = 0, 1, 2, \ldots$, is to augment the ecological model in Equation 7.1 with an equation that describes the dynamics of the strategy trajectory $\varepsilon_t, t = 1, 2, 3, \ldots$. In the context of the population model, in Equations 7.13, the standard augmentation (Vincent 1990; Getz & Kaitala 1993; Vincent *et al.* 1993) yields the system

$$
\begin{aligned}
x_{t+1} &= F(x_t, y_t)x_t \\
y_{t+1} &= G(x_t, y_t; \varepsilon_y)y_t \qquad (7.13) \\
\varepsilon_{t+1} &= \varepsilon_t + \sigma_t^2 \frac{\partial}{\partial \varepsilon}[\ln|G(x_t, y_t, \varepsilon_t)|]
\end{aligned}
$$

This augmentation of Equations 7.3 is an expression of Fisher's fundamental theorem of natural selection, if σ_t^2 has the interpretation of variance for the trait ε_t in the population (Roughgarden 1983), because $\ln|G(x_t, y_t, \varepsilon_t)|$ is a measure of the net reproductive fitness of the population at time t when the state of the system is (x_t, y_t) and all individuals are playing strategy ε_t. Note, that at an equilibrium $(\hat{x}, \hat{y}, \hat{z})$ of the system in Equations 7.13, we must have $\partial/\partial\varepsilon[\ln|G(\hat{x}, \hat{y}, \hat{\varepsilon})|] = 0$ given $\sigma_t^2 \neq 0$ for at least some $t \in [\tau, \infty]$ and all $\tau \geq 0$. If, in addition, $\partial^2/\partial^2\varepsilon[\ln|G(\hat{x}, \hat{y}, \hat{\varepsilon})|] = 0$, then $G(\hat{x}, \hat{y}, \varepsilon)$ is a maximum at $\hat{\varepsilon}$, so that $\hat{\varepsilon} = \varepsilon^*$ (i.e. it is an ESS) if this maximum is global. Hence, stable equilibria of the system in Equations 7.13 provide candidate ESS solutions that can be analysed further to check for global maximization of $G(\hat{x}, \hat{y}, \varepsilon)$ at $\varepsilon = \hat{\varepsilon}$.

If the system in Equations 7.3 has a chaotic solution for a particular value of ε, then the system in Equations 7.13 has no equilibrium solution either. In this case the fluctuations in ε_t depend on the value of σ_t^2. In general, ε_t can still be forced to some equilibrium value provided $\sigma_t^2 \to 0$ as $t \to \infty$. In this case, the equilibrium value $\hat{\varepsilon}$ may still be an ESS for the chaotic system if $I_\infty(\varepsilon)$ has a maximum at $\varepsilon = \hat{\varepsilon}$ and $I_\infty(\hat{\varepsilon}) = 0$. Possible choices for σ_t^2 as a function of time are $\sigma_t^2 = ae^{-bt}$ or $\sigma_t^2 = a/(b + t)$ (Schoombie & Getz 1998).

Of course, the augmented ecological model in Equations 7.13 can be extended through the use of apropriate notation to $n - m$ populations evolving within a background of m populations, where the strategy played by the $n - m$ evolving populations is represented by a vector ε of parameters.

The analysis can also be extended in the obvious way to continuous-time models of the type in Equation 7.2. In this case Equations 7.13 have the form

$$\frac{dx}{dt} = xf(x,y)$$

$$\frac{dy}{dt} = yg(x,y;\varepsilon) \tag{7.14}$$

$$\frac{d\varepsilon}{dt} = \sigma^2(t)\frac{\partial}{\partial\varepsilon}g(x,y,\varepsilon)$$

(For a discussion of trait evolution in continuous time models also see Taper & Case 1992; Abrams *et al.* 1993; Matsuda & Abrams 1994.) As in the discrete case, this model can be applied even if the solution to Equation 7.14 does not approach a stable equilibrium. In particular, for a periodic attractor $(\hat{x}(t),\hat{y}(t))$, of period T, candidate ESS solutions can be obtained by solving the equation $\int_0^T \partial/\partial\varepsilon g(x(t),y(t),\varepsilon)dt = 0$ for ε.

Discrete population models

Populations on a fixed resource

The simplest of all situations to model is a population with discrete nonoverlapping generations in which the density of individuals y_{t+1} in generation $t+1$ is determined in terms of y_t the density of individuals in generation t, and x_t the density of resources available to the population in generation t.

If the resources per unit consumer, i.e. *resources per capita* x_t/y_t, are sufficiently high in generation t so that, as a function of resource density x_t, each individual realizes its maximum fecundity b, and its maximum survival rate s (from birth to reproductive maturity), then the growth of the population from one generation to the next is given by $y_{t+1} = sby_t$. On the other hand, if the resources per capita x_t/y_t are close enough to zero in generation t so that either all individuals fail to survive from birth to reproductive maturity or a few do manage to survive but are in too poor a condition to reproduce, then the density of individuals in generation $t+1$ will be zero. For intermediate levels of resources per capita, we expect the maximum growth rate to be modified by a decreasing function $\theta(x_t/y_t)$, so that our model takes the form:

$$y_{t+1} = sb\theta(x_t/y_t)y_t \tag{7.15}$$

A number of forms have been proposed for θ in the context of constant background resource levels $x_t = \bar{x}$ for all t: e.g. $\theta(\bar{x}/y) = (\bar{x}/y)/(c + \bar{x}/y) = \bar{x}/(cy + \bar{x})$ is the so-called Beverton and Holt model and $\theta(\bar{x}/y) = e^{-cy/\bar{x}}$ is the so-called Ricker model. The two parameter model that provides that best fit to real insect population data (Bellows 1981), however, has the sigmoidal form

$$\theta(\bar{x}/y)=\frac{(\bar{x}/y)^{\gamma}}{c^{\gamma}+(\bar{x}/y)^{\gamma}} \tag{7.16}$$

Further, unlike the Ricker and Beverton and Holt models (Gatto 1993; Getz 1996), this sigmoidal form has the property that as long as $\gamma > 1$, the onset of density dependence at low values of y_t relative to $\bar{x}$ is gradual (i.e. its slope at the origin is zero; see Getz 1996).

Defining r to be the maximum per generational rate of increase (or maximum reproductive value of each female, i.e. putting $r = sb$, Equations 7.15 and 7.16 can be combined and terms rearranged to obtain

$$y_{t+1}=\frac{ry_t}{1+(cy_t/\bar{x})^{\gamma}} \quad \gamma \geq 1 \tag{7.17}$$

The dynamics of this system depend solely on the values of r and γ since the units of x can always be selected so that $c/\bar{x} = 1$.

The ecological and evolutionary dynamics of systems modelled by Equation 7.17 have been explored in considerable detail (Getz & Kaitala 1989; Doebeli 1995; Getz 1996; Schoombie & Getz 1998; for analysis of similar systems see Hansen 1992; Gatto 1993). The form of the density dependence in the model in Equation 7.17 is controlled by the parameter γ which is a density abruptness response parameter, because for increasing γ, the onset of the density-dependent response gets increasingly close to the density $c/\bar{x}$ and increasingly abrupt in the neighbourhood of density $c/\bar{x}$ (Fig. 7.1). In fact, one can think of interspecific competition becoming more interference or 'contest-like' as $\gamma \rightarrow 1$ from above, and more exploitative or 'scramble-like' as $\gamma \rightarrow \infty$ (also see Hassell 1978). Using standard stability analysis (Luenberger 1979; Lichtenberg & Liebermann 1991), it is easily shown that oscillatory solutions arise whenever r is sufficiently large to satisfy the inequality $r > \gamma/(\gamma - 2)$. Note we need to assume that $r > 1$ for the population to grow at even the lowest densities. Further, $\gamma/(\gamma - 2) \rightarrow 1$ as γ approaches infinity. As r is increased beyond the point where oscillations set in (Getz 1996), the system rapidly goes through the standard period doubling route to chaos (May & Oster 1976; Lichtenberg & Liebermann 1991).

Evolution of abrupt density dependence

A detailed ESS analysis of the model in Equation 7.17 reveals a number of interesting results (Getz 1996; Schoombie & Getz 1998). First, if we consider

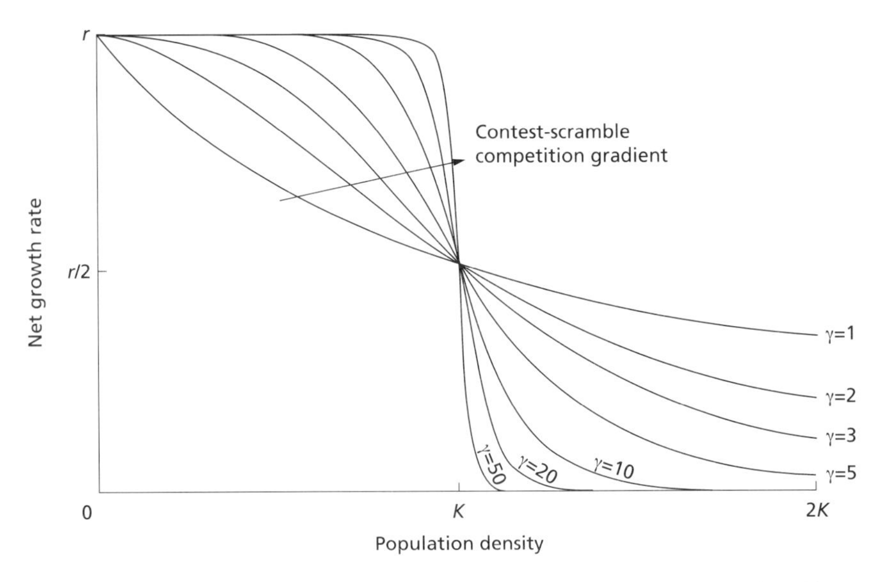

Fig. 7.1 The density-dependent sigmoidal response expressed in Equation 7.15 is plotted here as a family of curves that portray increasing levels of scramble-like competition as the parameter γ increases from 1 to 50.

the ESS value of the abruptness constant γ for different fixed values of r, then it turns out that the boundary value $\gamma = 1$ is an ESS whenever $r > 2$ but that γ assumes some relatively abrupt value whenever $r < 2$. For example, when $r = 1.5$, $\gamma^* = 9.58$, while $\gamma^* \rightarrow 4$ as $r \rightarrow 2$. This result has an interesting biological interpretation in the context of populations exploiting distributed patches of resources, where females lay eggs or larvaposit their young on these resource patches, and the developing young are not mobile enough to easily move from one patch to another. In particular, it suggests the following hypothesis (Getz 1996):

> In synchronous semelparous populations, competition among young for resources should be more contest-like if the population has the potential at low densities to more than double in each generation and scramble-like if the population does not have this growth potential. Further, contest-like competition may be avoided in slower growing populations by females investing the effort to over-disperse their eggs or young among available resources.

A related result was obtained by Hansen (1992), using a slightly different approach. From his analysis he concluded that stable equilibria (i.e. low γ) are favoured by selection at high population densities while oscillatory dynamics is favoured at low population densities. These results were taken a step further by considering the density-dependent or state-dependent strategy $\gamma(y) = \gamma_0 y + \gamma_1$. A strategy of this type assumes the behaviour or physiology of individuals is adjusted in each generation in response to the density of the population in that generation. This assumption is not unreasonable (Yashuda 1990; Guisande 1993; Gage 1995; also see Kaitala *et al.* 1997). In this case, it is no longer the evolution of γ itself that is investigated, but the evolution of the parameters γ_0 and γ_1 in the expanded model

$$y_{t+1} = \frac{ry_t}{1+(cy_t/\bar{x})^{\gamma_0 y_t + \gamma_1}} \qquad (7.18)$$

An ESS analysis of the parameters γ_0 and γ_1 demonstrates that if $r > 2$, the system evolves towards an equilibrium value for $y^* \to (r-1)$ and values of γ_0 and γ_1 that ensured $y^* \to (r-1)$. In the more interesting case of $1 < r < 2$, the ESS solution produces oscillating dynamics, very often cycles of period 2, but some times more complex oscillations. These oscillations are generally smaller than those produced by non-ESS values for γ_0 and γ_1. Further, the ESS value γ_0^* is generally as negative as it can be without leading to a violation of the constraint $\gamma_t = \gamma(y_t) \geq 1$ for all t. In the case $r = 1.5$, for example, γ_t^* oscillates between a value of 1 when γ_t is at its lowest density and a little less than 7 when γ_t is at its highest density. This results suggests the additional hypothesis:

> In synchronous semelparous populations where females are able to adjust the way they distribute their young on patchily distributed resources in response to the density of the population, these females should invest effort when the population density is high to over-disperse their eggs or young among the patches only if the maximum growth rate of the population is less than 100% per generation.

Evolutionary trade-off

Most populations are not free to evolve one trait at a time. The evolution of traits are typically linked through epigenetic and pleiotropic effects (Hoffman & Parsons 1991). Further, trade-offs in the evolution of traits might be expected since physiological resources are limited when it comes to improving fitness through any particular trait. Individuals that evolve to reduce the onset of density dependence through mechanisms that allow them to disperse and search out new resources are likely to have reduced fecundity or survivorship (the latter through increased exposure to predation). One way to analyse this in the context of Equation 7.17 is to make both the abruptness parameter γ and growth rate parameter r functions of 'investment' in the parameter ε. We can assume, for example, that as investment in some component μ increases (e.g. improvements in locomotion or some sensory system), the ability of individuals to realize a more abrupt level of density dependence increases linearly with μ, i.e. $\gamma = 1 + \varepsilon$, and the growth rate of the population decreases from a maximum level r_0 to 0. One expression that has the desired properties is

$$r(\varepsilon)=\frac{r_0}{1+(\varepsilon/b)^m} \tag{7.19}$$

where $r_0 > 1$ and the parameters m and b respectively are positive 'shape' and 'scaling' constants.

The more general state-dependent case $\varepsilon = \gamma_0 y + \gamma_1$ also has been studied (Schoombie & Getz 1998). The results for this case reinforce the results obtained for the less general cases in the sense that if $r_0 > 2$, then the ESS is $\gamma^* = 1$ and the population approaches its corresponding equilibrium. If $1 < r_0 < 2$, then typically the population evolves towards an ESS that produces a dynamical system with the following properties: solutions will approach an equilibrium state when b and m are relatively small, they will exhibit period 2 dynamics for most other values of b and m, but they will exhibit higher order oscillations for relatively large values of b when $1 \leq m \leq 2$ (Fig. 7.2).

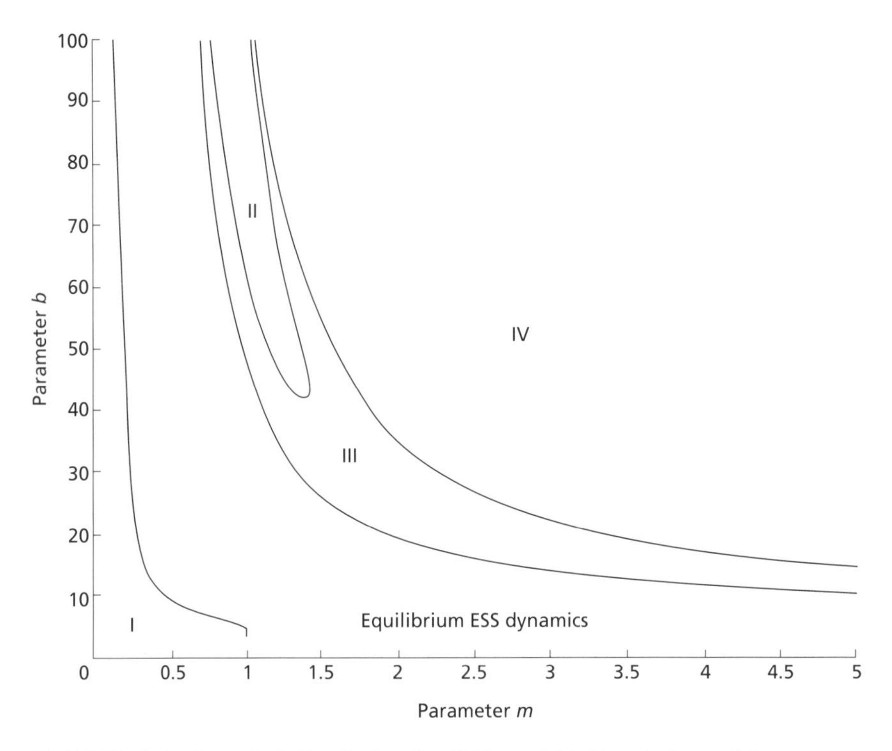

Fig. 7.2 The behaviour of solutions to Equation 7.17 are plotted here in terms of the parameters b and m in Expression 7.19 for the case $r_0 = 1.5$. Region I corresponds to boundary ESSs (i.e. the ESS value is actively constrained by $\gamma \geq 1$—for further details see Schoombie & Getz (1998), with equilibrium density states (i.e. $\gamma^* = 1$), and Region IV to boundary ESSs with period 2 dynamics. Region III corresponds to interior ESSs (i.e. the ESS value is not constrained by $\gamma \geq 1$) that produce period 2 solutions, and Region II to interior ESSs that produce period greater than 2 or chaotic solutions. The region where equilibrium dynamics occur is explicitly labelled.

Effects of harvesting

The natural mortality factor s that determines the growth rate parameter $r = sb$ in the model in Equation 7.17 must be modified to include the effects of human-induced mortality when it occurs. Specifically, if a constant proportion p of individuals in each generation are removed then the value of r must be reduced to $r = (1 - p)sb$ in Equation 7.17 or $r_0 = (1 - p)sb$ in the trade-off situation associated with Equation 7.19. Further, we might expect populations to evolve in response to this human-induced mortality.

In applying discrete population density-dependent models to investigate such questions, it is important to pay attention to the timing of events. For example, if the proportion p_t of the individuals that survive to maturity are harvested from a population modelled by Equation 7.15 then the dynamics

$$y_{t+1} = (1 - p_t)sb\theta(x_t/y_t)y_t \tag{7.20}$$

and the actual number h_t of harvested individuals is

$$h_t = p_t sb\theta(x_t/y_t)y_t \tag{7.21}$$

On the other hand, if the individuals are removed soon after birth, as may happen to the immature stage of anadromous fish diverted into irrigation channels as they make their way downstream to the sea, then the dynamics can be modelled by the equation

$$y_{t+1} = (1 - p_t)sb\theta\left(\frac{x_t}{(1 - p_t)y_t}\right)y_t \tag{7.22}$$

and the actual number h_t of harvested individuals is

$$h_t = p_t y_t \tag{7.23}$$

Finally, if individuals are not removed at a particular point in time—specifically after and before natural mortality has happened, as modelled by Equations 7.20 and 7.22 respectively—but are removed continuously over an interval of time, then one needs to decide what the best value of the exploited population might be to calculate the effects of density. This problem, of course, is moot if harvesting is not applied during the period when the effects of density dependence are most prominent as is the case for the class of age-structured models considered below.

In the context of Equations 7.20 and 7.22, the effect of removing a constant proportion of individuals in each generation is to reduce the value of $r = (1 - p)sb$ in Equation 7.17 (or $r_0 = (1 - p)sb$ in the trade-off situation associated with Equation 7.19). If r (or r_0) exceeds 2 prior to harvesting, and γ has evolved to its ESS value of 1, then by reducing r (or r_0) to a value less than 2 through harvesting we might expect γ to begin to evolve towards its new 'abrupt' ESS value, provided that sufficient variation and plasticity existed in the population to enable the parameter γ to evolve. In this case, harvesting would have the effect of destabilizing the population from its equilibrium state to an oscillating periodic or chaotic state.

Trophic interactions

The extension of Equation 7.17 to a situation where x is a biological resource that is itself a population growing off some other resource w depends on whether the trophic interaction is due to predation or parasitism, and whether the parasitism itself is due to a parasitoid, a macroparasite (helminths, nematodes, etc.) or a pathogen (fungal, bacterial, or viral). The dynamics of the density y_t of predators or macroparasites, for example, might be usefully described by the equation (Kaitala *et al.* 1997)

$$y_{t+1} = \frac{r_y y_t}{1+(c_y y_t / x_t)^{\gamma_y}} \quad \gamma_y \geq 1 \tag{7.24}$$

where x_t is the density of their prey or hosts at time and r_y, c_y, and γ_y are the predator or macroparasite specific parameters. Similarly, the density or dynamics of the prey (or host) population x_t might be usefully described by the equation

$$x_{t+1} = \frac{b_x s_x (y_t / x_t) y_t}{1+(c_x x_t / \overline{w})^{\gamma_x}} \quad \gamma_x \geq 1 \tag{7.25}$$

where $\overline{w}$ is a constant background resource for the prey or host, and b_x, c_x, and γ_x are the prey specific parameters. In the host or prey equation, however, we need to account for the effects of the parasite or predator by assuming that the survival rate of the prey is a decreasing function of the ratio (y_t/x_t).

Kaitala *et al.* (1997) recently considered a macroparasite–host problem where host individuals at an early developmental stage activated a switch with probability ε that lead to the development of an immune system: that is, a proportion of individuals ε in the population were immune to the macroparasites in question, while a proportion $(1 - \varepsilon)$ were susceptible to these macroparasites. Further they assumed that immunity was perfect in the sense that all the parasites attacking the host died without any ill effects on the host other than a fixed cost d to the reproductive rate of immune hosts, irrespective of whether or not the hosts were attacked by one or more parasites. Under these assumptions and the assumptions that in each generation the macroparasites attack infest all individuals at the same (uniform) rate that is directly proportional to the number of parasites per host (y_t/x_t), the parasite and host equations take the form (cf. Equations 7.24 & 7.25)

$$x_{t+1} = \frac{b_x s_x \left((1-\varepsilon)\mathrm{e}^{-ab_y y_t / x_t} + \varepsilon(1-d)\right) y_t}{1 + \left(c_x x_t / \overline{w}\right)^{\gamma_x}}$$
$$y_{t+1} = \frac{b_y s_y y_t}{1 + \left(c_y y_t / x_t\right)^{\gamma_y}} \tag{7.26}$$

where a can be interpreted as the virulence of the microparasites: a measures the proportional rate of reduction of host survivorship as a function of the severity of the infection (i.e. of the number of macroparasites per host). Alternatively, the parameter could represent reductions in the fecundity of hosts in cases where fecundity is known to decrease with the severity of the infection (Anderson & May 1978; May & Anderson 1978; Roberts *et al.* 1995).

In an ESS analysis of the host–macroparasite model (Equation 7.26), Kaitala *et al.* (1997), considered parasite fecundity b_y and the propensity ε of the host to develop immunity (or the proportion of hosts expressing immunity) as evolvable strategies. Their analysis indicated that the parasite fecundity parameter, b_y, should evolve towards its maximum possible value, even though at high parasite fecundity levels, the size of the parasite population may be lower than levels obtained at more moderate fecundity values. The reason for this is that at high parasite fecundity values the susceptible individuals experience high mortality rates leading to few resources for the parasite population to successfully exploit. In addition, Kaitala *et al.* (1997) were able to show that host population should be polymorphic in the proportion of hosts expressing immunity: that is, typically this proportion should not evolve to 0 or 1. They also obtained the expected result that the stability of the host—macroparasite interaction is strongly influenced by the values of density-dependent abruptness parameters γ_x and γ_y with equilibrium dynamics being more likely for low than high values of these parameters. Finally, the results obtained by Kaitala *et al.* (1997) suggest that the ESS proportion of hosts expressing immunity is lower when the dynamics of the host–macroparasite is chaotic than stable.

Age-structured populations

One class of discrete age-structured population models, which has had extensive application in fisheries management (Getz & Haight 1989), assumes that the density of births is proportional to a weighted sum of the density of females of different ages (the weighting factor is a measure of relative fecundity as a function of age), the mortality rate of the youngest age class is density-dependent, while mortality rates of the older age classes are density-independent. If no individuals survive beyond age n, then this model takes that form (Bergh & Getz 1988; Getz & Haight 1989; Kaitala & Getz 1995)

$$y_{t+1} = \theta\left(\frac{x_t}{B_t}\right)B_t \tag{7.27}$$

where y_t is the number of young produced at time t and

$$B_t = \sum_{j=1}^{n} c_j y_{t-j+1} \tag{7.28}$$

is the 'breeding density' of the population at time t (see Gatto 1993 for a simpler approach to modelling iteroparous fish stocks). The parameters c_j are themselves constructed from fecundity and survivorship parameters, b_i and s_i respectively, where b_i is the average number of females born to a female of age i, s_i is the probability of surviving from age i to age $i+1$, and hence $\ell_i = \prod_{j=0}^{i-1} s_j$ is the density-independent probability of surveying from birth to age i. As in Equation 7.15, the function $\theta(\cdot)$ is a density-dependent correction to survivorship that increases from a value of 0 to 1 as the resources per unit breeding density increases from 0 to ∞.

More specifically, if a proportion p_i of individuals is removed from the ith age class in each time period (in a fisheries model this proportion may be realized by applying a fishing effort v subject to an age-dependent catchability coefficient q_i to obtain $p_i = e^{-q_i^v}$) then the coefficients c_i are given by the expression

$$c_i = l_i b_i e^{-\left(\sum_{j=1}^{i-1} q_j\right)v} \tag{7.29}$$

As a consequence of the fact that breeding populations consist of individuals born at different times, the time-delays that arise in age-structured models, such as Equations 7.27 and 7.28, are known to destabilize populations (May 1981a). Harvesting, on the other hand leads to a proportionately greater reduction in the importance of the older age classes (see Expression 7.29), and hence to a de-emphasis of the longer time delay models such as Equations 7.27–7.29. Thus, in nonevolving systems, we might expect harvesting to lead to a stabilization of the dynamics of populations, although excessive levels of harvesting will ultimately cause a population to crash (Fig. 7.3). In evolving systems, however, harvesting or other anthropogenic sources of mortality tend to reduce r so that the growth rate of naturally fast growing populations ($r > 2$) is sufficiently reduced to the point $r < 2$ where the evolution of abrupt density dependence is facilitated and the equilibrium dynamics of the natural system disrupted to become oscillatory.

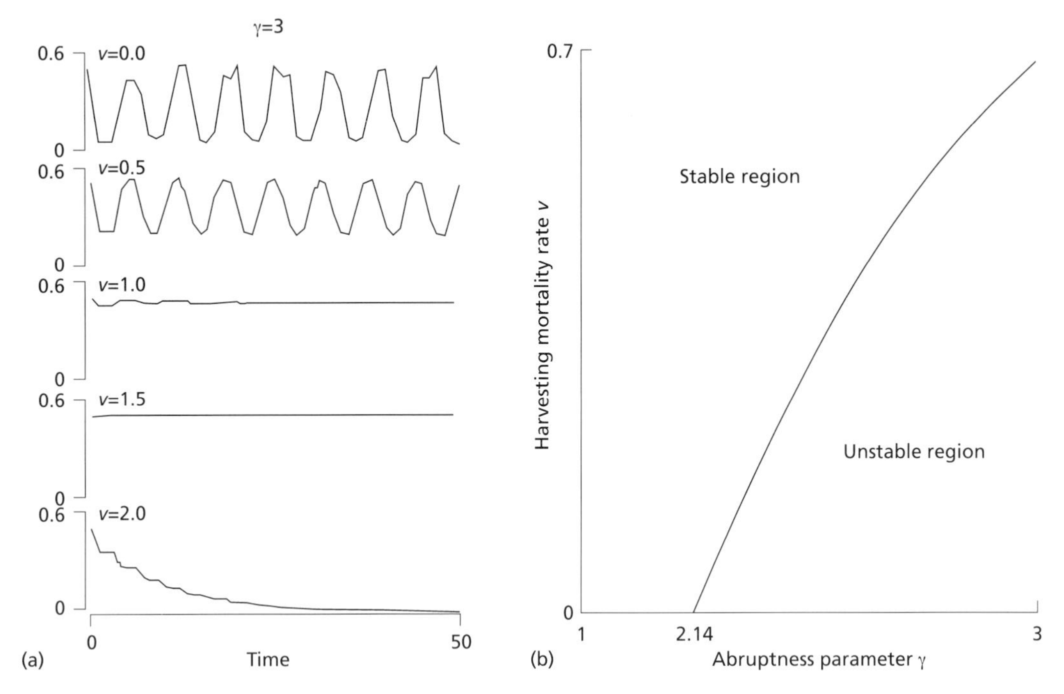

Fig. 7.3 (a) Solutions to Equations 7.27–7.29 are plotted for the initial condition $y_0 = 0.5$ when the parameters have the values $b_1 = 0$, $b_2 = b_3 = 10$, $q_1 = 0$, $q_2 = 1$, $q_3 = 2$, $l_2 = 0.5$, $l_3 = 0.3$, and $\gamma = 3$, and the harvesting intensity takes on the values $v = 0.0$, 0.5, 1.0, 1.5, and 2. The system bifurcates from oscillatory to stable solutions at $v = 0.683$, as plotted in (b). Note that the population ceases to have a positive equilibrium at $v = 1.72$ (correct to two decimal places). (b) The value of v at which a transition occurs from oscillatory to stable solutions, as v increases from 0 to 0.7, is plotted for γ ranging from 1 to 3. Note that the solutions are stable for all v when $\gamma < 2.14$ (correct to two decimal places).

Semelparous populations

Semelparous populations, such as Pacific salmon, where adults breed at some age n and then die, can be modelled by discrete-time scalar equations such as Equation 7.15, if the units of t are taken as n-years (i.e. a unit increase in t implies that n years have passed). Such a model, however, would be inadequate for addressing the problem of the evolution of breeding age n, since such a model would require that we look at the ESS value of n by analysing the conditions under which a mutant population playing breeding strategy of age n is able to invade a resident population playing a breeding strategy of age $n - 1$ or $n + 1$. In this case, an age-structured model such as that shown in Equations 7.27–7.29 is much more useful, especially if the effects of harvesting upon the ESS age at reproduction are to be evaluated.

The evolution of the breeding age in semelparous populations has been examined in some detail by Getz and Kaitala (Getz and Kaitala 1993; Kaitala & Getz 1995; also see Mylius & Diekmann 1995) in the context of

Pacific salmon. Kaitala & Getz (1995) focused their analysis on whether adult salmon should return to spawn at age 3 or age 4. They concluded that in random mating populations, polymorphisms for breeding age cannot be supported by the dynamics of the population. Rather, polymorphism would only be expected to arise in situations where assortative mating occurs. One mechanism that would support such matings is that individuals return to mate in different parts of the habitat (i.e. in different reaches or tributaries of the river) with older maturing fish travelling further up the river than the younger maturing fish. In this case, k-density-dependent functions θ_i apply in each of the k spawning areas, where the effects of density dependence are based on local rather than total population densities. In this case, breeding age polymorphisms can exist: the ESS strategy is characterized by a reproductive ideal free distribution in which the individuals in the different habitats are equally fit independent of whether they are early (age 3) or late (age 4) spawners.

On the other hand, in isolated panmictic populations, early spawning is favoured by natural selection whenever $c_3 > c_4$ (in this case Equations 7.27 and 7.28 apply to the early spawning phenotype with $c_1 = 0$ for all i except $i = 3$ and also applies to the late spawning phenotype with $c_1 = 0$ for all i except $i = 4$) and late spawning is favoured by natural selection whenever $c_4 > c_3$. If harvesting is now introduced into the marine environment so that all cohorts are vulnerable to harvesting prior to returning to their home rivers to spawn, then it follows from Expression 7.29 that older spawners are harvested more heavily than younger spawners because they have an extra year of vulnerability to fishing prior to spawning. Consider c_3 (v) and c_4 (v) as functions of fishing effort v, and suppose that initially $c_4 (0) > c_3 (0)$. Then it follows from Expression 7.29 that for sufficiently large v the inequality $c_3 (v) > c_4 (v)$ holds. In this case, sufficiently intense harvesting has the effect of causing the ESS spawning strategy to switch from late to early spawning, and we can expect a predominantly late spawning population to evolve into a predominantly early spawning population. A consequence of not taking this switch into account could be that the effort level calculated to provide the maximum sustainable yield may severely deplete or even destroy the fishery (Kaitala & Getz 1995).

Continuous population models

Consumer-resource stacks

Individuals in a trophic stack extract resources from the trophic level below and are extracted by individuals in the trophic level above. At the lowest trophic level, the resources are abiotic (light, nutrients, etc.), but organisms at all other levels except the top level are simultaneously both consumers of resources below and resource for the consumers above. Let $\phi(x_{i-1},x_i)$ represent the per-capita (or per unit biomass) rate at which consumers at density x_i extract resources at density x_{i-1}. Let x_0 represent the density of the abiotic resources at the lowest trophic level, and let x_n represent the density of the 'top predator'.

At the most fundamental level, the only other process we need to consider is the process $f_i(\phi_i)$ whereby consumers at the ith trophic level convert the mass (or energy) ϕ_i extracted at the $(i-1)$th trophic level into their own biomass. In this case, if $f_i(\phi_i)$ is interpreted as the per-capita (or per unit biomass) conversion rate, the model takes the form (Getz 1991, 1993, 1994)

$$\frac{\mathrm{d}x}{\mathrm{d}t}=x_i f_i(\phi_i)-x_{i+1}\phi_{i+1} \quad i=1,\ldots,n \tag{7.30}$$

where $\phi_{n+1}=0$ (since x_{n+1} does not exist) and the basal resource variable x_0 at the lowest trophic level is either a constant input (underlying resource flux) or satisfies an appropriate abiotic resource pool or detritus production equation (Getz 1994). Note that this formulation allows for the possibility that extraction from the top trophic level is due to harvesting or other anthropogenic sources, where f_n represents the process of turning the extracted biomass into a form useful to humans (e.g. filleted and gutted fish) and the variable x_n represents the accumulated processed yield.

The case where the extraction rate $\phi_i(x_{i-1},x_i)$ is simply proportional to the resource density, i.e. $\phi_i(x_{i-1},x_i)=bx_{i-1}$ for all i, and conversion is a linear function of extraction, i.e $f_i(\phi_i)=a_i\phi_i-c_i$ where c_i can be interpreted as a rate of decline in the absence of resources that is overcome in a linear fashion with the density of extracted resources, yields the standard Lotka–Volterra expression (Getz 1994; Hastings 1995) for the ith trophic level:

$$\frac{\mathrm{d}x_i}{\mathrm{d}t}=a_ib_ix_ix_{i-1}-c_ix_i+b_{i+1}x_ix_{i+1} \tag{7.31}$$

The problem with this approach is that it does not include satiation or interference competition in the extraction functions ϕ_i. Further, the assumption of a constant decline rate in the absence of resources implies that the population is able to draw at a constant rate upon its own internal resources in the absence of external resources. This allows the population to decline exponentially in the absence of resources, an unrealistic situation for most populations.

Other approaches to modelling multitrophic interactions can be, and have been, taken. They include the metabolic pool approach of Gutierrez (Gutierrez *et al.* 1984; Gutierrez 1992; Gutierrez *et al.* 1994), and the generalized functional response approach of Arditi & Michalski (1995).

We can focus on a pure consumer-resource interaction at the lowest trophic level (i.e. an autotroph growing on a resource flux) by setting $v=x_0$,

$x = x_1, x_2 \equiv 0$. Consider the case where $v = v(t)$ is a time dependent input flux, and the extraction rate $\phi(v,x)$ has the following form, which includes both the notion of satiation at a maximum extraction rate $\delta > 0$ as well as a notion of interference competition through a self-interference parameter $\alpha > 0$ (Beddington *et al.* 1975; DeAngelis *et al.* 1975; Getz 1991, 1993, 1994; Arditi & Akçakaya 1990; Szathmáry 1991; Arditi & Michalski 1995),

$$\phi(v,x) = \frac{\delta v}{\beta + v + \alpha x} \tag{7.32}$$

where $\beta \geq 0$ is the half-saturation parameter. When self-interference is zero, i.e. $\alpha = 0$, then Expression 7.32 is the so-called Holling disc equation (Holling 1959; Emlen 1984; Yodzis 1989). On the other hand, when $\beta = 0$ then Expression 7.32 represents pure ratio-dependent extraction with saturation (Getz 1984; Arditi & Ginsburg 1989; Ginsburg & Akakaya 1992; Arditi & Michalski 1995).

Extraction of resources

Expression 7.32 can be rederived in a more general form that provides clearer biological interpretation of the parameters involved, and the assumptions implicit in Expression 7.32 can be clarified. The resource density in Expression 7.31 is represented by a single 'lumped' variable that may be time dependent if we regard $v = v(t)$. Almost all resources, however, have spatial and temporal structures on scales that impact the ability of individuals to meet physiological demands for food. In developing a new expression to replace Expression 7.32, two temporal and two spatial scales will be considered: one that pertains to individuals, and one that pertains to population averages.

At a local or individual scale relating to the extraction of food over one cycle of activity—for example, a foraging bout or a daily scale—two extremes exist in terms of how individuals perceive a homogeneous resource: either the resource is perceived as a flux (or, equivalently, a non-depletable pool) or a nonrenewable pool that may be entirely consumed if the individual is sufficiently hungry. In the case of a flux, we assume the extraction rate is purely resource density dependent (Arditi & Saïah 1992). In the case of periodically replenished pool that is completely consumed by all individuals present, we assume that each individual obtains an equal share; that is, the rate is purely ratio dependent. Thus, assuming that resource levels are sufficiently low so that satiation does not occur, the average extraction rate over population time scales for pure density-dependent extraction is $\phi(v,x) = \beta_1 v$, where β_1 is an extraction efficiency parameter, and for pure ratio dependent it is $\phi(v,x) = \beta_0 x/v$, where β_0 is the amount of resource extracted per individual per unit time. Note that we do not assume that v is necessarily constant. To the contrary, over population time scales we expect v to depend on time t. It is only at local scales for the ratio-dependent case that we think of resources occurring in pulses (replenishable packages). At population scales, where only longer term averages are important, local resource pulses translate into a smoothed function $v(t)$ of time (cf. Getz & Schreiber, in press). If satiation does occur, and δ is the maximum extraction rate for each individual, then the two extremes are:

Spatially homogeneous pure resource dependence

$$\phi(v,x)=\begin{cases}\beta_1 v & \text{for } 0<\beta_1 v<\delta \\ \delta & \text{otherwise,}\end{cases} \tag{7.33}$$

Spatially homogeneous pure ratio dependence

$$\phi(v,x)=\begin{cases}\beta_0 v/x & \text{for } 0<\beta_0 v/x<\delta \\ \delta & \text{otherwise}\end{cases} \tag{7.34}$$

Now suppose the resource is no longer locally homogeneous but patchy instead and that the distribution of the resource in each local patch is given by a probability density function $\pi(\eta;\bar{v},\sigma)$, where $\bar{v}$ is the mean resource density, σ^2 the variance across patches, and η is a measure of the density of v. Then, in the case of pure resource-dependent extraction, the mean extraction rate for the population is (cf. Expression 7.33):

$$\phi(\bar{v},x)=\int_0^{\delta/\beta_1}\beta_1\eta\pi(\eta;\bar{v},\sigma)\mathrm{d}\eta+\int_{\delta/\beta_1}^{\infty}\delta\pi(\eta;\bar{v},\sigma)\mathrm{d}\eta \tag{7.35}$$

For $\bar{v}<\delta/\beta_1$ this is more illuminatingly written as

$$\begin{aligned}\phi(\bar{v},x)&=\beta_1\int_0^{\infty}\eta\pi(\eta;\bar{v},\sigma)\mathrm{d}\eta-\int_{\delta/\beta_1}^{\infty}(\beta_1\eta-\delta)\pi(\eta;\bar{v},\sigma)\mathrm{d}\eta\\ &<\beta_1\bar{v}\end{aligned} \tag{7.36}$$

since, by definition, $\int_0^{\infty}\eta\pi(\eta;\bar{v},\sigma)\mathrm{d}\eta=\bar{v}$ and $(\beta_1\eta-\delta)>0$ for $\eta>\delta/\beta_1$. On the other hand, for $\bar{v}>\delta/\beta_1$ Expression 7.35 is more illuminatingly written as

$$\begin{aligned}\phi(\bar{v},x)&=\delta\int_0^{\infty}\pi(\eta;\bar{v},\sigma)\mathrm{d}\eta-\int_0^{\delta/\beta_1}(\delta-\beta_1\eta)\pi(\eta;\bar{v},\sigma)\mathrm{d}\eta\\ &<\delta\end{aligned} \tag{7.37}$$

since, by definition, $\int_0^{\infty}\pi(\eta;\bar{v},\sigma)\mathrm{d}\eta=1$ and $(\delta-\beta_1\eta)>0$ for $\eta<\delta/\beta_1$.

Because equality is implied in Expressions 7.35 and 7.37 when $\sigma=0$ ($f(\eta;\bar{v},\sigma)=0$ for $\eta\neq\bar{v}$), it follows that local heterogeneity in the distribution of the resource reduces the average rate of intake of the population compared with the corresponding rate for homogeneously distributed resources. Further, from Expression 7.37, $\phi(\bar{v},x)\to\delta$ as $\bar{v}\to\infty$ provided the coefficient of variation remains bounded. Thus the effect of the heterogeneity is not to depress the ultimate consumption rate as a function of the average resource density $\bar{v}$, but to cut away the corner that occurs in Expression 7.33 at the point $\bar{v} = \delta/\beta_1$ (Fig. 7.4). The exact form of the expression $\phi(\bar{v},x)$ will depend not only on the variance σ which will change with mean density $\bar{v}$, but also on the distribution $f(\eta;\bar{v},\sigma)$ which may also take on different forms at different resource densities $\bar{v}$. The simplest approach to including the effects of a spatially heterogeneous distribution of resources with respect to the distribution of individuals is to fit a one-spatial-parameter or, if necessary, a two-spatial-parameter family of curves to data on the average per-capita extraction rates in a population for the different densities of the resource. One such 1-spatial-parameter family is the following:

Spatially heterogeneous pure resource dependence

$$\phi(v,x) = \frac{\delta\beta_1 v}{\left(\delta^\gamma + (\beta_1 v)^\gamma\right)^{1/\gamma}} \qquad (7.38)$$

where $\gamma > 0$ is a measure of the degree of spatial heterogeneity of the underlying resource level with respect to the distribution of individuals. Equation 7.33 is the special case of Equation 7.38 when $\gamma \to \infty$. For finite $\gamma > 0$, the 'sharpness of the corner' is determined by the size of γ. Note that the case $\gamma = 1$ yields Holling's 'disc' equation (Holling 1959; Emlen 1984), although the parameters here enter in a different way so that altering the value of δ does not affect the slope of the function at the lowest densities. The traditional representation of the disc equation (Expression 7.32) with $\gamma = 1$ does not have this property. Thus, irrespective of the value of γ in Expression 7.38,

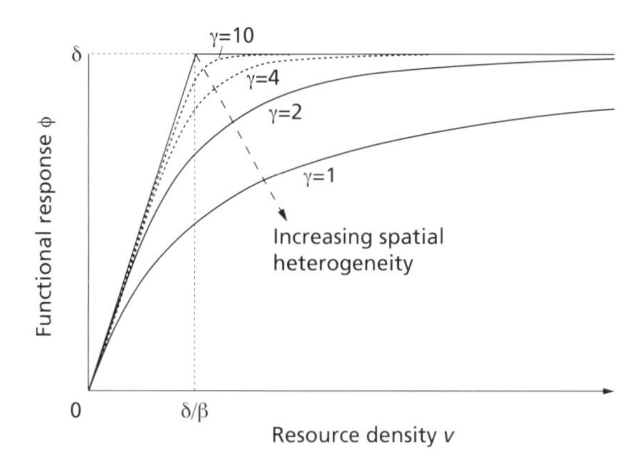

Fig. 7.4 A Holling type I or 'spatially homogeneous pure resource dependent' functional response (i.e. Expression 7.33) has a corner that is smoothed out by spatial heterogeneity in the distribution of the resource with respect to the distribution of the consumers. The parameter γ in Expression 7.38 controls the degree of smoothing or depression of the corner in the type I response. Expression 7.33 approaches a type I response as $\gamma \to \infty$.

if $\beta_1 v >> \delta$ then $\beta_1 v$ dominates the denominator in this expression so that $\phi(v,x) \approx \delta$ and when $\delta >> \beta_1 v$ then δ dominates the denominator in this expression so that $\phi(v,x) \approx \beta_1 v$. Also, note from Expression 7.38 that $\phi(v,x)$ has the same form as the extraction function used to model herbivore grazing systems for which two stable and two unstable equilibria exist (Thornley & Johnson 1990, Section 6.3).

Applying similar arguments the spatially homogeneous pure ratio-dependent extraction function, Equation 7.34, can be generalized to obtain:

Spatially heterogeneous pure ratio dependence

$$\phi(v,x) = \frac{\delta\beta_0 v}{\left((\delta x)^\gamma + (\beta_0 v)^\gamma\right)^{1/\gamma}} \tag{7.39}$$

Equation 7.34 is now the special case $\gamma \rightarrow \infty$ of Expression 7.39. The case γ = 1 yields the ratio-dependent form proposed elsewhere (Getz 1984; Arditi & Ginzburg 1989). Note, irrespective of the value of γ, if $\beta_0 v >> \delta x$ then $\beta_0 v$ dominates the denominator in Equation 7.39 so that $\phi(v,x) \approx \delta$; and when $\delta x >> \beta_0 v$ then δx dominates the denominator in Expression 7.39 so that $\phi(v,x) \approx \beta_0 v/x$.

Note that some ecologists regard the parameter β_1 in the Holling type II function $\phi(v,x) = \delta v/(\beta_1 + v)$ (cf. Equation 7.32 with $\alpha = 0$) as a shape parameter that determines the density of the resource at which the extraction is $\delta/2$ (i.e. it is referred to as the 'half-saturation' parameter; see Yodzis 1989). The problem with this interpretation is that β_1 also determines the slope of the extraction function $\phi(v,x)$ at low resource densities and is thus also an extraction 'efficiency' parameter. One way to decouple these interpretations, as is done in deriving Equation 7.37, is to introduce γ as the shape parameter and leave β_1 as the efficiency parameter that determines the slope of the response at low resource densities, irrespective of the value of γ: i.e. extraction rates at relatively low resource densities are determined purely by the extraction efficiency of the consumer, rather than any factors relating to the distribution of the resource itself. If this is not a reasonable assumption, then a different family of curves than Expressions 7.38 or 7.39 should be used to model the functional response.

From an empirical standpoint, β_1 and δ in Equation 7.38 should be estimated by evaluating extraction rates respectively under entirely limiting (all individuals are limited) and entirely nonlimiting (no individuals are limited) conditions. The parameter γ is then estimated by evaluating how much the corner has been depressed at the resource level $v = \delta/\beta_1$. A quick calculation indicates that the value of Expression 7.38 at $v = \delta/\beta_1$ is $\phi(v,x) = \delta/2^{1/\gamma}$. In this way we estimate β_1, γ and δ independently of one another. The resulting function can then be fitted to a set of data that estimates extraction rates over a range of values for the resource density v and verified for goodness of fit (which is legitimate since all the function parameters are estimated independently rather than being simultaneously fitted to the same set of data).

Finally, the parameters α and β in Equation 7.32 have been interpreted also in terms of the search efficiency of consumers and the time that it takes the consumer to handle a resource item relative to the total time available for searching for resources (Hassell 1978; Emlen 1984). If each resource item is relatively tiny, however, then handling time does not explain the fact that type II response curves have been observed to occur in practice for many classes of consumers (see Emlen 1984).

A general expression for extraction

Most real resource-limiting situations fall between the two ideals of pure resource dependence and pure ratio extraction, respectively represented by Expressions 7.38 and 7.39. First, the individuals extracting resources from the flux will, to some extent, either shade one another, or locally deplete a flux so that individuals in groups experience, on average, lower levels of the resource than isolated individuals. Second, individuals foraging for hidden resources, such as in the deer-mouse/sawfly-cocoon system studied by Holling (1959), may interfere with each other through aggressive interactions or avoidance behaviour. These examples imply that pure resource dependence should be modified to include a component of population density dependence. On the other hand, pure ratio-dependent extraction assumes that the extraction rate of resources is independent of resource density *per se*, and only dependent on how much resource is available per consumer. If individuals are required to actively forage or search for resources, then clearly resource density in its own right becomes a limiting factor when the densities are sufficiently low so that individuals cannot achieve the maximum rate of consumption because of time spent searching for resources.

At high resource-per-capita (consumer) concentrations, where extraction is limited by the capacity of individuals rather than competition for resources, the per-capita extraction rate will be δ. At resource-per-capita densities that are low enough to prevent many individuals from having access to sufficient resources to satiate, we may expect resource extraction to be some mix between pure resource and pure ratio dependence. Thus a general expression for resource extraction should be some appropriate interpolation of the pure resource and pure ratio-dependent Expressions 7.38 and 7.39. A linear interpolation has the drawback that no matter how large the population density x is, each individual will always have a fixed proportion of the resource exclusively available for exploitation. For many resources, however, it may be more reasonable to assume that when x is particularly large, the extraction rate becomes more ratio than pure resource dependent, in which case the following interpolation is appropriate:

$$\phi(v,x)=\begin{cases}\dfrac{\beta_0\beta_1 v}{\beta_1(1-\alpha)x+\alpha\beta_0} & \text{for } x>C(v)\\ \delta & \text{otherwise}\end{cases} \tag{7.40}$$

where

$$C(v)=\left(\frac{v}{\delta}-\frac{\alpha}{\beta_1}\right)\frac{\beta_0}{1-\alpha} \tag{7.41}$$

Note that the inequality $x > C(v)$ assumes that $0 \leq \alpha < 1$. Pure ratio dependence corresponds to $\alpha = 0$ and pure resource dependence corresponds to the limit as $\alpha \to 1$. If we now account for spatial heterogeneity of the resource and/or variation in extraction efficiency or satiation rates among consumers, then this function can be generalized to (cf. Expressions 7.38 and 7.39):

$$\phi(v,x)=\frac{\delta\beta_0\beta_1 v}{\left[\left((\beta_1(1-\alpha)x+\alpha\beta_0)\delta\right)^{\gamma}+(\beta_0\beta_1 v)^{\gamma}\right]^{1/\gamma}} \tag{7.42}$$

The special case $\gamma = 1$ of Expression 7.42 has the form

$$\phi(v,x)=\frac{\delta\beta_0\beta_1 v}{(\beta_1(1-\alpha)x+\alpha\beta_0)\delta+\beta_0\beta_1 v} \tag{7.43}$$

This is the same functional form as Expression 7.32, although the parameters are rearranged and are biologically more appealing. In particular, δ is the maximum extraction rate in both, while β_1 in Equation 7.43 is the extraction efficiency parameter that is estimated by measuring extraction rates at low resource densities. Contrast this with β in Equation 7.32, which is the resource level at which the extraction rate is $\delta/2$, and thus cannot be estimated directly. (In general, β must be interpolated from resource values that result in extraction rates both less than and greater than $\delta/2$.) Estimation of α in the application of Equations 7.32 or 7.43 to real data requires measurement of resource extraction rates over a range of population densities. In Expression 7.43, however, α has the interpretation of the relative degree to which extraction is limited purely by search rates versus competition, while in Expression 7.32 α has the more nebulous interpretation of somehow reflecting competition (but it is not clear how $\alpha > 0$ should be bounded or what it means to allow α to be unbounded). Finally, estimation of β_0 is less direct than β_1, but it can be obtained from estimates of the carrying capacity of the environment once the other three parameters are known (as illustrated in the next section, where it is shown for the case $\gamma = 1$ that the carrying capacity is directly proportional to the value of β_0).

Exponential and logistic growth revisited

Consider the special case of Equation 7.31 in which a population at density x exploits a constant resource v:

$$\frac{dx}{dt} = xf(\phi(v,x)) \tag{7.44}$$

If the conversion function $f(\cdot)$ is linear, we obtain a Lotka–Volterra-type consumer equation, as already mentioned above. Another approach is to assume that in the absence of resources, all individuals immediately die, at least to a reasonable approximation within the time-scales of interest. The simplest appropriate function with this property (Getz 1991, 1993, 1994) is the hyperbolic function

$$f(\phi) = \rho\left(1 - \frac{\mu}{\phi}\right), \tag{7.45}$$

where ρ is a growth rate parameter and μ is a metabolic break-even point: when $\phi > \mu$ the population will grow and when $\phi < \mu$ the population will decline. Small amounts of internal storage of resources can be modelled by incorporating a time delay τ so that $f = f(\phi(v,x(t-\tau)))$; or one could explicitly model the storage process, as discussed later in this chapter.

A more general expression for the function $f(\phi)$ can be derived from considerations of the underlying processes of conversion of ingested resources, basal metabolism, and mortality from senescence and stress induced by suboptimal intake of resources (Getz & Owen-Smith, in press). In this case, the hyperbolic function in Equation 7.45 generalizes to $f(\phi) = (c_g\phi - c_m - c_s/\phi)$, where c_g, c_m and c_s are positive constants and $f(\phi) \to c_g\phi - c_m$ as $\phi \to 0$. Below, however, we will only consider the model arising from Expression 7.45.

If we now substitute Equation 7.42 into Equation 7.45, Equation 7.43 yields the model

$$\frac{dx}{dt} = \rho x\left[1 - \frac{\mu\left(((\beta_1(1-\alpha)x + \alpha\beta_0)\delta)^\gamma + (\beta_0\beta_1 v)^\gamma\right)^{1/\gamma}}{\delta\beta_0\beta_1 v}\right] \tag{7.46}$$

The parameter γ complicates the analysis of this model, but the two extreme cases $\gamma = 1$ and $\gamma \to \infty$ are easily analysed.

For the case $\gamma \to \infty$ (i.e. Expression 7.40 applies) and the resource density v is large enough to ensure that $\phi(v,x) = \delta$ (i.e. $C(v) \geq x$; see Equation 7.41), it follows from Equations 7.44 and 7.45 that

$$\frac{dx}{dt} = r_s x, \quad x < C(v) \tag{7.47}$$

where

$$r_s = \rho\left(1 - \frac{\mu}{\delta}\right) \tag{7.48}$$

On the other hand, if $C(v) < x$ so that consumer satiation does not occur then it follows from Expression 7.38 that

$$\frac{dx}{dt} = r_\infty x\left(1 - \frac{x}{K_\infty}\right) \tag{7.49}$$

This is the logistic equation with intrinsic growth rate

$$r_\infty(v) = \rho\left(1 - \frac{\mu\alpha\beta_0}{\beta_1 v}\right) \tag{7.50}$$

and environmental carrying capacity

$$K_\infty(v) = \left(\frac{v}{\mu} - \frac{\alpha}{\beta_1}\right)\frac{\beta_0}{1-\alpha} \tag{7.51}$$

Since it is assumed that $\delta > \mu$, it follows from Expressions 7.41 and 7.51 that $K_\infty(v) > C(v)$. Further, if we define

$$v_{\min} = \frac{\mu\alpha}{\beta_1} \quad \text{and} \quad v_s = \frac{\delta\alpha}{\beta_1} \tag{7.52}$$

(note that $v_s > v_{\min}$ since $\delta > \mu$) it also follows from Equations 7.50 and 7.51 that $K_\infty(v)$ and r_∞ are positive if and only if $v > v_{\min}$ while $C(v)$ is positive if and only if $v > v_s$. Thus, if $v_{\min} < v \leq v_s$ the population is governed purely by the logistic process modelled by Equation 7.49. If, however, $v > v_s$, then very low initial population levels will increase exponentially at the rate r_s to $x_s = C(v)$ (see Equation 7.47), and then continue to grow logistically to asymptotically approach K_∞. Further, for $v > v_s$, it follows from Equation 7.52 that $1/\delta > \alpha/\beta_1 v$ which, after comparing Equations 7.48 and 7.49, implies that $r_s < r_\infty$. This inequality may seem paradoxical until one realizes that r_∞ is not the actual intrinsic per capita growth of the population, but the intrinsic growth rate that would have occurred if individuals had not become satiated.

For $v > v_s$, however, individuals do become satiated at low population densities so that the actual intrinsic growth rate is truncated to $r_s < r_\infty$.

For the case $\gamma = 1$ and $0 \le \alpha < 1$ it follows from Expression 7.46 that

$$\frac{dx}{dt} = r_1 x\left(1 - \frac{x}{K_1}\right) \tag{7.53}$$

where, after some substitutions involving Expressions 7.48 and 7.50, the intrinsic growth rate of the population is

$$r_1(v) = r_\infty(v) - \frac{\rho\mu}{\delta} \tag{7.54}$$

and its carrying capacity is

$$K_1(v) = K_\infty(v) - \frac{\beta_0 v}{\delta(1-\alpha)} \tag{7.55}$$

It now follows from the positivity of the resource level v and the parameters ρ, μ, and δ, and from the constraint $0 \le \alpha < 1$, that $r_s < r_\infty$ and $K_1 < K_\infty$. These relationships are a direct consequence of the inefficiencies expressed by the inequalities shown in Equations 7.36 and 7.37 that arise when populations exploit resources that are spatially heterogeneous. In these situations some consumers become satiated when local resource densities are relatively high, while other consumers are resource limited when local resource densities are relatively low, even though the density experienced by the average consumer is sufficient for satiation.

Finally, we note from Expressions 7.50, 7.51, 7.54 and 7.55 that r_s and K_1 are zero for the same set of values of $\alpha, \beta_1, \delta, \mu$ and v, and are in fact related to each other by the expression

$$r_1(v) = \chi(v) K_1(v) \tag{7.56}$$

where

$$\chi(v) = \frac{\rho\mu(1-\alpha)}{\beta_0 v} \tag{7.57}$$

is positive provided $\alpha \ne 1$.

Trade-off between search efficiency and maximum extraction rate

In populations that experience relatively low resource levels v, one might expect selection to favour individuals that are relatively efficient at extracting resources, that is individuals that have a relatively high value of β_1 in Expression 7.42. On the other hand, in populations that experience relatively high resource levels v, one might expect selection to favour individuals that have relatively large maximum rates of extraction δ in Expression 7.42. This problem can be formally studied by considering a trade-off between the values of β_1 and δ through an 'investment' parameter ε. In particular, if we define

$$\delta = d\varepsilon^p \quad \text{and} \quad \beta_1 = 1/\varepsilon \tag{7.58}$$

where d and p are positive constants, then by varying ε between 0 and ∞ we have the following trade-off between the two extremes of $\varepsilon = 0$ and $\varepsilon \to \infty$: when $\varepsilon = 0$ consumption is infinitely efficient but the maximal consumption rate is zero, while when $\varepsilon \to \infty$ satiation never occurs but consumption is completely inefficient.

For purposes of illustration, we present the results obtained for the logistic model Equation 7.53 with the parameters in Expressions 7.54 and 7.55 (also see Expressions 7.50 and 7.51) taking on the values $\rho = d = p = \beta_0 = 1$, $\alpha = 0.5$, and $\mu = 0.2$ (note that we can always rescale time so that setting $\rho = 1$ does not sacrifice generality). We also subtract the term hx from the logistic equation to allow us to explore the effects of harvesting at effort level h on the evolution of the population parameters in question. Further, we explicitly identify r_1 and K_1 as functions of the resource level v and the investment or evolutionary strategy parameter ε, because we want to explore how the level of v affects the ESS value ε^*. In this case the system shown in Equation 7.14, which describes the evolution of the parameter ε, has the specific form

$$\begin{aligned} \frac{dx}{dt} &= r_1(v;\varepsilon)x\left(1 - \frac{x}{K_1(v;\varepsilon)}\right) - hx \\ \frac{d\varepsilon}{dt} &= \sigma^2\left(\left(1 - \frac{x}{K_1(v;\varepsilon)}\right)\frac{\partial r}{\partial \varepsilon} + \frac{r_1(v;\varepsilon)x}{K_1(v;\varepsilon)^2}\frac{\partial K}{\partial \varepsilon}\right) \end{aligned} \tag{7.59}$$

where, for the parameters in question, we have

$$r_1(v;\varepsilon) = \left(1 - \frac{\varepsilon}{10v} - \frac{1}{5\varepsilon}\right) \quad \text{and} \quad K_1(v;\varepsilon) = \left(\frac{5v}{2} - \frac{\varepsilon}{4} - \frac{v}{2\varepsilon}\right) \tag{7.60}$$

It follows from the first expression in Equation 7.59 that the equilibrium solution $(\hat{x},\hat{\varepsilon})$ satisfies the equation (see also Table 7.1)

$$\hat{x} = K_t(v;\hat{\varepsilon})\frac{r_1(v;\hat{\varepsilon}) - h}{r_1(v;\varepsilon)} \tag{7.61}$$

If this expression is now substituted in the second expression in Equations 7.59 we obtain the second equilibrium condition

$$\frac{-h}{r_1(v;\hat{\varepsilon})}\frac{\partial r}{\partial \varepsilon} + \frac{(r_1(v;\hat{\varepsilon}) - h)}{K_1(v;\hat{\varepsilon})}\frac{\partial K}{\partial \varepsilon} = 0 \tag{7.62}$$

Resource level v	$\hat{\varepsilon}$	Consumer equilibrium
0.1	0.45	0.26
1	1.41	1.79
5.5	3.32	12.1
10	4.46	22.8
100	7.85	242
$5.5 + 4.5 \sin t$	2.42	3.2–15.1†

† Oscillates between these two values.

Table 7.1 ESS values $\varepsilon^* = \hat{\varepsilon}$ obtained from an analysis of the trade-off between search efficiency $\beta_1(\varepsilon)$ and maximum extraction rate $\delta(\varepsilon)$.

Because in the case considered here, $\chi(v)$ in Equation 7.56 (cf. Expression 7.57) is independent of ε so that $\partial r/\partial z = \chi(v)\partial k/\partial \varepsilon$ and Equation 7.62 reduces to the condition $\partial r/\partial \varepsilon = 0$ for all h. It follows from the second expression in Equation 7.60 that $\hat{\varepsilon} = \sqrt{2\hat{v}}$. An invasion analysis (cf. Equation 7.7) can be used to verify that the equilibrium values $\hat{\varepsilon}(v)$ resist invasion from mutants with values of $\varepsilon \neq \hat{\varepsilon}$. These values are listed in Table 7.1 and have been verified using an invasion analysis (cf. Equation 7.7) to be ESS solutions (i.e. $\hat{\varepsilon} = \hat{\varepsilon}$).

Since the equilibrium $\hat{\varepsilon}$ is independent of h, harvesting has no selective effect in this case. The density v of the underlying resources, however, do influence the value of $\hat{\varepsilon}$. As one would expect, low values of v result in relatively low values of $\hat{\varepsilon}$ which, from Expressions 7.58, favour search efficiency $\beta_1(\varepsilon)$ over the maximal intake rate $\delta(\varepsilon)$. As v increases, so $\delta(\varepsilon)$ gains at the expense of $\beta_1(\varepsilon)$. If the resource input is variable, then when the variability is symmetrical (e.g. sinusoidal) the low levels are more critical than the high levels in determining $\hat{\varepsilon}$. This is seen in Table 7.1 for the sinusoidal input $v(t)$ oscillating between $v = 1$ and $v = 10$. In this case $\hat{\varepsilon}$ is below the level corresponding to the average $v = 5.5$ (compare $\hat{\varepsilon} = 2.42$ vs $\hat{\varepsilon} = 3.32$ in Table 7.1).

Trade-off between resources density and resources-per-capita extraction rates

Individuals in a population can reduce the effects of shading or resource competition due to local depletion of resources by increasing their mobility or, in the case of plants, growing towards more favourable locations. In many cases, improved access may come at the expense of extraction efficiency itself. For example, sessile organisms may invest in support structures at the expense of resource extraction structures (e.g. plants investing in stems at the expense of leaves), while mobile organisms may invest in faster or more efficient locomotion at the expense of, for example, a larger or more efficient digestive system. In this case the trade-off is through the dependence of the value of the parameters α and β_1, say, on the investment parameter ε. Since $\alpha(\varepsilon)\in[0,1]$ and $\beta_1(\varepsilon)\geq 0$, and the value of $\alpha(\varepsilon)$ should decrease from 1 to 0 as the value of $\beta_1(\varepsilon)$ increases without bound, the following expressions are useful for evaluating the evolutionary trade-off between $\alpha(\varepsilon)$ and $\beta_1(\varepsilon)$.

$$\alpha(\varepsilon)=\frac{1}{1+\varepsilon} \quad \text{and} \quad \beta_1(\varepsilon)=c\varepsilon \tag{7.63}$$

where c is a positive scaling parameter.

Again, for purposes of illustration, consider the special case $\rho=\delta=\beta_1=1$, and $\mu=0.2$. In this case, it follows from Equations 7.50, 7.51, 7.54, 7.55 and 7.63 that

$$r_1(v;\varepsilon)=\frac{4}{5}-\frac{1}{5vc\varepsilon(1+\varepsilon)} \quad \text{and} \quad K_1(v;\varepsilon)=\frac{4v(1+\varepsilon)}{\varepsilon}-\frac{1}{c\varepsilon^2} \tag{7.64}$$

Equilibrium Equations 7.61 and 7.62 still apply. In the absence of harvesting (i.e. $h=0$), Equation 7.62 implies that $\partial K/\partial\varepsilon=0$ For the special case represented by Equation 7.64, this latter condition implies that

$$\hat{\varepsilon}=\frac{1}{2cv} \tag{7.65}$$

The values obtained from this expression were used in an invasion analysis (cf. Equation 7.7) to verify that they are indeed ESS values. This appears to be the case. Substituting Expression 7.65 in Expressions 7.63, the ESS values β_1^* and α_1^* now regarded as functions of v are:

$$\alpha^*(v)=\frac{2cv}{1+2cv} \quad \text{and} \quad \beta_1^*(v)=\frac{1}{2v} \tag{7.66}$$

Thus, not surprisingly, if resource levels v decrease then populations will respond by individuals evolving to be more efficient at extracting resources at the expense of adaptations for reducing the effects of shading. On the other hand, if resource levels increase then individuals will evolve to reduce the effects of resource shading at the expense of being efficient extractors of resources.

Unlike the previous case, harvesting now does have an effect on the values of $\hat{\varepsilon}$. The equilibrium values $(\hat{x},\hat{\varepsilon})$ can now be generated for this case by substituting Expressions 7.64 and their derivatives with respect to ε and in Equations 7.61 and 7.62 and then solving for $\hat{x}$ and $\hat{\varepsilon}$. Instead of taking this approach, however, we generated the values directly through numerical solution of Equation 7.59 over a sufficiently long time period so that equilibrium was attained to within six significant figures. The results for the special case $c=0.1$ and $v=10$ are given in Table 7.2 and have been

Harvest level h	ε^*	x^*	α^*	β_1^*
0	0.5	80	0.667	0.05
0.1	0.571	65.6	0.636	0.057
0.2	0.667	52.5	0.6	0.067
0.3	0.8	40.6	0.556	0.08
0.4	1	30	0.5	0.1
0.5	1.33	20.6	0.429	0.133
0.6	2	12.5	0.333	0.2
0.7	4	5.63	0.2	0.4
0.75	8	2.66	0.111	0.8
0.78	20	1.03	0.048	2
0.79	40	0.51	0.024	4
$\to 0.8$	$\to \infty$	$\to 0$	$\to 0$	$\to \infty$

Table 7.2 ESS values $\varepsilon^* = \hat{\varepsilon}$ of the investment parameter and corresponding values for the resource extraction parameters α^* and β_1^* (see Expressions 7.63) for the case $c = 0.1$ and $v = 10$.

verified using an invasion analysis (cf. Equation 7.7) to be ESS solutions (i.e. $\varepsilon^* = \hat{\varepsilon}$).

Harvesting appears to have a dramatic effect on the population. Since harvesting reduces population levels, it improves the resource-per-capita ratio. Thus, harvesting has the effect of shifting the balance of the trade-off from adaptations to reduce the effect of shading to adaptations to improve the efficiency of individuals extracting of the resource itself. The switch can be quite dramatic, as seen in Table 7.2, with the parameter α declining from a value of $\alpha^* = 2/3$ at $h = 0$ to $\alpha^* = 0$ at $h = 0.8$. Beyond $h = 0.8$, positive population levels are no longer sustainable, and the population will crash despite any evolutionary adaptations in the parameters α and β_1.

Storage in food webs

The hyperbolic growth function given by Expression 7.45 does not take into account the fact, as previously mentioned, that populations can exist for some time without having to extract any external resources, because they are able to draw upon internal resources stored in the form of carbohydrates, oils, and fats. As described in Getz & Owen-Smith (1998), the basic trophic Equations 7.30 can be augmented to include a storage component for each population. In this section, I provide an outline of this approach.

Let us assume that the ith population in a food-web is partitioned into 'active' biomass at density x_i and 'stored' biomass at density y_i. Let ϕ_{ei} represent the per unit x_i rate at which resources extracted are from the environment (i.e. $\phi_{ei} \equiv \phi_i$ in Equation 7.30) and let ϕ_{si} represent the per unit x_i rate at which biomass flows from the storage to active tissue component. For simplicity, we assume that the flow rate ϕ_{si} of biomass into and out of storage can be partitioned into a recipient-controlled (i.e. the flow is proportional to the amount of active biomass) 'buffering flow' ϕ_{si} that helps individuals maintain vital functions when resources are scarce and into a donor controlled (i.e. the flow is proportional to the amount of storage biomass) 'translocation flow' ϕ_{wi} that shunts storage to active tissue at some appropriate time of the year (e.g. the first flush of growth of perennial plants after the dormant season). Further, we assume that these two flows are controlled by the following three devices.

1 A maintenance metabolism device μ_i that is compared with the intake ϕ_{ei} in the case of Equation 7.45 the parameter is the device and it satisfies $f(\mu) = 0$, although for the case $f(\phi) = (c_g\phi - c_m - c_s/\phi)$ the situation is more complicated and $f(\mu) \neq 0$.

2 A capacity parameter κ_i, that is compared with the storage ratio

$$w_i = \frac{y_i}{x_i} \tag{7.67}$$

3 A translocation flow device τ_i, that corresponds to a certain time or, more realistically, phenological time (determined using degree-days; see Podolsky 1984; Gutierrez 1996) during the season.

Note, under these assumptions, the total rate of flow from storage is

$$\phi_{si} = \phi_{ei} + w_i\phi_{wi} \tag{7.68}$$

For simplicity, I will drop the subscript i and consider a single population, represented by active biomass density x and storage biomass density y, exploiting a constant underlying resource v that is itself not subject to exploitation by any other population (a more general treatment can be found in Getz & Owen-Smith in press). Based on the assumptions discussed above, we expect the growth rate of active biomass tissue to depend not only on the extraction of external resources but also on the flow of resources from storage; that is, we can still apply Equation 7.44 provided we interpret ϕ to be $\phi = \pi\phi_e + \phi_s$, where π is the proportion of resources extracted from the environment that are allocated to the growth of active tissue and $1 - \pi$ is the proportion allocated to storage tissue. In general, since it is inefficient to unnecessarily shunt resources through storage (external resources must be converted to storage and then reconverted to active tissue), we should expect that $\phi_s = 0$ when $\pi < 1$. This case should arise when the external resource extraction rate $\phi_e > \mu$, so that external resource flow more than covers basal metabolic needs. Conversely when $\phi_e < \mu$, we should expect $\phi_s > 0$ so that basal metabolic needs can be met. Also we should expect flow into storage to slow down as the density of storage y approaches its storage capacity κx. Further, for some reason $y(t) > \kappa x(t)$ (either active tissue is selectively removed by some predator, or $x(t)$ drops faster than $y(t)$ when external resources crash), then we should expect biomass to flow out of storage even though $\phi_e > \mu$ at that point t in time.

Under these assumptions, the net flow in and out of storage is $(1-\pi)\, c_1\phi_e - \phi_s$, where $c_1 < 1$ allows us to account for the fact that shunting resources through storage is less than 100% efficient. Further, since some storage tissue will be lost due to natural mortality, we need to account for this in any equation that is used to model the temporal dynamics of storage. In particular, it can be shown that the particular form given by Equation 7.45 for the growth function $f(\phi)$ in Equation 7.44 embodies the notion of a natural mortality rate that is proportional to $1/\phi$ (Getz & Owen-Smith in press); or, more generally, is given by $(c_2/\phi)^{c_3}$ where $c_2 > 0$ and, in the simplest case, $c_3 = 1$. If we assume some heterogeneity in the ratio ω of storage to active biomass and we assume that animals with a higher ratio are likely to die at a lower rate than those with a lower ratio then this can be accounted for by setting $c_3 = \omega/\kappa$ (Getz & Owen-Smith in press).

From the above discussion, it follows that with the addition of a storage biomass component, Equation 7.44 describing the biomass dynamics of an isolated population extends to

$$\begin{aligned} \frac{dx}{dt} &= xf(\pi\phi_e + \phi_s) \\ \frac{dy}{dt} &= xc_1(1-\pi)\phi_e - x\phi_s - y\left(\frac{c_2}{\pi\phi_e + \phi_s}\right)^{\omega/\kappa} \end{aligned} \tag{7.69}$$

The allocation function π and the buffering and translocation flow rate functions ϕ_e and ϕ_w respectively can be expressed in a number of different ways and still be compatible with the assumptions made above. A particular set of expressions has been derived by Getz and Owen-Smith (1998), while a general theory for deriving such expressions from mechanistic principles has been proposed (Michalski & Getz unpublished manuscript). I will not pursue these details here, other than to remark that the existence of a storage component is critical to promoting the persistence of populations in highly variable resource environments. Further, depending on the degree of variability, we can expect allocation and flow rates to evolve to levels that optimize a population's ability to store resources while still remaining as competitive as possible (Getz & Owen-Smith in press).

Conclusion

Over the past 10 years, the mathematical tools presented in this chapter have been used increasingly to investigate questions in theoretical evolutionary ecology. I have focused primarily on my own work in this area, carried out principally in collaboration with Kaitala and Schoombie (Getz & Kaitala 1989; Kaitala *et al.* 1989; Getz 1993; Getz & Kaitala 1993; Kaitala & Getz 1995; Getz 1996; Kaitala *et al.* 1997; Schoombie & Getz 1998), although I have provided references to some others who have used the same mathematical tools to study a variety of problems, including the evolution of traits in consumer-resource systems (e.g. Abrams 1992; Brown & Vincent 1992; Matsuda & Abrams 1994; Rand *et al.* 1994) and in simple populations that have applications to fisheries management (e.g. Gatto 1993). The material I have presented illustrates the potential that these methods have to provide insights into the structure of population assemblages and food webs that self-organize under the forces of natural selection.

For populations most appropriately modelled by discrete-time equations, significant insights have been obtained into the nature of the density-dependent response of populations as a function of the magnitude of their growth rates (Getz 1996; Schoombie & Getz 1998). The methods have also provided insights into the impacts of harvesting on populations in the context of both lumped and age-structured systems, as well as insights into the evolution of immunity in populations susceptible to exploitation by macroparasites (Kaitala *et al.* 1997). Finally, Brown & Vincent (1992) have used discrete-time models in an evolutionary ecology context to study questions regarding the assembly of predator–prey communities when both consumptive and competitive processes are considered. They found that although predators may ultimately disappear from communities, they can play a critical role in promoting the evolution of diversity that is maintained once the predators disappear (Brown and Vincent refer to this phenomenon as the 'ghost of predation past').

The application of continuous-time models, such as Equation 7.30, to the evolution of consumer-resource systems has received less attention than the application of discrete-time models (but see Abrams 1992; Getz 1993; Matsuda & Abrams 1994) although Vincent and colleagues have made extensive use of Lotka–Volterra continuous-time models to study the coevolution of competing populations (for a tutorial style review see Vincent *et al.* 1996). In this chapter, I have confined the discussion to consumers growing on a constant or time-varying resource input, rather than a full two-species trophic interaction. The ecological dynamics of trophic stacks have been discussed elsewhere (Getz 1991, 1993, 1994), but the evolutionary dynamics of such stacks has yet to be thoroughly investigated. Future investigation of questions such as how the traits of individuals may evolve in populations purely as a function of the position of those populations in a trophic stack should provide us with significant insight into fundamental differences between the evolutionary dynamics of plants, herbivores, and carnivores.

In essence, consumer-resource interactions are the primary ingredient of all ecological food webs, with resource extraction and consumption rates modified by competitive and mutualistic interactions. The ecological aspects of these kinds of interactions have been studied in great detail, including some analysis based on Equations 7.30 (Getz 1991, 1993, 1994). The evolutionary aspects, however, remain relatively unexplored (but see Brown & Vincent 1992). Future evolutionary investigations of more complex trophic interactions than those considered here, based on the methods and models presented, should yield substantial rewards, including a better understanding of how food webs self-assemble on evolutionary time scales and how the structure and connectness of food webs is influenced by the richness and variability of their nutrient and energy inputs.

Finally, the role of storage in food webs has received little attention in ecological and especially evolutionary settings. An application of the methods outlined in this chapter to food web models that include storage components remains one of the most exciting areas for future research into the dynamics of evolving ecological systems.

Acknowledgements

I would like to thank Veijo Kaitala and Schalk Schoombie for their collaboration on much of the work reviewed here.

References

Abrams, P. (1992) Adaptive foraging by predators as a cause of predator–prey cycles. *Evolutionary Ecology*, **6**, 56–72.

Abrams, P., Matsuda, H. & Harada, Y. (1993) Evolutionarily unstable fitness maxima and stable fitness minima of continuous traits. *Evolutionary Ecology*, **7**, 465–487.

Anderson, R.M. & May, R.M. (1978) Regulations and stability of host–parasite population interactions. I. Regulatory processes. *Journal of Animal Ecology*, **47**, 219–247.

Arditi, R. & Akçakaya, H.R. (1990) Underestimation of mutual interference of predators. *Oecologia*, **83**, 358–361.

Arditi, R. & Ginsburg, L.R. (1989) Coupling in predator–prey dynamics: ratio-dependence. *Journal of Theoretical Biology*, **139**, 311–326.

Arditi, R. & Michalski, J. (1995) Nonlinear food web models and their responses to increased basal productivity. In: *Food Webbs: Integration of Patterns and Dynamics* (eds G.A. Polis & K.O. Winemiller). Chapman & Hall, London.

Arditi, R. & Saïah, H. (1992) Empirical evidence of the role of heterogeneity in ratio-dependent consumption. *Ecology*, **73**, 1544–1551.

Beddington, J.R., Free, C.A. & Lawton, J.H. (1975) Dynamic complexity in predator–prey models framed in difference equations. *Nature*, **255**, 58–60.

Bellows, T.S. (1981) The descriptive properties of some models for density dependence. *Journal of Animal Ecology*, **50**, 139–156.

Bergh, M.O. & Getz, W.M. (1988) Stability of discrete age-structured and aggregated delay-difference population models. *Journal of Mathematical Biology*, **26**, 551–581.

Berryman, A.A., Michalski, J., Gutierrez, A.P. & Arditi, R. (1995) Logistic theory of food web dynamics. *Ecology*, **76**, 336–343.

Brown, J.S. & Vincent, T.L. (1987) A theory for the evolutionary game. *Theoretical Population Biology*, **31**, 140–166.

Brown, J.S. & Vincent, T.L. (1992) Organization of predator–prey communities as an evolutionary game. *Evolution*, **46**, 1269–1283.

Cressman, R. & Dash, A.T. (1987) Density dependence and evolutionary stable strategies. *Journal of Theoretical Biology*, **126**, 393–406.

Cressman, R., Dash, A.T. & Akin, E. (1986) Evolutionary games and the two species population dynamics. *Journal of Mathematical Biology*, **23**, 221–230.

DeAngelis, D.L., Goldstein, R.A. & O'Neill, R.V. (1975) A model for trophic interaction. *Ecology*, **56**, 881–892.

Doebeli, M. (1995) Updating Gillespie with controlled chaos. *American Naturalist*, **146**, 479–487.

Emlen, J.M. (1984) *Population Biology: The Coevolution of Population Dynamics and Behavior*. Macmillan, New York.

Ferrière, R. & Gatto, M. (1993) Chaotic population dynamics can result from natural selection. *Proceedings of the Royal Society of London*, **B251**, 33–38.

Ferrière, R. & Gatto, M. (1995) Lyapunov exponents and the mathematics of invasion in oscillatory or chaotic populations. *Theoretical Population Biology*, **48**, 126–171.

Gage, M.J.G. (1995) Continuous variation in reproductive strategy as an adaptive response to population density in the moth *Plodia interpunctella*. *Proceedings of the Royal Society of London*, **B261**, 25–30.

Gatto, M. (1993) The evolutionary optimality of oscillatory and chaotic dynamics in simple population models. *Theoretical Population Biology*, **43**, 310–336.

Getz, W.M. (1984) Population dynamics: a resource per-capita approach. *Journal of Theoretical Biology*, **108**, 623–644.

Getz, W.M. (1991) A unified approach to multispecies modelling. *Natural Resource Modelling*, **5**, 393–421.

Getz, W.M. (1993) Metaphysiological and evolutionary dynamics of populations exploiting constant and interactive resources: r-K selection revisited. *Evolutionary Ecology*, **7**, 287–305.

Getz, W.M. (1994) A metaphysiological approach to modelling ecological populations and communities. In: *Frontiers in Mathematical Biology* (ed. S.A. Levin). Springer-Verlag, New York.

Getz, W.M. (1996) A hypothesis regarding density dependence and the growth rate of populations. *Ecology*, **77**, 2014–2026.

Getz, W.M. (1998) An introspection on the art of modeling in population ecology. *Bioscience* (in press).

Getz, W.M. & Haight, R.G. (1989) *Population Harvesting: Demographic Models of Fish, Forests and Animal Resources*. Princeton University Press, Princeton, NJ.

Getz, W.M. & Kaitala, V. (1989) Ecogenetic models, competition, and heteropatry. *Theoretical Population Biology*, **36**, 34–58.

Getz, W.M. & Kaitala, V. (1993) Ecogenetic analysis and evolutionarily stable strategies in harvested populations. In: *The Exploitation of Evolving Resources* (eds K. Stokes, J. McGlade & R. Law). Springer-Verlag, Berlin.

Getz, W.M. & Owen-Smith, N. (1998) A metaphysical population model of storage in variable environments. *Natural Resource Modeling* (in press).

Getz, W.M. & Schreiber (in press) Multiple time scales in consumer-resource interactions. *Annales Zoologici Fennici*.

Ginsburg, L.R. & Akçakaya, H.R. (1992) Consequences of ratio-dependent predation for steady-state properties of ecosystems. *Ecology*, **73**, 1536–1543.

Guisande, C. (1993) Reproductive strategy as population density varies in *Daphnia magna* (Cladocera). *Freshwater Biology*, **29**, 463–467.

Gutierrez, A.P. (1992) Physiological basis of ratio-dependent prey–predator theory: the metabolic pool model as a paradigm. *Ecology*, **73**, 1552–1563.

Gutierrez, A.P. (1996) *Applied Population Ecology: a Supply–Demand Approach.* John Wiley & Sons, New York, NY.

Gutierrez, A.P., Baumgaertner, J.U. & Summer, C.G. (1984) Multitrophic models of predator–prey energetics. *Canadian Entomologist,* **116**, 923–963.

Gutierrez, A.P., Mills, N.J., Schrediber, S.J. & Ellis, C.K. (1994) A physiological based tritrophic perspective on bottom-up–top-down regulation in populations. *Ecology,* **75**, 2227–2242.

Hansen, T.F. (1992) Evolution of stability parameters in single-species population models: stability or chaos? *Theoretical Population Biology,* **31**, 195–273.

Hassell, M.P. (1978) *The Dynamics of Arthropod Predator–Prey Systems.* Princeton University Press, Princeton, NJ.

Hastings, A. (1995) What equilibirum behaviour of Lotka–Volterra models does not tell us about food webs. In: *Food Webs: Integration of Patterns and Dynamics* (eds G.A. Polis & K.O. Winemiller). Chapman & Hall, London.

Hines, W.G.S. (1987) Evolutionary stable strategies: a review of basic theory. *Theoretical Population Biology,* **31**, 195–273.

Hoffmann, A.A. & Parsons, P.A. (1991) *Evolutionary Genetics and Environmental Stress.* Oxford University Press, Oxford.

Holling, C.S. (1959) The components of predation as revealed by a study of small mammal predation of the European pine sawfly. *Canadian Entomologist,* **91**, 293–328.

Kaitala, V. & Getz, W.M. (1995) Population dynamics and harvesting of semelparous species with phenotypic and genotypic variability in reproductive age. *Journal of Mathematical Biology,* **33**, 521–556.

Kaitala, V., Kaitala, A. & Getz, W.M. (1989) Evolutionary stable dispersal of a water-strider in a temporally and spatially heterogeneous environment. *Evolutionary Ecology,* **3**, 283–298.

Kaitala, V., Heino, M. & Getz, W.M. (1997a) Host–parasite dynamics and the evolution of host immunization and parasite fecundity strategies. *Bulletin of Mathematical Biology,* **59**, 427–450.

Kaitala, V., Mappes, T. & Ylönen, H. (1997b) Delayed female reproduction in equilibrium and chaotic populations. *Evolutionary Ecology,* **11**, 105–126.

Lichtenberg, A.J. & Liebermann, M.A. (1991) *Regular and Chaotic Dynamics.* Springer Verlag, New York.

Luenberger, D.G. (1979) *Introduction to Dynamics Systems: Theory, Models and Applications.* Wiley, New York.

Matsuda, H. & Abrams, P.A. (1994) Timid consumers: self extinction due to adaptive change in foraging and anti-predator effort. *Theoretical Population Biology,* **45**, 76–91.

May, R.M. (1981a) Models for single populations. In: *Theoretical Ecology* (ed. R.M. May). Sinauer Associates, Sunderland, MA.

May, R.M. (1981b) Models for two interacting populations. In: *Theoretical Ecology* (ed. R.M. May). Sinauer Associates, Sunderland, MA.

May, R.M. & Anderson, R.M. (1978) Regulation and stability of host–parasite population interactions. II. Destabilizing processes. *Journal of Animal Ecology,* **47**, 249–267.

May, R.M. & Oster, G.F. (1976) Bifurcations and dynamics complexity in simple ecological models. *American Naturalist,* **110**, 573–590.

Maynard Smith, J. (1982) *Evolution and the Theory of Games.* Cambridge University Press, Cambridge.

Metz, J.A.J., Nisbet, R.M. & Geritz, S.A.H. (1992) How should we define 'fitness' for general ecological scenarios? *Trends in Evolution and Ecology,* **7**, 198–202.

Mylius, S.D. & Diekmann, O. (1995) On evolutionarily stable life histories, optimisation, and the need to be specific about density dependence. *Oikos* **74**, 218–224.

Podolsky, A.S. (1984) *New Phonology: Elements of Mathematical Forecasting in Ecology.* John Wiley & Sons, New York.

Rand, D.A., Wilson, H.B. & McGlade, J.M. (1994) Dynamics and evolution: evolutionarily stable attractors; invasion exponents and phenotype dynamics. *Philosophical Transactions of the Royal Society of London*, **B343**, 261–283.

Roberts, M.G., Smith, G. & Grenfell, B.T. (1995) Mathematical models for macroparasitic wildlife. In: *Ecology of Infectious Diseases in Natural Populations* (eds B.T. Grenfell & A.P. Dobson). Cambridge University Press, Cambridge.

Roughgarden, J. (1983) The theory of coevolution. In: *Coevolution* (eds D.J. Futuyma & M. Slatkin). Sinauer Associates, Sunderland.

Schoombie, S.W. & Getz, W.M. (1998) Evolutionary stable density-dependent strategies in a generalized Beverton and Holt growth model. *Theoretical Population Biology*, **53**, 216–235.

Szathmáry, E. (1991) Simple growth laws and selection consequences. *Trends in Evolution and Ecology*, **6**, 366–370.

Taper, M.L. & Case, T.L. (1992) Models of character displacement and theoretical robustness of taxon cycles. *Evolution*, **46**, 317–333.

Taylor, P.D. (1989) Evolutionary stability in one-parameter models under weak selection. *Theoretical Population Biology*, **36**, 125–143.

Thornley, J.H.M. & Johnson, I.R. (1990). *Plant and Crop Modelling: A Mathematical Approach to Plant and Crop Physiology*. Oxford University Press, New York.

Vincent, T.L. (1990) Strategy dynamics and the ESS. In: *Dynamics of Complex Interconnected Biological Systems* (eds. T.L. Vincent *et al.*). Birkhäuser, Boston.

Vincent, T.L. & Brown, J.S. (1984) Stability in an evolutionary game. *Theoretical Population Biology*, **26**, 408–427.

Vincent, T.L. & Brown, J.S. (1988) The evolution of ESS theory. *Annual Review of Ecology and Systematics*, **19**, 423–433.

Vincent, T.L. & Fisher, M.E. (1988) Evolutionarily stable strategies in differential and difference equation models. *Evolutionary Ecology*, **2**, 321–337.

Vincent, T.L., Cohen, Y. & Brown, J.S. (1993) Evolution via strategy dynamics. *Theoretical Population Biology*, **44**, 149–176.

Vincent, T.L., Van, M.V. & Goh, B.S. (1996) Ecological stability, evolutionary stability, and the ESS maximum principle. *Evolutionary Ecology*, **10**, 567–591.

Yashuda, H. (1990) Effect of population density on reproduction of two sympatric dung beetle species, *Aphodius haroldianus* and *A. elegans* (Coleoptera: Scarabaeidae). *Research into Population Ecology*, **32**, 99–111.

Yodzis, P. (1989) *Introduction to Theoretical Ecology*. Harper and Row, New York.

8: Understanding the ecological and evolutionary reasons for life history variation: mammals as a case study

P.H. Harvey and A. Purvis

Introduction

For ecologists, a good mathematical model provides two types of insight: it explains what we know and it predicts what we do not know. Models that place life history evolution into an ecological context can become very complicated very rapidly, yet we know that the patterns they are trying to explain are often relatively simple (see Chapter 6). Over the years, data have accumulated which describe growth, fecundity and mortality schedules for a wide range of species. When those data are analysed statistically, we find very clear relationships among variables: weights and rates tend to be highly correlated with each other. For example, among mammal species neonatal weight, adult weight, gestation length, litter size, ages at maturity and weaning, and life span are generally very highly intercorrelated, which leads us to believe that a properly parameterized model would contain very few key variables. In this chapter we focus on the main developments which by trial, error, and insight have led to a more straightforward understanding of life history variation in mammals than we had just 10 years ago. For more general reviews of life history theory, we refer the reader to Stearns (1992) or Roff (1992).

For convenience we recognize four key historical phases in the comparative study of mammalian life history variation. In the first phase, life histories were compared across species and the differences were correlated with, and attributed directly to, differences in body weight. In the second phase, when phylogenetically based comparative analyses were used to test the idea that body weight was central to explaining life history variation, it became apparent that the same significant correlations in life history variation among taxa remained even when body weight was controlled for statistically. At this stage piecemeal explanations were used to explain some of the patterns, with body size treated as more or less a confounding variable. The third phase was a comprehensive model developed by Charnov (1993) which claimed to explain the patterns that were known and which made several predictions about relationships between variables which had not previously been investigated. Charnov's model successfully explained many of the patterns that had been described during the first and second historical phases, but some of the new predictions were wrong. The question then was whether Charnov's model needed minor modifications or whether it was necessary to start from scratch with a new comprehensive model. The current or fourth phase of investigation centres around a more general model produced by Kozlowski & Weiner (1996) of which Charnov's model turns out, in large part, to be an unrealistic special case. Both models thrust body size once more into the limelight, but view the size of a mammal species more as an effect of its life history than as the cause. Little more than a decade of interplay between theoreticians and empiricists has brought enormous progress: the state-of-play has moved from description and piecemeal hypotheses to explicitly formulated, testable and comprehensive optimality models.

The first phase: allometry with body size as a key explanatory variable

Julian Huxley (1932) focused the attention of biologists on the importance of body size. He was interested in the question of the relative growth rate of different components of morphology. It was possible to plot the size of, say, an organ against time. He then went further by plotting the size or weight of organs against the adult weight of different species. When he did that he found that relationships were usually 'allometric', which was to say that:

$$y = \alpha w^{\beta} \quad (8.1)$$

where y is the weight or size of the organ and w is body weight. Such plots become linear when logarithmically scaled:

$$\log(y) = \log(\alpha) + \beta \log(w) \quad (8.2)$$

The exponent of the power equation, β, was estimated as the slope of the line from the logarithmic plot. Huxley took a hormonal perspective to the underlying cause of allometric relationships. He imagined a growth hormone being released into the body with different organs responding in different ways such that relative growth rates differed. When adult body size was reached, so the growth of the different organs stopped and the inter-species allometric growth relationship resulted because different species had what he referred to as a 'common growth mechanism'. Given this simple view of life, all the biologist needs to do is determine an animal's optimal adult body size and the growth laws to 'understand' the relative sizes of organs.

Huxley's approach became extraordinarily influential, particularly amongst that set of biologists who believed that much of life's diversity was nonadaptive. Take, for example, Lewontin's (1979, p. 13) claim that, because the size of deers' antlers varies interspecifically and is highly correlated with body size, 'it is then unnecessary to give a specifically adaptive reason for the extremely large antlers of deer'. The relative size of organs was only a start to the exercise. Life history timings, such as gestation length, age at maturity and life span are also allometrically related to body weight (e.g. Millar & Zammuto 1983). As a consequence, many authors interpreted this covariation too as an automatic consequence of common growth laws. Western & Ssemakula (1982, p. 287) provide a typical claim for the period in which it was written: 'Gestation time, postnatal growth rate and age at maturity [can be viewed] as a single growth continuum related to mature weight, which alone might be the target for selection'. One extreme variant of these views was the suggestion by Lindstedt & Swain (1988) that, because many physiological variables show a similar scaling with body size, there exists a physical *Periodengeber* that regulates all body processes around some internal metric set by body weight.

The belief that adult body size is a key variable determining an organism's life history was challenged by some who believed that there is indeed a key determinant of life, but that it is some other factor such as brain weight (Sacher 1959; Sacher & Staffeldt 1974) or metabolic rate (e.g. McNab 1983, 1986a,b), rather than body weight.

The second phase: phylogenetically controlled tests of allometric ideas

Statistical tests of the idea that adult body weight in some sense determines species' life histories needed to incorporate information on the phylogenetic relatedness of the species concerned. The reason for this is that more closely related species are more similar to each other in almost every way than are less closely related species. That is a very obvious statement but it means that cross-species correlations, which treat species values as statistically independent points, heavily inflate the degrees of freedom involved and thereby produce spurious significant correlations. By the mid-1980s, this problem had become acknowledged and Felsenstein (1985) had produced a solution: if the phylogenetic tree connecting the species under consideration is known, then it is possible to partition the variation among species into independent evolutionary events. Essentially, Felsenstein's method constructs ancestral character states for every node in a bifurcating phylogenetic tree, and then uses the differences that have evolved between sister nodes or taxa since they last shared a common ancestor. In a phylogenetic tree relating n species, there are $n - 1$ such independent differences, which have come to be known as independent contrasts. Unfortunately, however, practical ways of developing Felsenstein's test for use on real data took longer to achieve, partially because reliable phylogenies were rare (see Harvey & Pagel 1991).

In the meantime, research workers who recognized the need for reducing degrees of freedom so that data points were effectively independent generally performed analyses across the species-averaged data for higher level taxa. For mammals, generic values were produced as the average of constituent species values, and then family values were estimated from the generic averages, and so on.

Correlations across logarithmically transformed family or order values demonstrated a surprising series of results. It was known that larger-bodied species lived slower lives than smaller-bodied species: they have longer gestation lengths, they mature later, and can potentially live longer. The same highly significant finding held when family or order means were correlated with each other. However, when body weight was controlled for by partial correlation, the same relationships held, although different taxa lay at the fast and slow ends of the continuum (Harvey & Clutton-Brock 1985; Read & Harvey 1989). For example, families (or orders) whose constituent species had short gestation lengths for their body weights also had earlier ages at weaning and maturity for their body weights (see Fig. 8.1). This seemed to suggest that body weight was not necessarily the key to understanding life history variation in mammals.

If body size was not the key variable to be controlled for in comparative life history analyses, then what was? Attention turned to brain weight and metabolic rate. However, the same partial correlations held and body size was almost invariably a better predictor of life history variation than was brain weight or metabolic rate (see, for example, Trevelyan *et al.* 1990; Harvey & Pagel 1991; Harvey *et al.* 1991).

The variables that had been used in most earlier analyses could be measured from animals in zoos, and the time had come to incorporate extrinsic ecologically determined information. Harvey & Zammuto (1985) used mortality rates from natural populations of mammals to show that life expectation at birth or life expectation at maturity was correlated with age at maturity even when body size was controlled for statistically. Since mortality must on average balance fecundity in natural populations, the fact that mammals with higher mortality rates also have increased fecundity per unit time should not appear too surprising (Sutherland *et al.* 1986). Of interest, however, is that, on further analysis of more detailed mortality data from 48 species of placental mammals, those with high mortality rates for their body size in natural populations had the whole suite of life history characters associated with the fast end of the fast–slow continuum: for their body weights, they had short gestation lengths, early ages at weaning and maturity, short periods from weaning to maturity, and large litters (Promislow & Harvey 1990, 1991). Furthermore, and not surprisingly, mortality correlates with body variation: small-bodied species have higher mortality rates. Perhaps body size may have been acting in part as a surrogate for mortality in earlier

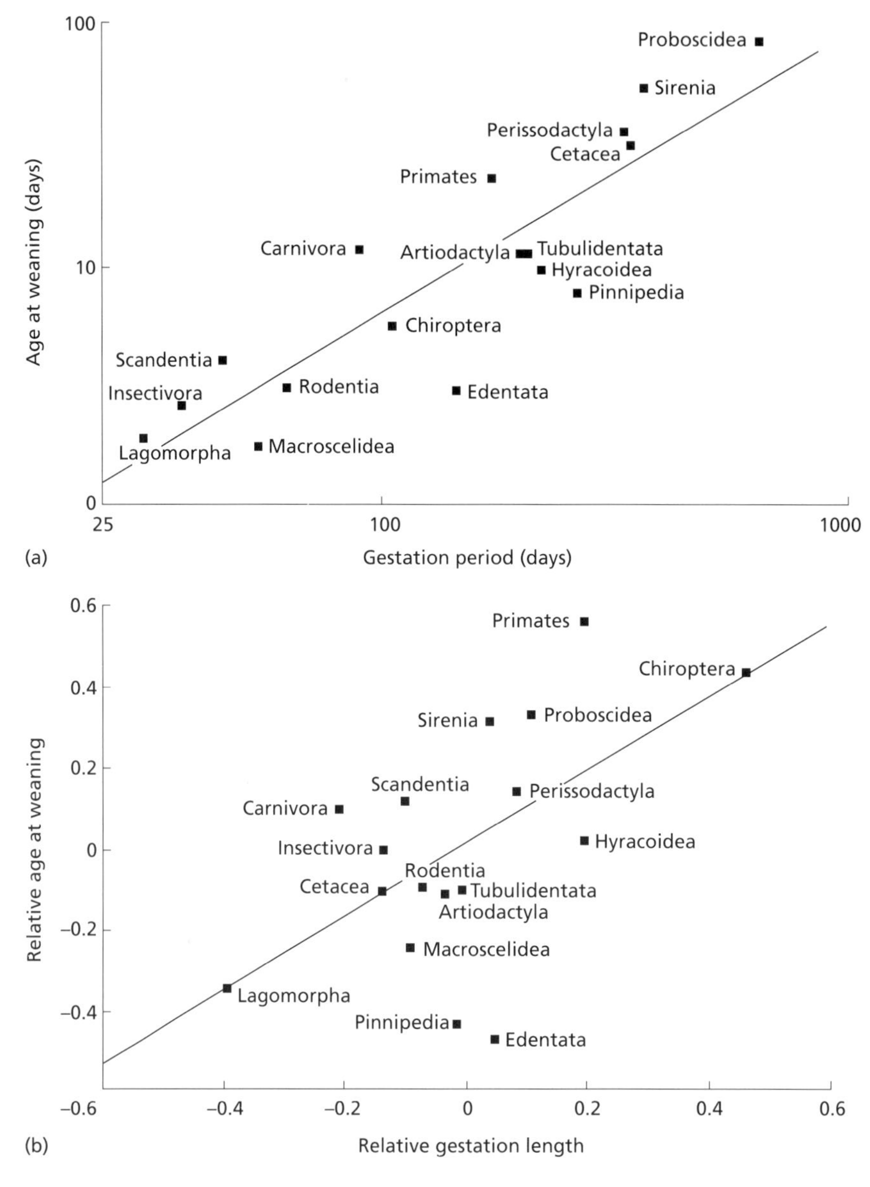

Fig. 8.1 (a) Comparison between gestation peiod and age of weaning across order means for placental mammals. Orders with large-bodied species have longer gestation periods and later ages at weaning. (b) Comparison with relative gestation period and relative age at weaning across order means for placental mammals. Relative values refer to deviations from logarithmic regressions of gestation period and age at weaning on body weight, so that relative values are not correlated with body weight. The positive correlation in (a) persists, but the orders are rearranged along the fast–slow axis.

analysis? If different ecologies impose similar mortality schedules, these findings may help to explain the general failure to find relationships between ecology and life history when size effects are held constant.

The new appreciation of the possible importance of mortality led to piecemeal explanations of particular correlations. For example, it was now known that mammals with long life expectations delay reproduction to beyond an age when other species of the same body size have started to reproduce. If there are mortality costs associated with reproduction, and if reproductive efficiency increases with age, then animals with lower non-reproductive mortality rates should be selected to delay reproduction. In contrast, species with high nonreproductive mortality rates should start breeding as soon as possible because they are less likely to survive to an age at which reproductive efficiency is high. The trade-off here is between delaying reproduction to a time when reproductive efficiency is high and dying while waiting (Harvey *et al.* 1989).

Other suspected trade-offs were refuted by the comparative data. For example, it had been argued that species with a short gestation length would wean their young at a later age than species of the same adult body size with longer gestation lengths. As we have seen, that trade-off between a longer gestation length and earlier age at weaning is totally contradicted by the comparative data. In fact, the relationship between the durations of prenatal and postnatal development is positive whether body size is held constant or not, and it occurs both among and within orders of mammals. It was now known that developmental timings correlated with mortality rates, and thought that this might hold the key to understanding why the periods of prenatal and postnatal development are positively correlated. If a mother dies before her offspring are weaned in the wild, the young will also perish, thus favouring shorter periods of parental investment. If long periods of investment lead to higher-quality offspring, where quality is measured in terms of survival and fecundity, then species with low rates of adult mortality will be selected to invest for longer in their young both prenatally and postnatally. The trade-off here is between slowing down the whole process of development to produce a higher quality young, and dying during the period of investment thus leaving no young at all (Harvey *et al.* 1989).

By 1991, Harvey and Pagel were able to write 'We suggest, then, that the scaling of life history variables with body weight may derive from more fundamental relationships of weight with mortality schedules . . . This approach, of attempting to explain allometric relationships as a function of demographic reality, is, we think, preferable to the largely descriptive role that allometry has played over the years' (Harvey & Pagel 1991, p. 198). It was time to produce an integrated model of mammal life history variation which could explain the correlations and partial correlations which were repeated time and again within independently evolved orders of mammals as well as across the orders as a whole.

The third phase: Charnov's comprehensive and testable model of mammalian life history variation

In the early 1990s, Charnov had developed his interest in life history invariants (Charnov 1993). One way of thinking about the importance of Charnov's approach is to realize that if invariant relationships between life history characteristics can be identified, then the number of parameters in a model of life histories can be reduced. For example, many timing variables such as life expectation at birth, age at maturity and life expectation at maturity each scales with approximately the +0.25 power of body weight while annual fecundity scales with the –0.25 power of body weight (as of course do the instantaneous mortality rates after birth and maturity because they are the inverse of life expectancy). This means that there are only two exponents to deal with, and they often cancel out in the algebra. Products of different variables with shared exponents of opposite sign are constant values (invariant) across species. If the dimensions are in equivalent but reciprocal units, their product will not only be invariant, it will also be dimensionless. For example, the products of age at maturity with juvenile mortality rates, adult mortality rates and annual fecundity are dimensionless invariants. In a series of papers, summarized and extended in a research monograph (Charnov 1993), Charnov brought the full force of his approach to produce a model which aimed to integrate all the various correlations, partial correlations, and power relationships that had hitherto been identified, by reproducing them from a model that had few but realistic assumptions. The basic structure of his model, which considers only female mammals, is straightforward and is summarized as a flow chart in Fig. 8.2 (from Harvey & Nee 1991).

Adult mortality is the driving force behind Charnov's model. The rate of adult mortality is assumed to be extrinsic, meaning that it is a characteristic of the species about which an individual can do nothing by altering its reproductive decisions; causes might be predators or lightning bolts. Different species have different adult mortalities which become a key to understanding variation in their life histories (Fig. 8.3a). If adult mortality is high, then females should start reproduction when they are younger or they might die before they start. What is the benefit to delaying reproduction? Charnov's answer lies in the way mammals grow after they are weaned. In his model, mammals grow rather like money in a savings account (Fig. 8.3b). Their body weight corresponds to the capital, while the interest is the amount of energy that is available for either growth or reproduction. Mammals reinvest the interest and so grow exponentially (Charnov uses an exponent of 0.75)—the

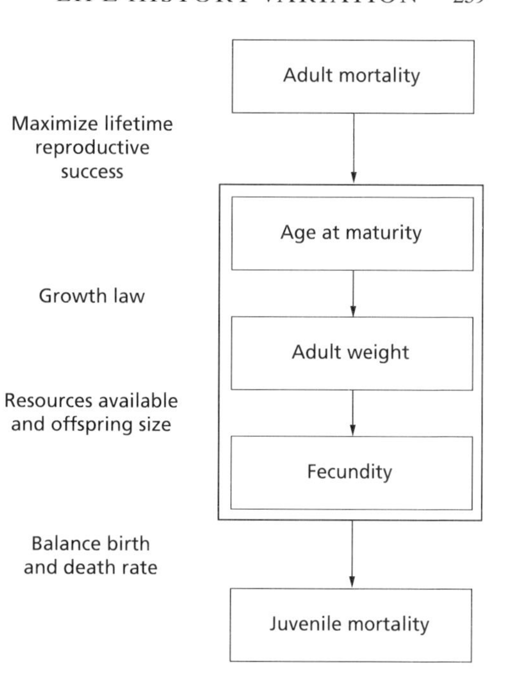

Fig. 8.2 Charnov's view of the causal chain in mammalian life history evolution. Adult mortality is determined by the environment and is not related to body weight. Mammals mature at the age which maximizes lifetime reproductive success. A growth law determines adult size as a function of age at maturity. At maturity, the energy which would have gone into growth is channelled into the production of offspring. In the long term, the birth rate must equal the death rate in natural populations. That equality is ensured by juvenile mortality being density-dependent.

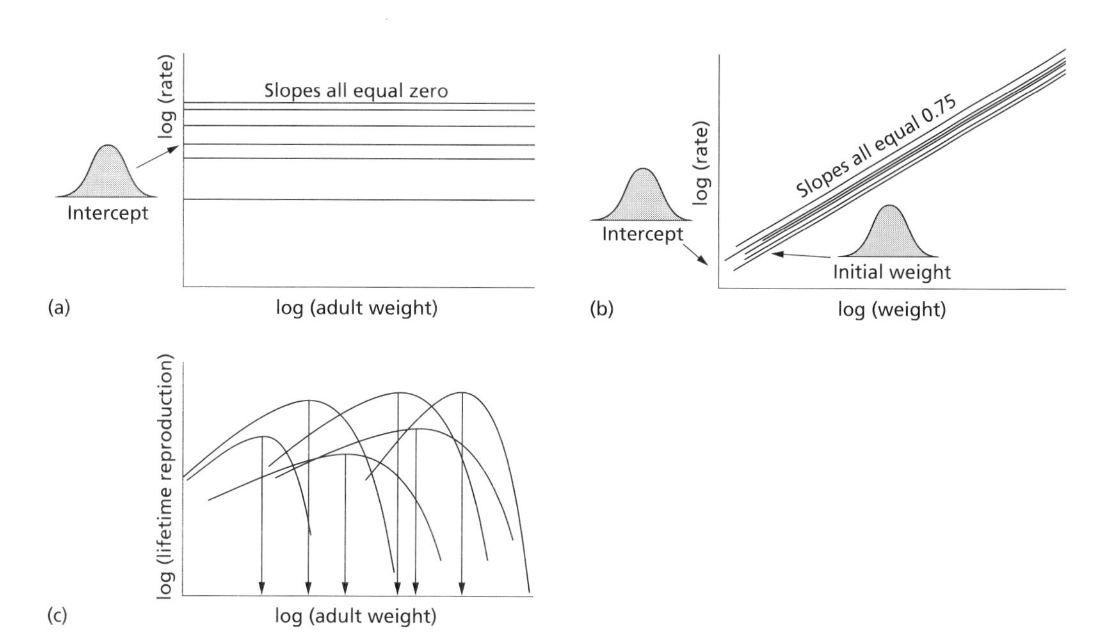

Fig. 8.3 Charnov's model of mammalian life history evolution. Lines represent within-species relationships. (a) Adult mortality rates are set by ecological factors beyond the animals' control, and are independent of body size. (b) Species grow after weaning according to the 'growth law'. (c) Natural selection trades production rate against mortality within each species, producing intermediate optimum sizes at maturity. For further discussion, see text. See also Fig. 8.4.

bigger the animal, the faster it grows—until they reach sexual maturity. At sexual maturity, they switch to spending all of the interest on reproduction, and so do not grow any further. The longer a female delays sexual maturity, the larger she will be, and the more energy she will be able to put into reproduction per unit time. Because growth is exponential, a late developer could be very many times larger than a female who matured early, and could spend correspondingly more on her offspring. Clearly, there is a trade-off between maturing early (high chance of surviving to reproduce, but inevitably low reproductive effort) and maturing late (high chance of dying before reproduction, but survivors can spend massively on reproduction). Natural selection acts to optimize age—and hence size—at maturity, given the extrinsic adult mortality rate (Fig. 8.3c). Whereas adult mortality drives the model, juvenile mortality takes up the slack: density-dependence must occur somewhere in the model if population sizes are to remain stable (which, on the average, they must), and the very youngest age classes are the obvious candidates to bear the brunt.

Extrinsic mortality, then, together with the growth law and the reproductive allocation decision allow Charnov, at least in principle, to define the optimal age at maturity and the optimal adult body weight for a species. In terms of cause and effect, Charnov (1993, p. 97) is explicit: 'This argument does not have adult body size determining mortality rates; quite the opposite—it is mortality rates which determine body size'.

Charnov's model was startlingly good at incorporating what we knew and predicting what we did not. By assuming that growth rates within taxonomic groups all scale with the 0.75 power of body weight, but that growth rates at a particular body weight differ among taxa, Charnov was able to reproduce the fast–slow continuum, the correlations among life history variables (mentioned in the previous section) that hold even after body size is held constant. Under this view, mammals that live slow lives for their body size, like primates, do so because they have relatively slow individual growth rates—these slow growth rates become the phenomenon requiring explanation (Charnov & Berrigan 1993). The first phylogenetically controlled test of one of Charnov's novel predictions (Berrigan *et al.* 1993) supported the model. Later, Purvis & Harvey (1995) performed a more detailed phylogenetic comparative test of many assumptions and predictions of the model, using data from 64 mammal species belonging to nine different orders. They found that the allometric relationships of time from independence to maturity, annual fecundity and adult and juvenile mortality rates were all in agreement with Charnov's model, as were the signs of the correlations between those traits when body size was controlled for. Furthermore, the nondimensional products of age at maturity and each of the previously mentioned traits were independent of adult body size, in line with Charnov's model. The proportion of neonates that survive to reproduce is, counter to intuition but again as predicted, independent of adult body size. However, there were some important discrepancies, all involving the same variable—the ratio of weight at weaning to adult weight (which Charnov terms δ). In order to make some of his predictions, Charnov assumed that δ is independent of adult weight, which turns out not to be the case; in fact it tends to be larger in heavier mammals. And δ was predicted to correlate with several other dimensionless numbers, but did not. Interestingly, δ was correlated with ecological population density independently of body size: taxa with high population densities for their body size also had relatively heavy weanlings. This finding perhaps suggests that the nature of mortality around weaning sets weaning weight (and hence δ): if mortality is largely density-dependent, as is likely where densities are relatively high, it pays mothers to prolong care and increase the size and competitive ability of their offspring. Aside from δ, Charnov's model passed all the comparative tests, leading (Purvis and Harvey (1995, p. 259) to conclude that 'the areas of agreement between Charnov's theory and the data are more impressive than the differences, indicating that it could be a major breakthrough in understanding the evolution of life histories in placental mammals'.

The fourth phase: Kozlowski and Weiner's comprehensive and testable model of mammalian life history evolution

Is Charnov's model the major breakthrough that Purvis and Harvey claimed? Charnov had assumed that weaning weight is proportional to adult weight in order to get many of the partial correlations found in the real data but, as we have seen, that assumption is wrong. Perhaps the model could be tinkered with a little to deal with this problem? But perhaps not. Kozlowski and Weiner (1996) argue that Charnov's model is built upon shaky foundations: they disagree with the existence of a 'growth law', which Charnov assumes, and show that δ must vary with adult weight—as it does, but contrary to Charnov's assumption—if adult weight is to be optimized.

Charnov's 'growth law' is an example of a long tradition in allometric studies: it assumes that the intraspecific allometric growth trajectory is also the interspecific line. Kozlowski and Weiner do not accept that proposition. Without making any assumptions about interspecific allometries, they produce a model which makes very similar predictions to Charnov's. Their model, however, has two clear advantages: it is less restrictive than Charnov's (which can be derived as a special case of theirs) and it makes an additional prediction about body size distributions. Figure 8.4 shows how Kozlowski and Weiner's model works; comparison with Fig. 8.3 shows both similarities to and differences from Charnov's formulation.

Kozlowski and Weiner's model is based on three intraspecific allometric relationships: mortality rate (Fig. 8.4a), assimilation rate (Fig. 8.4b), and

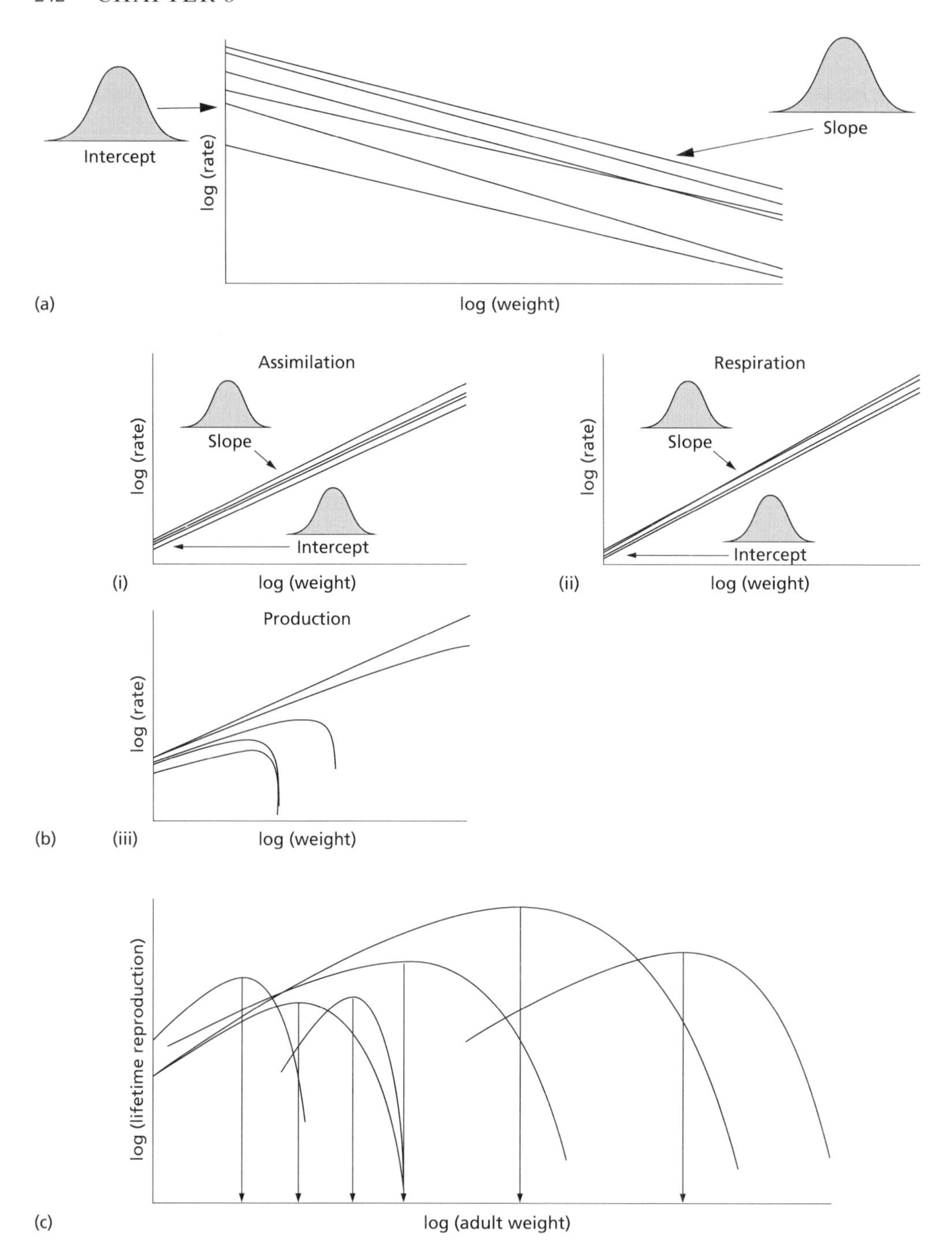

Fig. 8.4 Kozlowski and Weiner's model of mammalian life history evolution. Lines represent within-species relationships. (a) Mortality rate scales with body weight differently in each species. (b) Assimilation and respiration scale differently within each species, so the production rate is not a simple allometric function of body size. Contrary to Charnov's model, all species are weaned at the same size. (c) Natural selection optimizes size at maturity within each species. For further discussion, see text. See also Fig. 8.3.

respiration rate (Fig. 8.4b). Each relationship is described by two parameters: the allometric constant (α) and the exponent (β), in the formula $Y = \alpha W^{\beta}$, where Y is mortality, assimilation or respiration and W is body weight. These three allometric constants and three allometric exponents reflect the species' ecology; like adult mortality rate in Charnov's model, they are not affected by allocation decisions. Given these six parameters, it is possible to estimate the optimal weight at maturity (Fig. 8.4c) using a previous derivation (Kozlowski & Wiegert 1987; Perrin & Sibley 1993). In summary, lifetime reproductive allocation is maximized if an animal stops growing and starts reproducing when the following condition is met:

$$\frac{dP(w)}{dw} E(w) + \frac{dE(w)}{dw} P(w) = 1 \qquad (8.3)$$

where $P(w)$ is production and $E(w)$ is life expectation for an animal that has reached weight w. The derivatives $dP(w)/dw$ and $dE(w)/dw$ measure how fast production rate and life expectation change with body size. In terms of the three original allometric equations, production is measured as assimilation minus respiration, and life expectation is the reciprocal of mortality rate. Biologically, this means that an animal should allocate a particular calorie to growth only if that means future expected reproductive success is increased by more than one calorie. The trade-off is very similar to Charnov's: small animals inevitably have low production rates, so would have low reproductive allocation were they to mature, but time spent growing is time spent flirting with mortality.

At this point it is possible to see two components of Charnov's model that become special cases of Kozlowski and Weiner's. First, Charnov's allometric relationship for production (growth) follows if the allometric exponents for assimilation and respiration are both 0.75. If assimilation and respiration are each allometric functions of body weight with different exponents, then production will not scale as a power of body weight in the manner of Equation 8.1. If, for instance, respiration rate rises more rapidly with body size than does assimilation rate, production rate will not go on increasing indefinitely as an animal grows, but will show a maximum (Fig. 8.4b). Second, whereas Kozlowski and Weiner model adult mortality as an allometric function of body weight (Fig. 8.4a), Charnov assumes that the allometric exponent is zero (Fig. 8.3a) with the result that $dE(w)/dw$ must also equal zero.

Kozlowski and Weiner's model is too complex for analytical solution, so they examined its properties under a series of computer simulations. They considered 100 'species' from a 'taxon'. The simulations set each of the three allometric constants and the three allometric exponents to plausible intraspecific values which are identical for all species except that a separate random, normally distributed variable with a specified variance and a mean of zero is added to each parameter. This procedure mimics the case where phylogenetically related species have similar physiological properties and live in similar environments, but slight differences have accumulated among the species. Those differences arise from many causes—morphological, physiological, ecological or behavioural—and the accumulation of many small differences derived from different causes is well described by a normal distribution. Once the simulations had been completed using Equation 8.3 above to determine when a species should stop growing and start reproducing, each of the 100 species had become characterized by initial and adult body weights, age at maturity, life expectation at maturity, adult assimilation rate, and adult production rate. Given the optimal adult body weight and the known intraspecific allometric constants and exponents, it is possible to estimate all the other variables in that list except initial weight (weight at weaning) which was set at the same value for each species.

Some of the more important initial outcomes of Kozlowski and Weiner's simulations were that:

1 strong interspecific allometric relationships appeared for the rates of assimilation, respiration and mortality with exponents that differed from the intraspecific values which went into the model;

2 production rate was related allometrically to body weight among species, even though the within-species scaling of production rate was not allometric;

3 age at maturity was allometrically related to body weight;

4 life expectation at maturity and age at maturity were positively correlated with each other when body size was held constant—there is a fast–slow continuum; and

5 the interspecific frequency distribution of adult body weights was right skewed under a logarithmic transformation.

These results immediately imply that the search for the meaning of interspecific allometric exponents in terms of functional equivalence may be a misguided endeavour; that strong interspecific correlations among life history variables can occur when body size is held constant, even when body size has been the specific focus of the optimization procedure; and that the skewed lognormal interspecific frequency distribution of body sizes, itself the subject of a burgeoning literature (reviewed by Brown 1995), can follow from a fairly straightforward model in which the key parameters that differ among species are themselves normally distributed.

By holding some parameters constant while varying others, Kozlowski and Weiner were able to produce some analytical results and carry out further simulations that gave insight into some of the underlying causes for the findings.

Why should interspecific allometric exponents differ from the intraspecific ones from which they were derived? Consider a group of species that are identical except that they differ in their allometric exponents for assimilation. Species having larger exponents will, at any given body size, have higher assimilation rates and therefore higher production rates as well. According to Equation 8.3 above, such species will therefore also have larger optimal body sizes. This means that when we plot production rate against body size *across* species, the interspecific line (linking the species' optima) will be steeper than the within-species line. If the differences among species in their assimilation exponent are made greater, then the interspecific line steepens even further. Therefore, the interspecific slopes yielded by Kozlowski and Weiner's model depend not only on the average values that the six parameters take within species, but also on their variances. Recall that in Charnov's model, in contrast, the intraspecific growth (production) exponent was assumed to be identical to the interspecific exponent.

Why do some correlations exist between life history variables when body size is held constant by partial correlation? Consider the known relationship between life expectation at maturity and age at maturity which is known to be positive in natural populations when body size is held constant. The same relationship results from the simulation studies. Why should that be? Size is, of course, normally free to vary as a result of changes in the six parameters that specify the model. However, it is possible for animals with different life histories to have the same optimal size. If size is held constant in Equation 8.3 then a species has a higher life expectation for its size ($E(w)$) and will also have a low productivity ($P(w)$) for its size. A species with a lower productivity has either a lower assimilation rate or a higher respiration rate and, in either event, takes longer to reach the same weight as a species with higher productivity. As a consequence, there will be a positive relationship between life expectation at maturity and time to reach maturity when body size is held constant. This result demonstrates that there is expected to be a 'fast–slow' continuum existing independently of size when the allometric parameter values vary with a random component among species.

Interestingly, one of the few correlations between mammalian life histories and ecology that is found when size is controlled for relates clearly to the fast–slow continuum. Primate species living in tropical rain forests have lower potential reproductive rates for their size than species living in other habitats such as savannah and secondary growth forest (Ross 1987). This was originally interpreted as a consequence of r versus K selection: tropical rain forests are taken to be particularly stable environments inhabited by K-selected populations in which mortality is primarily density-dependent, whereas the other habitats are presumed less stable, normally containing populations that are beneath the carrying capacity for the environment. Kozlowski and Weiner's model assumes a stable population size and shows that the fast–slow continuum can evolve under such circumstances, and so the *r*/*K* perspective may be unnecessary. It remains to be determined what the driving force for lower productivity and presumed lower mortality is in tropical rain forest species of primates, but the new model provides a useful framework within which to approach the problem.

Finally, we turn to a possible error in Charnov's model. He had assumed that weaning weight was a constant proportion of adult weight across species, whereas Kozlowski and Weiner assume that weaning weight is a constant value across species. Neither is a sensible assumption, because both are known to be wrong. While admitting this problem, Kozlowski and Weiner argue that Charnov's assumption is logically flawed. The reason is that a female that delays reproduction will grow larger and have more energy to put into reproduction per unit time, but she will also have to produce larger offspring. As a consequence, the number of offspring she can produce may be reduced if she delays maturity while her chances of producing any at all (because she may die while waiting) will also be reduced. Optimal adult sizes can exist only if weaning size increases with adult size more slowly than the production rate does. In Charnov's model, weaning age increases at the same rate as adult size while production rate increases exponentially with adult size, but the exponent is less than 1. The consequent claim is that no optimal size can be derived under Charnov's model.

Summary

Selection operating on species living in different environments has resulted in the evolution of different life history patterns. Mammals are taken here as a case study to show how theory can be developed and applied to help identify the selective forces operating. The many variables that comprise life histories are highly correlated across species, suggesting that very few general processes have been at work. Initially, body size was seen as a key determinant of life history evolution, although for the wrong reasons. The view that life history differences were adaptations to body size differences took a serious blow when phylogenetically controlled statistical tests revealed that many life history traits were intercorrelated independently of body size (the 'fast–slow' continuum); as a result, body size was for a while viewed almost as a confounding variable and was relegated to the margins. The recent mathematical models focus once more on body size, but they view size as an adaptation to life history, rather than the reverse. These models not only explain the 'fast–slow' continuum, but they can also explain other biological phenomena including body size distributions. The between-species allometries, which first attracted workers to the field, are ironically viewed by the most recent model as mere side-effects; they are neither the lines of functional equivalence, nor the evolutionary trajectories, that they have often been assumed to be.

Acknowledgements

We thank the BBSRC, the Royal Society and the Wellcome Trust for support.

References

Berrigan, D., Charnov, E.L., Purvis, A. & Harvey, P.H. (1993) Phylogenetic contrasts and the evolution of mammalian life histories. *Journal of Evolutionary Biology*, **7**, 270–278.

Brown, J.H. (1995) *Macroecology.* Chicago University Press, Chicago.

Charnov, E.L. (1993) *Life History Invariants: Some Explorations of Symmetry in Evolutionary Ecology.* Oxford University Press, Oxford.

Charnov, E.L. & Berrigan, D. (1993) Why do female primates have such long lifespans and so few babies? or, life in the slow lane. *Evolutionary Anthropology*, **1**, 191–194.

Felsenstein, J. (1985) Phylogenies and the comparative method. *American Naturalist*, **125**, 1–15.

Harvey, P.H. & Clutton-Brock, T.H. (1985) Life history variation in primates. *Evolution*, **39**, 559–581.

Harvey, P.H. & Nee, S. (1991) How to live like a mammal. *Nature*, **350**, 23–24.

Harvey, P.H. & Pagel, M.D. (1991) *The Comparative Method in Evolutionary Biology.* Oxford University Press, Oxford.

Harvey, P.H. & Zammuto, R.M. (1985) Patterns of mortality and age at first reproduction in natural populations of mammals. *Nature*, **315**, 319–320.

Harvey, P.H., Read, A.F. & Promislow, D.E.L. (1989) Life history variation in placental mammals: unifying the data with the theory. In: *Oxford Surveys in Evolutionary Biology* (eds P.H. Harvey & L. Partridge), Vol. 6. Oxford University Press, Oxford.

Harvey, P.H., Pagel, M.D. & Rees, J.A. (1991) Mammalian metabolism and life histories. *American Naturalist*, **137**, 556–566.

Huxley, J.S. (1932) *Problems of Relative Growth.* Methuen, London.

Kozlowski, J. & Weiner, J. (1996) Interspecific allometries are byproducts of body size optimization. *American Naturalist*, **149**, 352–380.

Kozlowski, J. & Wiegert, R.G. (1987) Optimal age and size at maturity in annuals and perennials with determinate growth. *Evolutionary Ecology*, **1**, 231–244.

Lewontin, R.C. (1979) Sociobiology as an adaptationist program. *Behavioral Science*, **24**, 5–14.

Lindstedt, S.L. & Swain, S.D. (1988) Body size as a constraint of design and function. In: *Evolution of Life Histories of Mammals: Theory and Pattern* (ed. M.S. Boyce), pp. 93–106. Yale University Press, New Haven, CN.

McNab, B.K. (1983) Energetics, body size, and the limits to endothermy. *Journal of Zoology*, **199**, 1–29.

McNab, B.K. (1986a) Food habits, energetics and reproduction of marsupials. *Journal of Zoology*, **208**, 595–614.

McNab, B.K. (1986b) The influence of food habits on the energetics of eutherian mammals. *Ecological Monographs*, **56**, 1–19.

Millar, J.S. & Zammuto, R.M. (1983) Life histories of mammals: a life table analysis. *Ecology*, **64**, 631–635.

Perrin, N. & Sibley, R.M. (1993) Dynamic models of energy allocation and investment. *Annual Review of Ecology and Systematics*, **24**, 379–410.

Promislow, D.E.L. & Harvey, P.H. (1990) Living fast and dying young: a comparative analysis of life history variation among mammals. *Journal of Zoology*, **220**, 417–437.

Promislow, D.E.L. & Harvey, P.H. (1991) Mortality rates and the evolution of mammal life histories. *Acta Oecologica*, **12**, 1–19.

Purvis, A. & Harvey, P.H. (1995) Mammal life history evolution: a comparative test of Charnov's model. *Journal of Zoology*, **237**, 259–283.

Read, A.F. & Harvey, P.H. (1989) Life history differences among the eutherian radiations. *Journal of Zoology*, **219**, 329–353.

Roff, D.A. (1992) *The Evolution of Life Histories.* Chapman & Hall, London.

Ross, C.R. (1987) The intrinsic rate of natural increase and reproductive effort in primates. *Journal of Zoology*, **214**, 199–220.

Sacher, G.A. (1959) Relationship of lifespan to brain weight and body weight in mammals. In: *C. I. B. A. Foundation Symposium on the Lifespan of Animals* (eds G.E.W. Wolstenholme & M. O'Connor), pp. 115–133. Little, Brown and Co., Boston, MA.

Sacher, G.A. & Staffeldt, E.F. (1974) Relationship of gestation time to brain weight for placental mammals. *American Naturalist*, **108**, 593–616.

Stearns, S.C. (1992) *The Evolution of Life Histories.* Oxford University Press, Oxford.

Sutherland, W.J., Grafen, A. & Harvey, P.H. (1986) Life history correlations and demography. *Nature*, **320**, 88.

Trevelyan, R., Harvey, P.H. & Pagel, M.D. (1990) Metabolic rates and life histories in birds. *Functional Ecology*, **4**, 135–141.

Western, D. & Ssemakula, J. (1982) Life history patterns in birds and mammals and their evolutionary interpretation. *Oecologia*, **54**, 281–290.

9: Species diversity

M.L. Rosenzweig

Measurement

Most biologists ask questions that deal with objects too small to be sensed directly by humans. So, biologists invent and employ sophisticated devices like microscopes and chromatographs. But students of species diversity did not worry about the adequacy of their senses. They simply watched or plucked or netted or trapped or shot their study objects. They tabulated their results, reported them and went on to the next place, time and taxon. Those days are long over.

Today's diversity investigator, even one dealing with accessible life forms, must confront the fact that diversity is not always what it seems to be. Some species are rare, unlikely to be included in a sample until an infinite amount of work is done. Some are so common and widespread that they obscure diversity differences from place to place. In addition, today we face more sophisticated questions than merely supplying lists. Comparing diversities from place to place and from time to time means confronting and correcting for the unequal sampling conditions of different environments.

The first theoretical challenge of diversity study is thus measuring it reliably. Can we develop methods that will allow us to compare say, the mammal diversity of an open semiarid grassland in the dry season with the moth diversity of a wet–dry tropical forest in the rainy season? How about the grasses of the first with the trees of the second?

The last half-century has provided many answers to this question, although problems do remain. In this section, I shall introduce and evaluate some of these. I shall not treat the issues involved in obtaining better samples; those are issues of instrumentation, not theory. Similarly, I shall not treat the question of 'What is a species?' or whether an investigator has correctly classified all observed individuals. I will assume instead that we have trustworthy data and that they reflect up-to-date, reasonable solutions to such nontheoretical questions. The question that then remains is: after samples are collected and tabulated, what mathematical procedures can we use to determine their true, underlying diversity? I shall symbolize this diversity with the Greek letter zeta (ζ) which resembles an S enough to be mnemonically useful.

Estimators of diversity

In many cases, a flora or fauna is so well worked that we need not worry about sampling problems. For such cases we use the simplest measure of diversity: count the known species. We dignify this simple procedure with the term 'enumeration'.

We do not know when we may use enumeration to estimate ζ without difficulty. Clearly, if samples of a place or time or taxon grow and diversity changes little, we should choose enumeration. But what does the phrase 'diversity changes little' mean? Meanwhile, let us examine some of the more sophisticated measures of diversity, keeping in mind that these measures present some of their own problems. Thus, as our knowledge converges to ζ, enumeration may become the best alternative.

Estimators to reduce bias

When sample sizes are small, S (the diversity of the sample) will also be small regardless of ζ (the underlying diversity). In the limit, a sample size of 1 has one species and yields no information about diversity. But very large sample sizes may also tell us little about ζ when ζ is very large itself and most species are represented by only one or two individuals. All measures of diversity share the need to continue collecting until many species have sample abundances >1.

Although we cannot entirely eliminate the problem of relatively small samples, we can ameliorate it. A number of diversity measures rapidly become unbiased as sample size grows. I will present four: Fisher's α; Simpson's index: Chao's estimator; and capture–recapture. Three of these have variants that I will discuss. An additional diversity measure, rarefaction, equalizes the bias among samples of varying size.

First Fisher's α: by assuming that real distributions of abundances include many rare species unlikely all to occur in any finite sample, Fisher *et al.* (1943) showed that sample abundances should fit a log-series distribution. That is, if the total number of individuals is Υ, and p is a constant proportion, then the most common species has $p\Upsilon$ individuals; the next most common, $p(1-p)\Upsilon$; the next, $p(1-p)^2\Upsilon$; etc. The value of α is fairly robust with respect to the log-series because small, incomplete samples of other distributions also fit the log-series fairly well (Boswell & Patil 1971).

Fisher showed that if abundances do fit the log-series, then S, the number of species in a sample, obeys

$$S = -\alpha \ln(1-x) \quad (9.1)$$

where α is a constant that depends on diversity alone, and x is a variable that depends on sample size. The variable x satisfies:

$$S/N = [(x-1)/x]\ln(1-x) \quad (9.2)$$

where N is the number of individuals in the sample. Because α depends solely on ζ, it provides an unbiased index of it. So, Fisher achieved a partition of sample diversity into a moiety that absorbs the sample size dependency, x, and one that reflects ζ.

To determine α, first solve Equation 9.2 iteratively. Then solve Equation 9.1 algebraically. Use Newton's method. (A computer program to do that is listed in Rosenzweig 1996; or use modern general mathematical software. Note: the listing in Rosenzweig 1995 lacks its subroutines. Old methods using nomograms are obsolete.)

Calculating α requires abundance data. You must know the total number of individuals in the sample. So α cannot be extracted from a mere list of species. But all methods for dealing with the sampling error in measuring ζ require more than a mere list. In fact, in utilizing only total sample size, N, α is the least demanding.

Fisher's α does not return an estimate of ζ. Instead, it is an index of ζ. Nevertheless, it is useful for comparisons. Where α is greater, ζ is greater. Taylor *et al.* (1976) have demonstrated its superior stability and reliability even in cases where the log-series assumption is false.

However, Fisher's α can fail. For example, once a complete list of species exists, α will decline as sample size grows (Rosenzweig 1995). In that case, the same ζ produces unequal α-values. The larger the sample size, the smaller the α. The problem is that the log-series is infinite, never allowing for the possibility of a fully known biota. In other words, at large sample size, the assumption of a log-series abundance distribution matters. Unless it is met, α will not be unbiased. Fisher was aware of this problem and included a test to determine if the data do not fit a log-series well. His test simultaneously calculates the parameter needed to correct α when data do not fit the log-series well (although I know of no case where this has been employed).

Fisher's α has another problem. When sample sizes are very tiny so that most or all species contribute only one individual to N, α has a negative bias. For example, suppose $N = S = 100$. Then $\alpha = 129.5$. For such an assemblage, it is likely that if $N = 25$ (say), $S = 25$ too. Then $\alpha = 32.4$ although the true ζ-value has not changed.

Second Simpson's index of concentration (*SI*): this measure makes no assumption about relative abundances. Instead, for its calculation, it requires the relative abundances of all species (Simpson 1949).

$$SI = \sum((n_i^2 - n_i)/(N^2 - N)) \tag{9.3}$$

where n_i is the abundance of species i.

However, Simpson's index departs even more from ζ than does α. In addition to not returning ζ, it scales inversely to ζ. The higher ζ, the lower SI. That is why Simpson termed it an index of concentration.

The usual remedy to the nature of SI is to calculate its inverse. Thus, Simpson's index of diversity would be $1/SI$. This transformation does have one advantage: it may be read as the 'number of equally common species'. (If all S species in a sample have the same abundance, $n_i = n_j$, then $SI = 1/S$.) But SI exhibits pathological behaviour because of its very high variance when N is low. In fact, when all species have a single individual in the sample, $SI = 0$ and diversity would seem to be infinite.

Kemp (cited in Pielou 1974) invented a better transformation. He defines Simpson's diversity index as $-\ln(SI)$. Pielou shows that Kemp's transformation behaves well. Although it still does not return ζ and lacks the advantage of easy interpretability, Kemp's transformation provides a reliable, unbiased measure of comparative diversity for relatively small samples. Perhaps its biggest disadvantage is that it requires the list of all species abundances, data rarely reported and often not even accurately recorded.

Rarefaction (Sanders 1968), like SI, requires the sample abundances of every species. It admits the effect of sample size on S and calculates the expected S in subsamples of smaller size. Hurlbert's (1971) formula based on the hypergeometric distribution is correct. It permits one to take a set of unequal samples, mathematically reduce them all to the size of the smallest one, and compare their diversities. Heck *et al.* (1975) derive the formula for the variance of the rarefaction estimate. I do not reproduce either the estimator or the formula for its variance here. Although rarefaction was a substantial improvement compared with using raw data, and although far too many have not yet absorbed its vital lesson regarding the sample size issue, I believe that, today, better methods have appeared. Briefly, here are my reasons.

Gotelli & Graves (1996), relying on Smith and Grassle (1977), report that rarefaction is the only measure of diversity that is free of sample-size bias. But rarefaction does not eliminate sample-size biases. It equalizes them by increasing all biases to match the worst bias, i.e. that of the smallest sample.

Rarefaction also assumes an abundance distribution—the multinomial—for all taxa at all sites, but the probability of detecting a species does not scale linearly with sample size. That is why Leitner & Rosenzweig (1997) found that the shape of sample abundance distributions varied with sample size. That is also why the numerical dominance pattern appears to vary regularly with sample size: dominance in a sample decreases as sample size grows. It follows that rarefaction overestimates the number of species in small samples compared to larger ones. The degree of the overestimate depends on the evenness of the whole sample compared with the subsample.

For the same reason, rarefaction does not permit us to extrapolate the diversity from a finite sample to the asymptotic value it would have in a complete sample (Gotelli & Graves, 1996). The extrapolated value would apply to an infinitely larger data set in which the numerical dominance of the most common species is the same as that of the smaller data set in hand. However, in reality, that numerical dominance will most probably be smaller in the complete sample.

Third Chao's estimators (Chao 1984): these measures greatly simplify the task of obtaining abundances because it requires them in only three classes: one, two and many. Chao's first estimator uses the number of species represented in the sample by only one (a), or by two (b), or by more than two individuals:

$$E[\zeta]=S+(a^2/2b) \tag{9.4}$$

A second Chao estimator requires a sequence of samples. It has exactly the form of Equation 9.4, but replaces a with L (the number of species found in only one sample), and b with M (the number of species found in two samples). Chao (1987) derives a formula for estimating the variance of $E[\zeta]$. The Chao estimators depend strongly on sample size until the collection contains about $\sqrt{2\zeta}$ species (Colwell & Coddington 1994).

Colwell & Coddington (1994) compared several methods for unbiased estimation of ζ and found the Chao estimator based on the sequence of samples to be slightly superior to all others. But the data on which they relied include some habitat heterogeneity and thus ζ grows with the number of samples; an ideal test would be based on a fixed ζ.

The fourth diversity measure is capture–recapture methodology: this is usually used to estimate population sizes. However, Rosenzweig & Taylor (1980) following the procedure of Burnham & Overton (1979), applied it to the problem of estimating ζ. Nichols & Pollock (1983) explored another procedure for doing the same. One can note that Chao's second estimator relies on recapture data, but the methods of this section focus on the classical machinery of capture–recapture estimates.

The method (Nichols 1992) requires a sequence of samples. A species observed during the first sample is imagined to be 'marked'. The fraction of species marked is determined during a second independent sample. That already yields an estimate of ζ, but not an unbiased one. If species are not equally easy to record, then the second sample will overestimate the proportion marked. That leads to a likely underestimate of ζ. On the other hand, if some species have left and others have entered the sampling arena, simple capture–recapture will overestimate ζ. The solutions to these biases have split recapture theory into two branches: closed-model and open-model recapture estimation.

Closed-model theory (Burnham & Overton 1979) does not allow for emigration or immigration. However, by using jack-knifing, it does permit each species to have a unique 'capture' probability. Open-model theory, often termed Jolly–Seber technique, assumes each species is equally easy to sample, but allows for dynamics in diversity composition. Each branch has its uses and its experts (Nichols 1992). The Burnham and Overton estimator that most nearly resembles Chao's sequence estimator is given by:

$$E[\zeta]=S+\left[\frac{L(2y-3)}{y}-\frac{M(y-2)^2}{y(y-1)}\right] \tag{9.5}$$

where y is the number of samples in the sequence. Colwell & Coddington (1994) showed that Equation 9.5 does virtually as well as the Chao estimator in estimating diversity, although unfortunately that conclusion, as before, depends on a data set whose ζ-value grows with y. The estimator given by Equation 9.5 depends strongly on sample size until the collection contains about $\zeta/2$ species (Colwell & Coddington 1994).

Capture–recapture methods do not utilize abundances. That reduces the cost of surveys to obtain their data. That also makes them practical for those who work with taxa such as plants or microbes, where it is often difficult to discern where one individual stops and another begins. These methods do not even require that all samples be taken with the same effort or skill (although such is certainly helpful).

Both closed- and open-model recapture estimators make unreasonable assumptions. However, Pollock (1982) has combined them so that each contributes its strength while its weakness is suppressed. 'Pollock's robust method' (for that is what it is called) begins with a set of samples taken at the same time and place. For these, we need not worry about species entering or leaving. Thus, we apply the closed model and obtain an unbiased estimate of ζ. Next, we move or wait awhile and collect a second set of samples. We repeat the closed model procedure on it. We continue, taking new sets of samples in time or space and, using the closed model, estimating ζ for each one. Then we insert our sequential estimates of ζ into the open model. We do not ask it to return ζ, but to estimate additions and deletions. Thus, we obtain the dynamical parameters associated with ζ, parameters that we shall shortly find very valuable.

Nevertheless, Pollock's robust method requires careful sampling design and considerable replicate sampling. I know of few cases where that sort of data has been collected at the species level (and none yet analysed with the method). Perhaps its requirement for so much advance planning and careful sampling will usually make it a poor choice for estimation of ζ. Yet it is tempting to imagine an ecology in which Pollock's method was used commonly. Meanwhile the theoretician can hope that the collective search for methods to better determine ζ will help the empiricist to design better data-gathering schemes.

Diversity and equability

Many diversity indices combine a measure of diversity with another attribute of natural populations, their equability, i.e. the degree of evenness with which their total abundance is subdivided amongst them. One example is Simpson's index. Another is the Shannon–Weiner index, H, or H', formerly very popular, but biased and very insensitive to scarce species:

$$H = -\sum_i p_i \ln p_i \tag{9.6}$$

where $p_i = n_i/N$.

Indices that incorporate equability increase when the number of species increases or when the species become more similar in abundance. All such measures achieve their maxima when their species are equally abundant. The effect of unequal abundances can be dramatic, causing diversity estimates to exchange rank order.

Fisher's α certainly relies on a measure of equability: a key parameter determining α is p, the proportion of abundance in the most abundant species. The higher that proportion, the less equable the species abundances. But equability in α is cryptic because α is not calculated from p; p works its influence on S via the assumption of a log-series distribution. For a given N, the more equable the abundance distribution, the greater is S.

Should we desire an index that incorporates both diversity and equability? Simpson more or less backed into his, whereas MacArthur wanted to incorporate both (MacArthur & MacArthur 1961). He argued that one would certainly not wish to call a place with 99 house sparrows and one robin as diverse as a place with 55 Carolina chickdees and 45 white-breasted nuthatches. Why? Because, in any given sample, we are more likely to observe two species in the more equable place. Notice here that the tail wags the dog. Our observations have sampling biases, yes, but we can virtually remove them. Besides, most indices that incorporate equability do not even correct the sampling bias (e.g. Shannon–Weiner). Ecologists will no doubt wish to study equability in its own right, and the review of Smith & Wilson (1996) will permit just that. But we have no reason to allow equability to adulterate our measures of diversity (Rosenzweig 1995; Gotelli and Graves 1996).

Division into dimensions: point, space and temporal components of species diversity

A combination of three sorts of heterogeneity maintain assemblages of species (MacArthur 1964): some comes from spatial variation, some from temporal variation (such as seasonality) and some inheres at a single point in space-time (such as variation among potential victims). Let us symbolize those components: ζ_3 for space (with its three dimensions), ζ_t for time, and ζ_p for point diversity. The problem of separating ζ into its three components presents fascinating theoretical questions. If we could resolve them, we would open the door to a virtually unexplored set of higher level diversity patterns.

In conceiving the problem, MacArthur used H to define a measure of ζ_p which is useful mostly for territorial species, so I do not review it here. Then he measured H in 2.2 ha plots and compared the former H to the latter H. Their difference measured spatial diversity. Other than mentioning it, he ignored time.

Whittaker (1972) also approached the problem without dealing with time. That may seem curious because Whittaker, like most plant ecologists, was acutely aware of the influence of time in such matters as plant succession.

However, Chesson & Huntly (1993) have focused on models that partition the sources of diversity maintenance into contributions from temporal heterogeneity and the rest. They are less interested in teasing pattern out of data than in understanding what sorts of responses to temporal variation can help to maintain diversity. Thus, their equations sweep all nontemporal components of coexistence into a single term while finely subdividing the temporal one into different scales and two roots. I shall return to their work in the section on diversity and temporal heterogeneity (p. 269).

Disentangling point diversity

MacArthur's method was not popular, but Whittaker's inspired considerable effort. Whittaker named the components of diversity α (which is ζ_p), β (which is ζ_3) and γ (which is ζ). In measuring α, he simply established very small plot sizes and hoped that they would be spatially homogeneous. (His standard small plots were 1 m^2, and I have seen some field data of his that had him experimenting with 0.01 m^2!) He ignored the problem of small sample sizes that perforce accompany such tiny areas.

Cleverly, however, instead of defining β as an amount of diversity, he treated it as a rate, that is, the rate of accumulation of species as one moves horizontally (and instantaneously) across a landscape. That suggested the equation $\alpha\beta = \gamma$ (Whittaker 1972). But this equation presents problems of spatial dimensionality. Point diversity has no dimensions, whereas β has one, two or even three. β could be 'rate of diversity increase as we sample more points in a straight line'; it could also be 'rate as we sample more space'. However, it usually involves increase as area is added. So, it usually has two dimensions.

Its very dimensionality suggests a better form for the relationship:

$$\gamma = \alpha + \beta A \tag{9.7}$$

where A is the area being considered.

Now, α is a true point diversity, i.e. the number of species to which diversity declines as area vanishes. Meanwhile, the units of β will be (new) species per unit area. As divisor, the units of area produce the required units in the product, βA, i.e. 'species'. We could accomplish the same thing with either distance or space, but then both the actual value of β and its units would change. So, we see that β must be carefully defined.

In fact, the literature contains many definitions of β, and most are not simple to relate to β as a rate of increase. Perhaps the most prevalent are definitions that calculate the average change in species composition from place to place. Cody (e.g. in 1993) has provided continually interesting examples of attempts to disentangle the components of diversity using this concept. Naturally, the value of β will vary with the distance chosen—see Cody (1993, Fig. 13.9) for a powerful, albeit differently interpreted example. Considering their variation with distance, one might suspect that such values cannot have much general significance. Nevertheless, values of β that vary with distance may come closer to the truth and be more informative than any presumed constant value, as we shall shortly see.

Equation 9.7 is a straight line with intercept α and slope β. But it also resembles a large number of ecological patterns called 'species–area curves' or SPARs (Rosenzweig 1995). SPARs that start from small areas are never linear, but they can be approximated with a power equation and thus linearized in logarithmic space:

$$\log S = c + z \log A \tag{9.8}$$

Equation 9.8 looks temptingly like Equation 9.7, but the logarithms do make a difference. In fact, z is not exactly the slope of the power equation, but its exponent. And c has more to do with the slope of the power equation than anything else; c is certainly not the intercept of the power equation, and thus it is not α.

So far, I have used Whittaker's symbols for historical reasons. But for clarity, I must now return to the symbols involving ζ. We write ζ_p for Whittaker's α, lest we confuse it with Fisher's α (which, being older, takes precedence). We write ζ_3 for β, because β has too many definitions to be clear enough. And we write ζ for γ, because γ does not incorporate temporal diversity.

Now we rewrite Equation 9.7 as a power equation. This will fit vast amounts of data and give us something like our desired theoretical curve (still without its temporal component however):

$$\zeta = \zeta_p + CA^z \qquad (9.9)$$

Based on that equation, I suggested (Rosenzweig 1995) that determining ζ_p would be a simple matter of performing nonlinear regression on the data. I was wrong. As the data descend to relevantly small areas, sample size becomes a very serious problem. Indicated diversity plunges and imposes a convex upward shape to known SPARs in logarithmic space. Instead, they should become concave upward and, over very tiny areas, practically horizontal (Leitner & Rosenzweig 1997).

We can address the problem by combining methods. Obtain replicates of tiny areas until jack-knifed capture–recapture or Chao's estimator is free of its sample size bias. Repeat on larger and larger areas until sample knowledge from simple lists of flora and fauna take over the function of the estimators with apparent reliability. That requires a subjective decision, but it is of little consequence. As areas grow even larger, their true diversities will more and more likely be revealed by such lists, and have more and more influence on the overall regression. Finally, perform the regression using Equation 9.9.

An alternative combination would use Fisher's α to assess the area that could safely represent a single point. Over areas smaller than that, α does not rise with area (Williams 1964; Rosenzweig 1995). Estimate α for a series of growing areas and judge the area at which α first begins to rise. That is the critical area. Now obtain ζ for a series of larger areas. Regress the data with Equation 9.8. The regression's estimate of ζ over the critical area is ζ_p. In following this method one must be sure that enough individuals appear in the smallest area so that many of the species have more than one individual.

Area, A, and the coefficients of the fitted Equation 9.8 or Equation 9.9 measure the contribution of spatial heterogeneity to ζ; $\zeta_3 = CA^z$. Thus, we see that we cannot achieve our goal of determining a single value of ζ_3 for a taxon in a province. The pattern is not linear and operation of a single ζ_3 would require that it would be. Instead, MacArthur had it right from the first. The contribution of spatial heterogeneity to diversity can be reduced to a single number for the expansion of space from a single point to a fixed area, but not to a variable one. Proceeding to a variable area requires two coefficients. Would this conclusion hold if we expanded the sample along a line instead of by increasing its area?

The temporal component

Measuring the temporal contribution to diversity is trivial, as long as we do not worry about actually doing it. Just measure ζ_p and sit. Record new species as they appear. The result will be a species–time curve analogous to a SPAR (Rosenzweig 1998). If it is linear, its slope will be ζ_t. If not, we will at least have measured it and know the pattern of its diminishing return, as we do now for spatial heterogeneity.

Still, it takes time to measure ζ_p. The longer it takes, the more likely we will have included a moiety of temporal heterogeneity in our results. In fact, we can be sure that the measure of ζ_p obtained above does include species that depend on temporal heterogeneity, heterogeneity as fine as variation within the day. Moreover, even if we knew ζ_p exactly, how would we know whether newly observed species would be the result of newly encountered temporal habitats, or just additions to the list of the species that already belonged in ζ_p but had not previously been recorded?

The answer seems to be to search for ζ_t by applying the same methods used to obtain ζ_3. Measure diversity during a sequence of times in a fixed location. Use the jack-knife or Chao estimator or α on small time segments. Then perform the appropriate regression. The small time intervals must measure the effect of diurnal variability. But the problem is the same as that associated with wringing out the droplets of spatial heterogeneity from ζ_p. A good natural historian should be able to do it.

It is now apparent that our estimate of ζ_p using different areas was not a true ζ_p. That ζ_p includes a component of temporal heterogeneity. Likewise, the ζ_p estimate using different time intervals will include a component of spatial heterogeneity. Perhaps we can again combine methods. The problem is inherently three-dimensional:

$$\zeta = \zeta_p + \zeta_3 + \zeta_t \tag{9.10a}$$

$$\zeta = \zeta_p + CA^z + Dt^w \tag{9.10b}$$

where t is time interval and D and w are the parameters of time's effect analogous to C and z for space. To use Equation 9.10b, we could establish a series of sampling quadrats of varying area, keep lists for each, applying good ζ-estimators where needed, and watch for an extended period. The multiple nonlinear regression of these data with Equation 9.10b would yield the answers, but that is a very expensive, impractical solution. Instead, we express ζ_p as a difference:

$$\zeta_p = \zeta - CA^z - Dt^w \tag{9.11}$$

and estimate it from data taken to measure the effects of area and time separately. These data will be easier to obtain.

Preston (1960), however, did not wait for them. He boldly conjectured a sort of ergodic principle with respect to diversity: increase of diversity in both space and time will fit power equations, and $z = w$. He did not offer a proof. At one scale, Preston must be correct. If small sample sizes dominate our measure of diversity, then it matters little whether we have collected them in time or space. Perhaps that is all Preston meant. Yet, in his discourse, he imagines very large amounts of space, and time extended for evolutionary intervals so that species have been entirely replaced. That is why previously, I have taken Preston at his word (Rosenzweig 1998). I looked for data sets to test his conjecture. I even analysed one suitably large period (Phanerozoic time, i.e. the past 530 million years) of time and space (the entire world's marine invertebrates) with nonlinear multiple regression models to attempt to disentangle time from space. It is too soon to be sure, but a few things do suggest themselves.

First, a power regression fits the data well in every case. But this is hardly impressive theoretically: many processes result in power equations, and even processes that do not, often produce patterns that are well fitted by power equations.

Second, the z-value characterizing space comes rather close to being the same as that of time at large scales of either one. In each case, z is near 1.0. This is easy to predict for time (Rosenzweig 1998), but no one yet knows how to predict it for space. Meanwhile, there are not enough data to measure the temporal pattern at intermediate scales (such as a few centuries to a few million years)—palaeobiologists please note; theoreticians, too, because we need more than data. We need the models that can help us disentangle time from space and suggest the likely processes that have produced the patterns.

Dynamic processes

The magnitude of species diversity, like that of any other state variable, may depend on three rates: the rate of species production; the rate of extinction; and, perhaps, the amount of time those rates have been in effect. Reducing a variable to rate processes is the classic device by which scientists turn capricious history into science. Einstein pointed that out, and though he himself overturned Newton's whole paradigm of the universe, he revered Newton for being the first scientist to model natural processes as differential equations.

The glory of being first to see that species diversity must reflect the differential equations that produce and destroy it belongs to MacArthur & Wilson (1963). But their theory dealt with island diversities—a special case—and so I will postpone its consideration until later in this section. I begin instead with the largest scale, the biogeographical province. Such a province may be defined as a region whose species arose internally. That is, none immigrated after originating in some other place. This definition is idealized; it allows us to construct theory, but no actual region has zero immigrant originations.

Do we need to know diversity's initial conditions and how long it has been evolving? Probably not. Data teach us that steady states often, perhaps usually, characterize diversity (Rosenzweig 1995). Some palaeobiological examples show diversities that wobble along trendlessly for 10^5 to 10^8 years. Palaeobiological data also show the rapid return of typical diversities after mass extinctions. After losses of 65–95%, diversity bounces back within a few million years (or less). This pattern of rapid recovery, repeated in taxon after taxon and era after era, suggests that diversity must be self-regulating. So, we should ask what features of speciation and extinction might be sensitive to the number of species itself. Such features will serve as feedback variables. If they turn out to be negative feedback variables, they can explain what brings diversity under control.

The negative feedback variables: population size and geographical range

Imagine a newly colonized province or one not yet recovered from a mass extinction. As diversity grows within it, many variables will change. But two in particular, population size and geographical range, have strong effects on speciation and extinction rates. Population size is the product of population density and geographical range, but we shall see that the latter should have effects of its own, separate from those of population size itself. I will first explain why population density and geographical range should change. Then I will examine how such changes ought to impact speciation and extinction rates.

Diversity influences average population and geographical range

Because energy flow and the resource base of ecosystems are limited, populations cannot continue to grow exponentially for long. Eventually they begin to fluctuate (whether regularly, chaotically or randomly) and no longer exhibit a trend of long-term increase. Similarly, they cannot forever expand their geographical ranges—eventually they occupy all parts of a province that their anatomies and physiologies will allow them to.

When a new species is added to a province, it may interact with an already established population. (If does not, we can ignore it.) If it does interact, that interaction could be mutualistic or negative. Mutualistic interactions will lead to increases of the population density or the range size of the established species. On the other hand, the negative interactions, competition and predation, have the opposite effects. Each of these may reduce an established species' density. Each may also reduce its geographical range.

When provincial diversity is very low and various resources are unexploited, the effects of mutualism on populations and ranges may actually dominate those of competition and predation. However, as new species enter and exploit those resources, the negative interactions must rule. A new species is most likely to reduce the population density and range of an established one. As we will see, were this not so, the mutualism would lead to positive feedback in speciation and extinction rates. Thus, we could not explain the steady states in the fossil record. So, we must focus on the consequences of competition and predation: beyond some threshold value of ζ (which may be zero), increasing diversity tends to shrink populations and ranges.

Population size and extinction rate

The relationship between population and extinction depends on a stochastic notion (see Chapter 2): if one individual, before reproducing, accidentally dies with probability p, then—assuming independence—all individuals will die with probability p^N. Since $p < 1$, the probability of extinction gets vanishingly small fairly quickly (MacArthur 1972). Richter-Dyn & Goel (1972) incorporated negative density-dependence into the birth and death rates, but that did not qualitatively change the outcome. Rare species should experience terminal accidents more readily than common species.

Leigh (1981) reformulated the entire question by considering not individual statistics, but those of the population as a whole, i.e. its mean growth rate and the variance of that mean. (Both those statistics can be modelled as density-dependent functions.) Goodman (1987) divided that variance into two components. One is intrinsic to the individual and so is independent among individuals. The other, environmental variation, is not independent; what happens to one individual, tends to happen to all.

Whereas individual variation does little to change the dependence of extinction probability on N, environmental variation weakens it considerably. Large population sizes contribute much less protection when all are subjected simultaneously to some major, debilitating environmental anomaly. Nevertheless, environmental variation does not reverse or eliminate the relationship. Larger N means smaller extinction rates. Data tend to support this idea, especially data from islands (which are characterized by speeded up rates making observation practical). A commonly studied example is that of the birds of small islands off the British coast (Rosenzweig & Clark 1994).

Range size and extinction rate

Suppose the only idea behind the theory of extinction was the notion of the simultaneous independent death of all individuals in a population. Then the only influence we would need to build into our theories would be the effect of total population size. Geographical range itself would have no independent effect at all. But once we add the idea that accidents can happen to entire populations, range becomes a crucial factor.

Most species constitute a metapopulation: they are loosely subdivided into demes, i.e. semiseparate subpopulations. The effect of metapopulation structure on extinction rates is a popular topic among ecologists. Gotelli (1995) crisply summarizes its essence. Gotelli's general model:

$$df/dt = p_i(1-f) - p_e f \tag{9.12}$$

where f is the fraction of occupied deme-sites, p_i is the immigration rate to each empty deme-site, and p_e is the rate of extinction of an occupied deme (as opposed to global species' extinction). The latter terms may be elaborated to fit various biological processes (e.g. Hanski & Gilpin 1991; Gotelli & Kelley 1993). A particularly sophisticated investigation (Allen *et al.* 1993; Rosko *et al.* 1994) finds that chaotic population dynamics reduces extinction rate in a metapopulation even if the chaos leads to frequent periods of rarity for individual demes. The chaos reduces synchrony among demes and thus the probability that all will vanish simultaneously.

Despite their complexity and variety, one conclusion of metapopulation theories at first seems robust enough to handle any elaboration: the more demes, the less global extinction. If deme size, deme density in the landscape, and the terms of Equation 9.12 do not vary with range size, then larger ranges will have more demes and lower rates of global extinction. That conclusion is what this chapter requires. Yet I am uneasy about the assumptions. For example, it seems quite likely that species with larger ranges would be less subdivided but have larger demes with smaller p_e. Thus, as ζ grows, ranges would shrink and add demes. However, these demes would have lower f and higher p_e values tending to counteract the increase in subdivision. Under what conditions would the smaller f-values and larger p_e-values dominate the effect of more demes? That is the nub of the relationship between diversity and the role of metapopulation structure in modifying extinction rates.

Some evidence already supports the conjectures of the previous paragraph. Maurer & Nott (1998) tested the correlation between range size and fragmentation in 242 species of insectivorous North American birds. They used an ingenious measure of the fractal dimension of species' ranges. They maintain that the more jagged the outer boundary of a range, the more fragmented it is. Certainly, larger ranges turned out to be much more jagged.

There is further work to be done. Deme extinctions may not be independent events. A weather disturbance that exterminates one deme is very likely to get neighbouring ones too. But because weather anomalies also have a frequency distribution of geographical extents, a distant deme undoubtedly suffers a much smaller risk of extermination from that same weather disturbance. I know no models of that process.

Whatever theories we eventually settle on, some palaeobiological data

(Jablonski 1986) already indicate that species with larger ranges have smaller extinction rates. We cannot be sure that this empirical relationship is independent of population size, because we cannot disentangle the effects of the range from those of population size. But these data say that even if large ranges *per se* tend contrarily to promote extinction, the effect is not sufficiently large as to overwhelm the general relationship.

Population size and speciation rate

Traditional population genetics works at a scale that may not be very useful for the modelling of speciation rates. It concerns itself with the replacement of one genetic entity by another within a species. More useful is the concept of 'genostasis'. Genostasis (Bradshaw 1984) describes a population that has run out of good new genetic ideas. Such a population stops evolving. What does this have to do with speciation rates?

When a geographical barrier isolates a population, successful divergence depends, in part, on whether the isolates can speciate before the barrier vanishes. In genostatic populations, the speed of the divergence corresponds to the rate at which the population develops new beneficial mutations. Total beneficial mutation rate correlates positively with population size.

A model by Walsh (1995) suggests the great importance of that correlation. Walsh relies on a widely held position originated by Haldane (1932): most novel genes result from modifications to duplications. But duplicate genes tend to damage fitness. That sets up an evolutionary race: will the duplicate gene be neutralized so that it does no harm? Or will it first evolve a new function and be available to improve fitness and promote speciation? The answer depends crucially on population size.

Walsh stipulates a population size for the species. He gives the inactivation process a mutation rate. He does the same for mutation to a new function with a small selective advantage, but sets this rate at only a tiny fraction of the previous one (10^{-5} or 10^{-6}). In one model, the advantage is fixed; in another, it comes from a distribution of values. The results for both these models are similar: small populations have virtually no chance of producing new genes. But with beneficial mutations only 10^{-5} as common as neutralizing mutations and producing benefits with only a 1% selective advantage (say), a population of 5×10^6 has a 0.5 probability of turning a duplicate into a new beneficial gene, and a population of 10^7 has a probability of 0.75 to do so. Moreover, the probability is a sharply sigmoid function of population size: only a narrow range of population sizes leads to intermediate probabilities.

So, populations much less than 10^5 or 10^6 may have a hard time evolving new genes. Their evolution is likely to be limited to reshuffling their gene frequencies or producing a rare new beneficial allele at an old locus. Perhaps that tells us why some small isolated populations appear to have such relictual gene pools, while their widespread, common sisters—which have had no more time to evolve—have diverged at a rapid rate (Rosenzweig 1995).

Range size and speciation rate

The geographical barriers competent to initiate allopatric speciation are like the poet's arrows: shot into the air to fall to Earth one knows not where. A barrier must fall properly across the range of a species to begin the process. The larger the range, the bigger the target and the more likely a 'random arrow' will encounter it.

Topologically, geographical barriers comprise only two groups: open curves and closed curves. The latter ('moats') constitute the outline of some two-dimensional figure, regular or not. The former ('knives') have two ends. Should a moat happen upon a range, it splits it. Thus, the probability of a random moat producing isolates is a monotonically increasing function of range size. Knives, however, are not so simple.

To produce isolates, a knife must find a range and transect its perimeter at two points, or more. Otherwise, individuals can maintain panmixia by moving around an end of the barrier. The larger a range, the more improbable its transection by a random knife. So, very large ranges are easy to find but difficult to split. Because the joint probability is a product, both large and small ranges should have relatively low rates of isolate production. Thus, knife barriers should generate a unimodal probability function that peaks over geographical ranges of intermediate size.

Theory suggests that, at least on land and probably also in the oceans, only the rising arm of the unimodal probability function matters (Rosenzweig 1978). If knives are segments of great circles with a uniform distribution of lengths and orientations, and range perimeters are small circles, the probability function peaks over a range amounting to the area of all the continents combined. Therefore, even for knives, the probability of a random barrier producing isolates is a monotonically increasing function of realistic range sizes.

Evolutionists know of several speciation processes besides geographical speciation. However, I believe these have rates governed by those of the geographical process (e.g. allopolyploidy) or else that their rates approach zero once diversities exceed a very small fraction of their steady states (e.g. competitive speciation) (Rosenzweig 1995).

Synthesis: the roots of steady states

All the elements needed for negative feedback are present. Diversity increases should tend to reduce both population sizes and geographical ranges. Smaller populations become extinct more rapidly and are quite unlikely to produce the novel genes needed for speciation. Smaller ranges probably also subject a species to a higher extinction probability and a reduced isolate formation rate. All in all, as diversity grows, its production slows and its losses grow. No wonder it is characterized by steady states.

Island biogeography

Compared to provincial theory, island biogeography dwells at a rapid temporal scale. We must carefully define that scale and the theory's variables. Its dynamics involve the arrival and disappearance of species that have already evolved. Thus its time intervals: on the order of 10^1 to 10^5 years (depending on the taxon). Thus also its assumption of a fixed overall diversity called the species pool. All members of the pool, ζ, live in a location called the 'mainland'. The questions asked include: How many species will also be found on an island? What is the dynamic status of an island's diversity?

In addition to the state variable itself, S_i (i.e. the number of species on island i) there are two rate variables. The immigration rate of species not now on an island is I_i. The extinction rate of species from the island is E_i. Now come some easy deductions.

1 Both I_i and E_i must be feedback variables connected to the state variable.
(a) When $S_i = \zeta$, I_i must be zero. When $S_i < \zeta$, I_i must be greater than zero. Thus, $I_i = I_i(S_i)$, and $dI_i(S_i)/dS_i$ must tend to be negative.
(b) On the other hand, when $S_i = 0$, E_i must be zero. When $S_i > \zeta 0$, E_i must be greater than zero. Thus, $E_i = E_i(S_i)$, and $dE_i(S_i)/dS_i$ must tend to be positive.

2 Hence, the two rate functions must intersect over at least one value of S_i. At that value, S^*, the island's diversity is at equilibrium. Further, at S^*, $dI_i/dS_i < dE_i/dS_i$; thus S^* is a steady state.

Notice that neither population interactions nor density dependence yet appear. MacArthur & Wilson (1963) incorporated them as second derivatives of the functions $I_i(S_i)$ and $E_i(S_i)$. They reasoned that the second derivative of $I_i(S_i)$ should be positive because species with higher immigration rates should tend to precede others. They reasoned that the second derivative of $E_i(S_i)$ should also be positive because new species should tend to reduce the average population size of species on the island, thus increasing the average extinction rate.

Diversity differences among islands

With their theory, MacArthur and Wilson explained why larger islands have more species than smaller ones: at any value of S_i, population sizes should be larger on a larger island. So, the $E_i(S_i)$ of a larger island should be less than that of a smaller one. Furthermore, the larger island is likely to comprise more habitat types. Absence of a habitat leads to reduction of $I_i(S_i)$ because an immigration event is defined as the arrival of a propagule of a new species. However, a propagule implies successful reproduction of the species, so it must have found suitable habitat (Rosenzweig 1995).

MacArthur and Wilson also predicted that islands closer to the mainland should have more species than more remote islands: at any value of S_i the $I_i(S_i)$ of a closer island should exceed that of a more remote one. This prediction has generated considerable controversy. No one doubts its mathematical accuracy, but some data challenge its prediction (e.g. Schoener 1976). However, a new analysis (Rosenzweig 1995) shows that even these data sets agree with the prediction.

Island theory also predicts the relative turnover rates of diversity: turnover rate should be inversely proportional to A_i, where A_i is the area covered by island *i*. But data appear to contradict this prediction. Unfortunately, however, the data are collected at discrete intervals (regular or irregular) whereas the prediction concerns instantaneous rates. The difference between them can be substantive and decisive (Rosenzweig 1995). Nevertheless, the data do support the theory's assumption that larger islands should have lower $E_i(S_i)$ curves. From there, the prediction of lower turnover is a rather trivial step.

Small-scale dynamics

Within small patches of a mainland, processes of origination outpace even those of islands. Individuals of many species move among patches and provide a ready source of replacement for locally extinct species. Such individuals can even rescue a local population otherwise doomed. We have already visited metapopulation theory, which is one outgrowth of these processes. In this section, I will summarize two other types of theory that are even more associated with issues of small-scale diversity. The first of these is the combined role of disturbance and population interactions (Levin & Paine 1974; Paine & Levin 1981; Caswell & Cohen 1993). The second is termed 'mass effects' (Shmida & Ellner 1984).

Disturbance

Community ecologists often study population dynamics to ask whether various combinations of competitors, predators and victims can coexist (see Chapter 5). That use remains valid. Nevertheless, using population dynamics by itself to predict diversity assumes that the dynamics have vanished, i.e. that all the state variables have reached equilibrium. No ecologist is happy with that assumption.

Ecosystems are usually interrupted on their way toward equilibrium.

Few may ever reach it. Even those that will occasionally reach it may spend most of their time away from it. Whether biotic or abiotic, such an interruption is called a 'disturbance'. The theory of the disturbance pattern has three component processes, two negative and one positive.

1 Species may disappear slowly from a patch by competition and predation.

2 Species may disappear suddenly from a patch because a disturbance sweeps them away. (Note: the disturbance may be a predator; predators that can coexist stably with some of their victims act as slow agents in process **1**, above; predators that cannot, and must constantly move from a patch after decimating it, act as disturbances.)

3 Species also move into the patch and replenish it.

These dynamic processes produce a distribution of diversities among patches of a fixed habitat type. The mean diversity of the distribution and its variance depend upon the frequency of disturbances in the habitat type.

1 In a habitat with frequent disturbances, the average patch accumulates only a few species before a new disturbance interrupts the process of accumulation. So, most such patches are depauperate.

2 In a habitat with an intermediate disturbance frequency, many species accumulate before disaster strikes. However, disturbance still comes often enough to interrupt the process of competitive exclusion. The average patch is quite diverse.

3 In a habitat with infrequent disturbances, competition and predation often drive systems close to local equilibria; diversity is lost. So, the average patch will be mature and depauperate.

Combining **1–3** above, we obtain a unimodal relationship between disturbance and diversity. Disturbance theory has enjoyed some powerful field confirmations (Lubchenco 1978; Sousa 1979; Petraitis *et al.* 1989). Nevertheless, it must be carefully approached. Its assumptions and goals direct it toward rapid-scale processes, not evolutionary ones. Therefore, it must not be extended to longer time scales without careful justification.

Mass effects

The presence of a species in a provincial patch depends—even at steady state—on the balance between births, deaths, departures and arrivals. Classical theories ignore departures and arrivals, however. They predict coexistence without taking into consideration the heterogeneity of the landscape. In particular, they do not allow a patch to have a species unless its births therein balance or exceed its deaths.

More realistically, Shmida & Ellner (1984) allow a species to occupy a patch if an excess of arrivals compensates for a birth–death imbalance. They term such species 'sink species' in that patch. Species whose births balance or exceed their deaths in the patch are its 'source species'. Of course, the excess arrivals of a sink species come from some place where it is a source species. In other words, some species may have no sink populations, but all must have source populations.

Temporal heterogeneity may also produce sink species, temporal sink species. Such a species' population would be dwindling everywhere now, but, at some other time—at least somewhere—it would grow. Migrants between temporal populations travel only to the future, which may make it easier to model their effects and detect them.

Source–sink dynamics can only add to a patch's diversity. That extra diversity is called the mass effect (Shmida & Ellner 1984; Shmida & Wilson 1985), but purely spatial sink populations cannot exist at the levels of whole provinces or whole islands. Somewhere the species must achieve a positive steady-state population size. Of course, temporal sink populations may well exist in whole provinces or in islands. A species now present may not be able to sustain itself in the current set of habitats. It owes its existence to a time gone by. It owes its future to the reoccurrence of that time (Rosenzweig 1998).

The difficulty of obtaining good local population dynamics has prevented widespread confirmation of the existence of sink populations. However, Keddy (1981) has demonstrated them for *Cakile edentula*, a dune annual; Kadmon & Shmida (1990) did the same for a grass, *Stipa capensis*.

Most plants have little vagility and are thus somewhat constrained to accept existence in a sink population. Perhaps then, only species with fairly immobile individuals have mass effects. However, Pulliam (1988) and Pulliam & Danielson (1991) conclude that individuals of mobile species such as birds may actually improve their fitness by settling in a sink population if source populations are scarce. Therefore, mobile taxa should also experience mass effects.

Diversity and temporal heterogeneity

Though we have long suspected that temporal habitat differences could support diversity, we had very little formal theory to help us understand how. Pioneering contributions include Stewart & Levin (1973), Armstrong & McGehee (1976) and Abrams (1984). Chiefly owing to the influence of Chesson, the past decade has seen a marked increase in attention to this question.

Chesson's theories depend on the concept of *nonlinear averaging*, first introduced to ecology by MacArthur (1968). Imagine any function $f(x)$. If it is linear, then $[f(x_1) + f(x_2)]/2 = f[x_1 + x_2)/2]$. In other words, the average value of a function equals the function of its average argument. However, if the function is nonlinear, those averages will differ.

Now imagine that the function is the reproductive rate of a species with density x in environment e: $f(x_e) = r_e$. Since such functions are nonlinear, we cannot determine the average rate by calculating the rate at the average population size or in the average environment. Instead, the average will have two components: the rate at the average condition and the deviation contributed by the nonlinear response to the temporal variation. That deviation is the source of temporal variation's power to add to diversity.

Two forms of comparative nonlinearity can help to prevent loss of diversity (Chesson & Huntly 1993): *subadditivity* and *relative nonlinearity*. Each requires the construction and comparison of reproductive rate functions that depend on competition with other species and on the state of the abiotic environment.

To determine subadditivity, we compare the effect of r on competitors in different temporal environments. If that effect is severe when the environment is good, but weak when the environment is poor, we term the combined effects subadditive. To enhance diversity, subadditivity must occur with two other relationships: substantial covariance between competition and environmental quality; but less covariance for species when they are at low density compared with high density. The combination of elements leads to the *storage effect*: gains in a good environment at low density carry over to times of high competition or poor environmental conditions. The greater the storage effect, the more the nonlinear averaging helps support diversity.

To determine relative nonlinearity, we compare two species' r-responses to changing conditions. If they differ in shape (i.e. the signs of their second derivatives differ) or in magnitude of curvature, the two are relatively nonlinear. Temporal variation may help or hinder the coexistence of relatively nonlinear species (Chesson 1994).

Comparative theory

A perfectly general, process-driven theory of species diversity would operate at all scales and make precise quantitative predictions, but such a theory remains a distant goal. Instead, we have a suite of pattern-driven theories (i.e. phenomenological theories). They are comparative, predicting the trends in diversity as conditions change in space and time. They identify the processes responsible for the trends. And they also help us advise conservation biology.

Species–area curves

Diversity grows with the area of a sample. Once, theory thought it had successfully predicted this pattern: a curve of the form $S = CA^z$ with an exponent of about 0.26 (Preston 1962; May 1975). But the prediction fell before the challenge of data (Connor & McCoy 1979; Connor *et al.* 1983). Although the form of the curve works rather well, real z-values run the gamut along the unit interval. Before focusing on z-values, let us turn to the processes that produce species–area curves in the first place.

Species–area curves compiled in two ways

In a nested species–area curve, a core area is sampled and then built upon. As the area of the sample grows, so, of course, must the species it contains. This merely shows that most species do not live everywhere. Yet, the exact form of this curve is not trivial. We shall revisit it below.

Most species–area curves come from separate provinces or separate islands or separate, nonoverlapping samples of a province. It is not so trivial to wonder what causes their positive slopes.

The theories on pp. 273–274 provide the explanations for both island and interprovincial species–area curves. As outlined above, larger islands should have lower extinction functions and higher immigration functions. And, as discussed, larger areas lead to lower extinction rates and higher speciation rates for each species. It is but a small and logical step from there to the realization that at the same ζ-value as a smaller province, the species of a larger province will generate more speciation and less extinction because they have larger ranges and populations sizes.

In the case of patches of a single province, however, the theoretical explanation for species–area curves is not yet well recognized. We do have a tested and true empirical explanation: larger samples of a province have more habitats (Williams 1943; Rosenzweig 1995 and the references therein). But if we theoreticians try to adopt this explanation uncritically, we run straight into a quantitative contradiction.

Suppose we take a sample area, A_q, from two provinces with a similar mixture of climate types and geological conditions. If area merely samples habitats, then we expect the same number of habitats in the two samples. That ought to lead to the same number of species. It does not. The A_q sample from the larger province has more species than the A_q sample from the smaller province. We can even go so far as to match habitats or take A_q from a single macrohabitat in each province. Still, there will be more species in the sample from the larger province. The explanation lies in the nature of habitats and the hierarchy of the processes determining diversity.

To a great extent, ecologists recognize habitats by looking at species distributions. If the list of plant species did not change as one ascends from 1000 m to 3000 m in the nearby mountains, I would not teach my students that the Catalina Mountains have four life zones. I would not know how to distinguish one elevational habitat from another. We even name most of those habitats after the species most common in them: pinyon–juniper; ponderosa pine; spruce–fir. How do those habitats come to be? The answer lies in understanding what limits species' distributions along any environmental axis. Why are species' niches circumscribed? We do not entirely know. Yet, we do recognize some general principles connected with spatial heterogeneity, and these suggest pathways toward an answer.

Some edges of distribution develop because a species falls prey to interactions with other species in parts of its fundamental niche. From this alone, we predict that if diversity were to decline, some surviving species would spread beyond their current ranges.

Physiological limits cause other edges. These would not change immediately after a diversity reduction. But we suspect that many limits evolved when a species was prevented by another from using some part of niche space, and thus prevented from being subject to natural selection in it. Take away the interactive constraints and, in evolutionary time, some of the physiological constraints may also disappear. The argument is even easier to appreciate in the reverse direction: add interactive species which reduce a distribution, and natural selection may force the species with the reduced distribution to abandon any costly physiological adaptations that helped to suit it to the abandoned parts of its niche. Whether considering physiology or interactions, we can see that the number of habitats life recognizes must correlate with species diversity.

Now we can explain the difference between S_q, the number of species, in equal areas, A_q, of unequal provinces. We view the evolution of diversity as a hierarchy of processes. Each province with a similar mixture of environments evolves a steady-state diversity depending on its size. At this diversity, extinction and speciation work equally slowly. In particular, extinction works too slowly to relieve the interactive pressures of competitors and predators. Species specialize and occupy fractions of niche space that correlate inversely with the steady state. In other words, the more species, the more habitats in the whole province. Because a given point in real space corresponds to only one point in niche space, habitat extents also must vary with diversity. In short, the more species, the less area each habitat tends to occupy. It is as if the province were paved with tiles called habitats, and larger provinces had smaller tiles. Thus, a sample area, A_q, from a province will have a diversity that correlates with the number of habitats it contains. But a large province will contain more habitats in A_q than will a small province.

Yet, the answer points to an unresolved question. Granted that the tiles should get smaller as diversity grows within a province, but why, at steady state, should a larger province have smaller tiles than a smaller province? What is the logical difficulty in predicting that the steady-state tiles of the larger province will be larger than those of the smaller province? If they are larger, then we would have predicted more species in the smaller province within the area A_q. We shall just have to leave this issue for the future. There is, however, a main point we can decide on: tile size should depend on total diversity within a province. So, theory does not contradict the data that ascribe the provincial species–area relationship to habitat sampling.

A gaggle of scales

Though z-values vary, they are far from random. The essential clue to their variation comes from noticing that those connected with mainland samples of varying area have values less than predicted while those that describe the curvature of interprovincial comparisons have values above 0.6. In between, we see values associated with islands of different size.

When we think about the different processes that control diversity, such variation seems less surprising. In retrospect, it is difficult to understand how we could have expected that a state variable determined by several scales of rate processes operating simultaneously should produce the same quantitative result despite dramatic differences in the influence of those processes. How can a diversity determined to a great extent by infrequent immigrants be expected to match one full of the mass effects implicit in the rapid relocation of individuals among semi-isolated patches of a region? And how could one whose new species come only from the slow processes of speciation match one that gets it 'new' species ready-made from a large pool on its donor mainland? The questions already suggest their solution: consider a fixed area whose diversity is set by one of the several temporal scales providing originations.

Consider a province with diversity ζ, a sample of it with area A_q, and an island of area A_i to which the province donates species. $A_q = A_i$ and they have the same mixture of habitats.

Now imagine a differential equation for the species S for each of the areas that incorporates all the rate processes:

$$dS/dt = O(S) + I(S_{old}) + I(S_{new}) - E(S) \tag{9.13}$$

where $O(S)$ is speciation rate, $I(S_{old})$ is immigration of species already present ('old' species), $I(S_{new})$ is immigration of locally new species, and $E(S)$ is extinction rate.

Although Equation 9.13 applies at all scales of time and space, its terms vary greatly in importance. For example, on both islands and in mainland patches, we may rely on $O(S)$ being negligible compared with immigration. Origination will be dominated by its immigration terms. This helps us compare islands with their donors.

In A_q, the sample of the donor area, extinction is local and rapid, but so is immigration. That allows some species with locally deficient demographics (average $R_0 < 1.0$) to persist. These are the spatial sink species. We may expect that the predominating immigration term will be $I(S_{old})$.

However, one can define a pure biogeographic island formally as an area with rare immigrations and no speciations (Rosenzweig 1995). So, on the island of area A_i, $I(S_{old})$ will be too low to compensate for $R_0 < 1$. That is why islands have no spatial sink species. Somewhere on the island, each of its species must achieve a positive steady-state population size owing solely to its local demography. Because all its species have average $R_0 = 1.0$, extinction also slows. The result is lower steady-state turnover.

However, the most interesting effect of these rate considerations is on the size of the steady state itself. At steady state, the island must support fewer species. It has all the source species of A_q but none of its spatial sink species. That explains the often noted fact that an island of area A_i does indeed have fewer species than a same-area patch of its mainland source pool (Rosenzweig 1995). It also tells us that the z-value of the island must be greater than that of the donor sample.

I have presented islandness as a discrete state where no spatial sink species live. But of course, some geographical islands are so near their donor pools that many spatial sink species of highly mobile taxa can usually be found there. Furthermore, some taxa are so mobile (e.g. ferns with wind-borne spores) that, for them, no place is an island. Holt (1993) presents a detailed model that treats the transition from mainland sample to island as a continuum.

We can also compare our island to another area lacking spatial sink species. We will compare it with a distant province that does not donate species to the island nor receive species from any source pool. Again, we assume the two have the same size, A_i, and the same mixture of environment types. But on the island, immigration dominates originations, whereas on the distant province, speciation does. So, we cannot expect the little province to maintain as much diversity as the island. The time scale of speciation is slow compared to immigration and we have assumed nothing that would lead to a change in the extinction function, $E(S)$.

Having a depressed diversity means that a line from the province's point in $\log A$–$\log S$ space to that of the island's donor province will have a steeper slope than that of the line joining the island's point in $\log A$–$\log S$ space to that of its donor province (Fig. 9.1). In other words, the province has a higher z-value than the island; interprovincial species–area curves ought to be steeper than those from islands.

Now recall from the previous section that, compared with nearby islands, more remote islands have lower steady-state S and higher z-values than nearby islands. Thus we expect a continuum in z-values among islands as immigration rates decline. Islands thus supply a bridge between mainland z-values and interprovincial z-values. As origination rates descend from those characterizing the demes in a metapopulation to those characterizing provinces, z should increase continuously. The data agree. The coefficient, z thus links the several scales of origination rate as no other parameter has succeeded in doing.

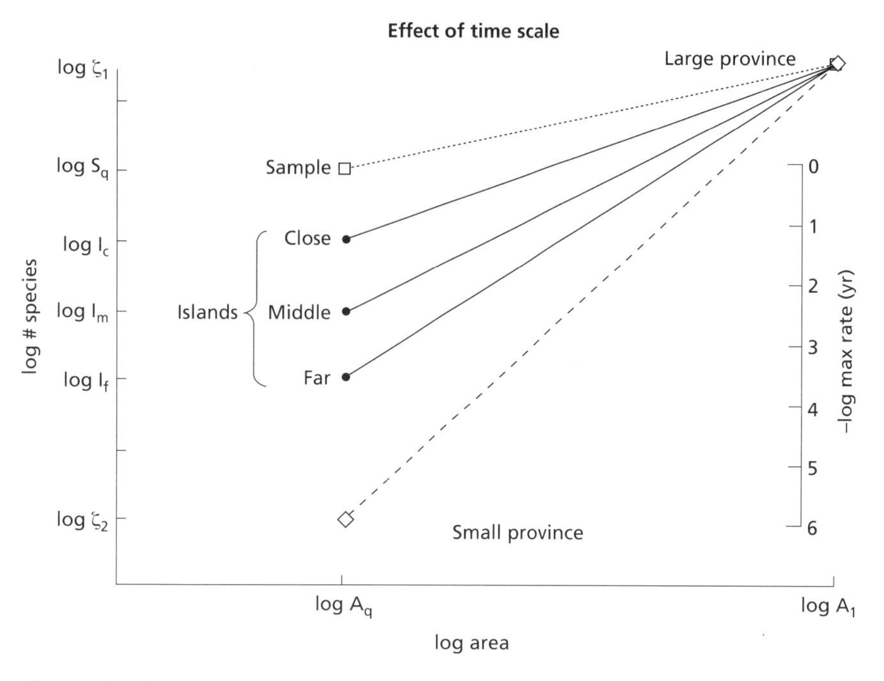

Fig. 9.1 The effect of time scale on species diversity. As the maximum origination rate rises from the order of 10^{-6} to 1, the diversity of an area, A_q, also rises. Sample quadrats of a large province are replenished annually (sometimes even faster), contain sink species and have the highest diversities. Isolated small provinces are almost never replenished; their origination rates come from internal speciation. They have the least diversity. Islands fall in between.

Productivity and local species diversity

The diversity contained within a sample patch of area A_q varies widely. Partly this must arise from noise in the number of habitats it contains. But at least one regular local pattern discloses that more than noise is at work: diversity is often unimodal with respect to local energy flow rates (Tilman & Pacala 1993). Is there a theoretical reason why patches of either low or high productivity should contain fewer habitats than one of intermediate productivity? One dynamic model suggests a pervasive reason (Tilman 1982, 1988). Tilman makes several assumptions.

1 All species require a set of resources; none can survive on only one. These resources may be nutrients, or a combination of light intensities and nutrients, or temperature and nutrients.

2 Competing species vary in the segments of niche space to which they are adapted.

3 All species have the same niche breadth (or ecological tolerance, see Rosenzweig 1991).

4 Tolerances of species do not depend on productivity.

5 Regardless of their average productivity, all equal-area local samples have the same habitat variance measured objectively. That is, they span the same amount of niche space.

From these assumptions, Tilman produces the well-known graphical theory that predicts the pattern. One may complain that the assumption set is very restrictive. In fact, it is. But if species that correspond to the assumption set are the only ones biasing the relative diversity of a local patch on the basis of its productivity, then those species drive the signal. Other species only attenuate the relationship and diminish its R^2.

Note that this section treats a pattern different from, although superficially similar to, the regional unimodal pattern (Rosenweig & Abramsky 1993). I know of no theory to handle that pattern that I would wish to include in here.

Abundance curves and diversity: the reduction of two mysteries to one

R.A. Fisher thought that we could reduce the problem of diversity to the problem of the species abundance distribution. As we have seen in the first section, he suggested that species abundances are well fitted by a log-series distribution. If we could characterize the parameters of that distribution, the total diversity, ζ, would be a simple consequence of that distribution and the total number of individuals in the area being studied.

Of course, one might complain that this approach solves nothing. Even if we accept the assumption that total abundance scales nicely with total area or energy or their product (Wright *et al.* 1993), Fisher's idea merely reduces two mysteries to one. We should still have to predict the abundance distribution. But science often advances by showing that two seemingly unconnected questions are really one. So, we should press ahead.

Log-normal distributions and diversity

Preston (1948) agreed with Fisher's basic idea, but not with the specific suggestion he made for the underlying distribution. Preston put forward a substitute. He suggested that abundance distributions are log-normal. Indeed, starting from the log-normal distribution, first Preston (1962) and then May (1975) developed a theory that claimed to predict both the form and the z-value of species–area curves. May went so far as to aver that log-normal abundance distributions are among the best confirmed patterns in ecology. But we now know that serious problems confront log-normal theory.

First, there are empirical problems. We have already seen that the z-values of species–area curves run a logical, theoretically explicable gamut along the unit interval. Moreover, both Hughes (1986) and Gaston (1994) show that the fit of the log-normal to real abundance distributions has never been adequately tested. Hughes (1986) looked at 222 different sets of abundances and found that only 65 (28%) showed a mode. Of course, that absence also characterized most of the data Preston worked with. In the absence of a mode, one cannot even calculate a log-normal distribution, let alone test it.

Second, Leitner & Rosenzweig (1997) built a simulation engine to determine whether the log-normal does in fact make the predictions claimed for it by Preston and by May. It does not. Leitner and Rosenzweig imagined a province in which species abundances do fit a log-normal. Its species were assigned square ranges in proportion to their abundances and those ranges were distributed at random in the province. Then they dropped square sampling quadrats of varying size on the virtual province to count the number of species in different areas A_q. The result was a z-value of 0.77.

Finally, Leitner and Rosenzweig approached the problem analytically. They predicted that the expected number of species S_q in an area A_q should be:

$$S_q = \xi_Q + 2\sqrt{q}\,l_Q + qp_Q + 1 - p_Q \quad (9.14)$$

where Q is a threshold range size such that species with ranges as large as Q are certain to be sampled, p_Q is the proportion of species with ranges smaller than Q, ζ_Q is the average range size of such species and l_Q is the average length of the ranges of such species. Equation 9.14 duplicated the result of the simulation exactly, indicating that the older theories had concealed some heretofore unnoticed problem. There are actually two.

Both Preston (1962) and May (1975) assumed that the form of the abundance distribution remains log-normal regardless of the proportion of the province sampled. But the simulations show that the abundance distribution depends on A_q. It gets more left skewed as A_q decreases.

The second problem comes from the use of a Taylor-series expansion for log(S) in terms of log(A) to estimate z about the point (A_T,ζ) where A_T is total provincial area (May 1975). But Taylor-series expansion requires the uniqueness of the expansion at the point it is made. Unfortunately, there is no unique expansion about $\{\log(A_T), \log(\zeta)\}$.

Both older theories relied on the assumption of canonicity. Preston conceived this assumption as follows. He invented a companion curve to the log-normal called the individuals curve. This shows the frequency of individuals in species of various abundances. Then he proposed that the individuals curve peaks over the most abundant species in the species curve. May improved the theory's robustness by inventing a parameter γ which allows the position of the individuals curve to vary with respect to the species abundance distribution. Leitner and Rosenzweig wondered if they could salvage the utility of the log-normal by varying γ. Unfortunately, over a broad range of reasonable γ-values, the value of γ has little to do with the value of z.

Hanski distributions

Hanski (1982) suggested that widespread species tend also to be common where they are found, whereas narrowly distributed species tend also to be rare where they are found. In contrast, implicit in previous theory was the assumption of no correlation between local abundance and range size. (Leitner and Rosenzweig had maintained this implicit assumption in their first simulations.) Data support the Hanski hypothesis (e.g. Maurer 1994), especially the scarcity of dense and narrowly distributed species. So, Leitner and Rosenzweig decided to add it to their model to see if it allowed the z-values predicted by the older theories.

No equation has yet been empirically fitted to real data on ranges and population densities. So, Leitner and Rosenzweig chose a flexible one for the Hanski distribution:

$$\xi_i = (\rho_i / M)^c \tag{9.15}$$

where ξ_i is the range size of species i; ρ_i is its population density; and M is the density of the densest species. As its parameter, c, varies from 0 to 1, Equation 9.15 will match any curvature from severely convex upward, to linear, to severely concave upward. When $c = 1$, density does not depend on range size; so Equation 9.15 can also match theory that does not incorporate an explicit Hanski distribution. (NB The c of Equation 9.15 is not the same as the c of Equation 9.8.)

A Hanski distribution with $c = 0.12$ combined with the log-normal abundance distribution proved sufficient to yield z-values in the range of 0.26. But this still does not prove that log-normal abundance distributions are necessary. In fact, Schoen and Rosenzweig (unpublished) have combined Hanski distributions with several other abundance distributions (e.g. log-series, Poisson, uniform) to show that, with the right choice of c, these also yield z-values in the range of 0.26. Hence, we must doubt the existence of either a necessary or a sufficient connection between abundance distributions and species diversity. Perhaps, once we have some knowledge of real c-values, a connection will become more apparent.

References

Abrams, P. (1984) Variability in resource consumption rates and the coexistence of competing species. *Theoretical Population Biology*, **25**, 106–124.

Allen, J.C., Schaffer, W.M. & Rosko, D.J. (1993) Chaos reduces species extinction by amplifying local population noise. *Nature*, **364**, 229–232.

Armstrong, R.A. & McGehee, R. (1976) Coexistence of species competing for shared resources. *Theoretical Population Biology*, **9**, 317–328.

Boswell, M.T. & Patil, G.P. (1971) Chance mechanisms generating the logarithmic series distribution used in the analysis of number of species and individuals. In: *Statistical Ecology* (eds G.P. Patil, E.C. Pielou & W.E. Waters), **vol. 3**. Pennsylvania State University Press, University Park, PA.

Bradshaw, A.D. (1984) The importance of evolutionary ideas in ecology—and vice-versa. In: *Evolutionary Ecology* (ed. B. Shorrocks). Blackwell Scientific, Oxford.

Burnham, K.P. & Overton, W.S. (1979) Robust estimation of population size when capture probabilities vary among animals. *Ecology*, **60**, 927–936.

Caswell, H. & Cohen, J.E. (1993) Local and regional regulation of species–area relations: a patch occupancy model. In: *Species Diversity in Ecological Communities* (eds R.E. Ricklefs & D. Schluter). University of Chicago Press, Chicago, IL.

Chao, A. (1984) Non-parametric estimation of the number of classes in a population. *Scandinavian Journal of Statistics*, **11**, 265–270.

Chao, A. (1987) Estimating the population size for capture–recapture data with unequal catchability. *Biometrics*, **43**, 783–791.

Chesson, P.L. (1994) Multispecies competition in variable environments. *Theoretical Population Biology*, **45**, 227–276.

Chesson, P. & Huntly, N. (1993) Temporal hierarchies of variation and the maintenance of diversity. *Plant Species Biology*, **8**, 195–206.

Cody, M.L. (1993) Bird diversity components within and between habitats in Australia. In: *Species Diversity in Ecological Communities* (eds R.E. Ricklefs & D. Schluter). University of Chicago Press, Chicago, IL.

Colwell, R.K. & Coddington, J.A. (1994) Estimating terrestrial biodiversity through extrapolation. *Philosophical Transactions of the Royal Society of London*, **B345**, 101–118.

Connor, E.F. & McCoy, E.D. (1979) The statistics and biology of the species–area relationship. *American Naturalist*, **113**, 791–833.

Connor, E.F., McCoy, E.D. & Cosby, B.J. (1983) Model discrimination and expected slope values in species–area studies. *American Naturalist*, **122**, 789–796.

Fisher, R.A., Corbet, A.S. & Williams, C.B. (1943) The relation between the number of species and the number of individuals in a random sample of an animal population. *Journal of Animal Ecology*, **12**, 42–58.

Gaston, K.J. (1994) *Rarity*. Chapman & Hall, London.

Goodman, D. (1987) The demography of chance extinction. In: *Viable Populations for Conservation* (ed. M.E. Soule). Cambridge University Press, Cambridge.

Gotelli, N.J. (1995) *A Primer of Ecology*. Sinauer Associates, Sunderland, MA.

Gotelli, N.J. & Graves, G.R. (1996) *Null Models in Ecology*. Smithsonian Institution Press, Washington, DC.

Gotelli, N.J. & Kelley, W.G. (1993) A general model of metapopulation dynamics. *Oikos*, **68**, 36–44.

Haldane, J.B.S. (1932) *The Causes of Evolution*. Cornell University Press, Ithaca, NY.

Hanski, I. (1982) Dynamics of regional distribution: the core and satellite species hypothesis. *Oikos*, **38**, 210–221.

Hanski, I. & Gilpin, M. (1991) Metapopulation dynamics: brief history and conceptual domain. *Biological Journal of the Linnaean Society*, **42**, 3–16.

Heck, K.L. Jr, Van Belle, G. & Simberloff, D. (1975) Explicit calculation of the rarefaction diversity measurement and the determination of sufficient sample size. *Ecology*, **56**, 1459–1461.

Holt, R.D. (1993) Ecology at the mesoscale: the influence of regional processes on local communities. In: *Species Diversity in Ecological Communities* (eds R.E. Ricklefs & D. Schluter). University of Chicago Press, Chicago, IL.

Hughes, R.G. (1986) Theories and models of species abundance. *American Naturalist*, **128**, 879–899.

Hurlbert, S.H. (1971) The nonconcept of species diversity: a critique and alternative parameters. *Ecology*, **52**, 577–585.

Jablonski, D. (1986) Background and mass extinctions: the alternation of macroevolutionary regimes. *Science*, **231**, 129–133.

Kadmon, R. & Shmida, A. (1990) Spatiotemporal demographic processes in plant populations: an approach and a case study. *American Naturalist*, **135**, 382–397.

Keddy, P.A. (1981) Experimental demography of the sand-dune annual, *Cakile edentula*, growing along an environmental gradient in Nova Scotia. *Journal of Ecology*, **69**, 615–630.

Leigh, E.G. Jr (1981) The average lifetime of a population in a varying environment. *Journal of Theoretical Biology*, **90**, 213–239.

Leitner, W.A. & Rosenzweig, M.L. (1997) Nested species–area curves and stochastic sampling: a new theory. *Oikos*, **79**, 503–512.

Levin, S.A. & Paine, R.T. (1974) Disturbance, patch formation, and community structure. *Proceedings of the National Academy of Sciences USA*, **71**, 2744–2747.

Lubchenco, J. (1978) Plant species diversity in a marine intertidal community: importance of herbivore food preference and algal competitive abilities. *American Naturalist*, **112**, 23–39.

MacArthur, R.H. (1964) Environmental factors affecting bird species diversity. *American Naturalist*, **98**, 387–397.

MacArthur, R.H. (1968) Selection for life tables in periodic environments. *American Naturalist*, **102**, 381–383.

MacArthur, R.M. (1972) *Geographical Ecology*. Harper & Row, New York.

MacArthur, R.H. & MacArthur, J.W. (1961) On bird species diversity. *Ecology*, **42**, 594–598.

MacArthur, R.H. & Wilson, E.O. (1963) An equilibrium theory of insular zoogeography. *Evolution*, **17**, 373–387.

Maurer, B.A. (1994) *Geographical Population Analysis*. Blackwell Science Ltd, Oxford.

Maurer, B.A. & Nott, M.P. (1998) Geographic range fragmentation and the evolution of biological diversity. In: *Biodiversity Dynamics* (ed. M.L. McKinney). Columbia University Press, New York.

May, R.M. (1975) Patterns of species abundance and diversity. In: *Ecology and Evolution of Communities* (eds M.L. Cody & J.M. Diamond). Belknap Press of Harvard University Press, Cambridge, MA.

Nichols, J.D. (1992) Capture–recapture models. *Bioscience*, **42**, 94–102.

Nichols, J.D. & Pollock, K.H. (1983) Estimating taxonomic diversity, extinction rates, and speciation rates from fossil data using capture–recapture models. *Paleobiology*, **9**, 150–163.

Paine, R. & Levin, S. (1981) Intertidal landscapes: disturbance and the dynamics of pattern. *Ecological Monographs*, **51**, 145–178.

Petraitis, P.S., Latham, R.E. & Niesenbaum, R.A. (1989) The maintenance of species diversity by disturbance. *Quarterly Review of Biology*, **6**, 393–418.

Pielou, E.C. (1974) *Population and Community Ecology: Principles and Methods*. Gordon & Breach Science, New York.

Pollock, K.H. (1982) A capture–recapture design robust to unequal probability of capture. *Journal of Wildlife Management*, **46**, 752–757.

Preston, F.W. (1948) The commonness, and rarity, of species. *Ecology*, **29**, 254–283.

Preston, F.W. (1960) Time and space and the variation of species. *Ecology*, **41**, 785–790.

Preston, F.W. (1962) The canonical distribution of commonness and rarity. *Ecology*, **43**, 185–215.

Pulliam, H.R. (1988) Sources, sinks and population regulation. *American Naturalist*, **132**, 652–661.

Pulliam, H.R. & Danielson, B.J. (1991) Sources, sinks, and habitat selection: a landscape perspective on population dynamics. *American Naturalist*, **137**, S50–S66.

Richter-Dyn, N. & Goel, N.S. (1972) On the extinction of a colonizing species. *Theoretical Population Biology*, **3**, 406–433.

Rosenzweig, M.L. (1978) Geographical speciation: on range size and the probability of isolate formation. In: *Proceedings of the Washington State University Conference on Biomathematics and Biostatistics* (ed. D. Wollkind). Department of Mathematics, Washington State University, Pullman, WA.

Rosenzweig, M.L. (1991) Habitat selection and population interactions: the search for mechanism. *American Naturalist*, **137**, S5–S5.

Rosenzweig, M.L. (1995) *Species Diversity in Space and Time*. Cambridge University Press, Cambridge.

Rosenzweig, M.L. (1996) *Species Diversity in Space and Time*, revised edn. Cambridge University Press, Cambridge.

Rosenzweig, M.L. (1998) Preston's ergodic conjecture: the accumulation of species in space and time. In: *Biodiversity Dynamics* (ed. M.L. McKinney). Columbia University Press, New York.

Rosenzweig, M.L. & Abramsky, Z. (1993) How are diversity and productivity related? In: *Species Diversity in Ecological Communities* (eds R.E. Ricklefs & D. Schluter). University of Chicago Press, Chicago, IL.

Rosenzweig, M.L. & Clark, C.W. (1994) Island extinction rates from regular censuses. *Conservation Biology*, **8**, 491–494.

Rosenzweig, M.L. & Taylor, J.A. (1980) Speciation and diversity in Ordovician invertebrates: filling niches quickly and carefully. *Oikos*, **35**, 236–243.

Rosko, D.J., Schaffer, W.M. & Allen, J.C. (1994) Chaos and metapopulation persistence. In: *Towards the Harnessing of Chaos* (ed. M. Yamaguti), pp. 115–126. Elsevier, Amsterdam.

Sanders, H.L. (1968) Marine benthic diversity: a comparative study. *American Naturalist*, **102**, 243–282.

Schoener, T.W. (1976) The species–area relation within archipelagoes: models and evidence from island land birds. *Proceedings of the XVI International Ornithological Congress*, pp. 629–642.

Shmida, A. & Ellner, S. (1984) Coexistence of plant species with similar niches. *Vegetatio*, **58**, 29–55.

Shmida, A. & Wilson, M.V. (1985) Biological determinants of species diversity. *Journal of Biogeography*, **12**, 1–20.

Simpson, E.H. (1949) Measurement of diversity. *Nature*, **163**, 688.

Smith, W. & Grassle, F. (1977) Sampling properties of a family of diversity measures. *Biometrics*, **33**, 283–292.

Smith, B. & Wilson, J.B. (1996) A consumer's guide to evenness indices. *Oikos*, **76**, 70–82.

Sousa, W.P. (1979) Disturbance in marine intertidal boulder fields: the nonequilibrium maintenance of species diversity. *Ecology*, **60**, 1225–1239.

Stewart, F.M. & Levin, B.R. (1973) Partitioning of resources and the outcome of interspecific competition: a model and some general conclusions. *American Naturalist*, **107**, 171–198.

Taylor, L.R., Kempton, R.A. & Woiwod, I.P. (1976) Diversity statistics and the log-series model. *Journal of Animal Ecology*, **45**, 255–272.

Tilman, D. (1982) *Resource Competition and Community Structure*. Princeton University Press, Princeton.

Tilman, D. (1988) *Plant Strategies and the Dynamics and Structure of Plant Communities*. Princeton University Press, Princeton.

Tilman, D. & Pacala, S. (1993) The maintenance of species richness in plant communities. In: *Species Diversity in Ecological Communities* (eds R.E. Ricklefs & D. Schluter). University of Chicago Press, Chicago, IL.

Walsh, J.B. (1995) How often do duplicated genes evolve new functions? *Genetics*, **139**, 421–428.

Whittaker, R.H. (1972) Evolution and measurement of species diversity. *Taxon*, **21**, 213–251.

Williams, C.B. (1943) Area and the number of species. *Nature*, **152**, 264–267.

Williams, C.B. (1964) *Patterns in the Balance of Nature*. Academic Press, London.

Wright, D.H., Currie, D.J. & Maurer, B.A. (1993) Energy supply and patterns of species richness on local and regional scales. In: *Species Diversity in Ecological Communities* (eds R.E. Ricklefs & D. Schluter). University of Chicago Press, Chicago, IL.

10: Ecological economics

E.J. Milner-Gulland

Introduction

Ecological economics is the study of the interactions between economics and ecology. Over the last decade, this area has grown in stature from an obscure discipline of economics to a field with major inputs into current debates about the conservation and management of the environment. So much so that if ecologists are concerned about the outcome of these debates, or wish to participate in them, a knowledge of economics is now essential in order to be aware of the assumptions lying behind the economic arguments. However, ecological economics is more than simply the branch of economics concerned with the use of natural resources. As a theoretical discipline, it overlaps with ecology, both in the methodologies it employs and the content of its models. Thus, for example, the treatment of decision-making is often identical in behavioural ecology and in the analysis of wildlife use by humans. In terms of model content, the now unpopular descriptor of 'bioeconomic' clearly identifies the symbiosis of biological and economic models that is inherent in the subject of ecological economics.

Ecological economics is usually divided into *environmental economics* and *resource economics*, the former concerned with the use of the environment as a sink for the waste products of the human economy, the latter with the use of natural resources as inputs into the economy. Resource economics is further subdivided into renewable resources and exhaustible (or nonrenewable) resources. This latter division is of more practical than conceptual use, since the definition of a resource as renewable or exhaustible is based on the renewability of the resource relative to time-scales of interest to humans.

The environmental/resource economics divide is even more clearly a traditional division, since the use of the environment as a sink for waste products can just as easily be seen as the use of a natural resource—the buffering capacity of the environment. The analytical techniques that apply to one area are usually applicable to the other as well, but there is still a clear division in the literature. The tradition of renewable resource economics is characteristically based on the use of single species, developed principally in the disciplines of fisheries science and forestry. It relies heavily on mathematical treatment of the problem, and is perhaps best exemplified by the work of C.W. Clark (Clark 1990). Practitioners tend to come from biology or mathematics, and to work from the perspective of the exploited stock. Environmental economics, on the other hand, is more the preserve of researchers trained in neoclassical economics, relies more on graphical treatments, and tends to start from the perspective of the exploiter rather than the exploited, partly due to the fact that the exploited resource is usually a whole ecosystem and so less tractable to mathematical treatment than a single species. The work of economists such as D. Pearce and R.K. Turner typify this approach (Pearce & Turner 1990). Exhaustible resources are treated in a separate and specialized branch of economics, and as they have little direct connection with ecology I shall not discuss them further. However, despite this generalization, various authors working in this area (such as Dasgupta & Heal 1979) have been influential throughout the field of ecological economics.

The position of ecological economics within the neoclassical economic paradigm is illustrated in Fig. 10.1. Neoclassical economics sees the economy as an open system, with inputs (labour, land, man-made and natural capital) entering the economy, and being transformed into useful products and waste products to be discarded. Ecological economics, by contrast, sees the economy and environment as linked together in a closed system. Inputs come from the environment, and waste products are returned there. A key linkage is the effect of the use of the environment as a sink for waste products on its productivity as a source of inputs for the economy. For example, acid rain caused by the release of industrial waste into the atmosphere has effects on the productivity of the timber and fishing industries.

The difference between the neoclassical and ecological views of economics shown in Fig. 10.1 has a further correlate, in that ecologists tend to have a strong sense that resources are limited, reflected in the ecological concept of carrying capacity. This concords with the paradigm of the closed economy. Neoclassical economics, with its view of the economy as open, has the sense of a system that is capable of indefinite growth so long as inputs are forthcoming. These inputs need not be of all the listed types—a scarcity of natural capital will lead to substitution into some other input. These contrasting paradigms, strongly engrained in each discipline, are one cause of many of the profound disagreements that arise within ecological economics.

The development of ecological economics has had a long, if patchy history. Many people have pointed out the interdependence of the economy and the environment, with Malthus as just one example. However, modern ecological economics can probably be traced most clearly to the work of Hardin (1968) and his contemporaries. The flowering of ecological economics has continued through the 1980s and 1990s, with many graduate level courses starting up. Sadly, though, it is still far outside the curricula of main-

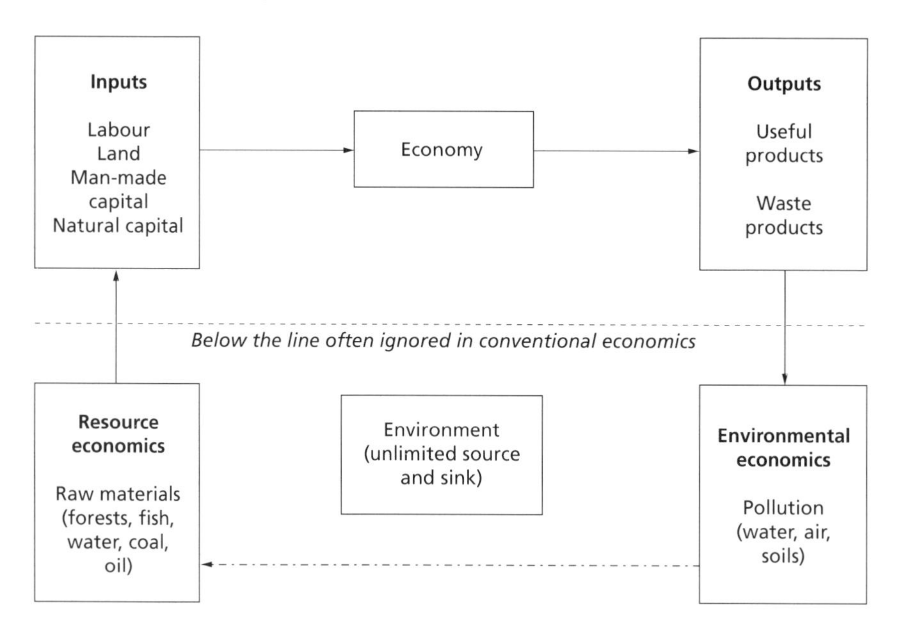

Fig. 10.1 The economy–environment linkage. Neoclassical economics tends to ignore processes below the dashed line. Of particular interest is the interaction between pollution and the ability of the environment to supply inputs to the economy (represented by the dot-dashed arrow).

stream undergraduate economics and ecology courses. In this chapter, I shall review some of the major areas of advance in ecological economics, the major current controversies, and a few of the questions that remain unsolved. As would be expected in a new and interdisciplinary subject, these are many, and each practitioner would probably pick a different set. My personal bias is as strong as anyone else's, and arises from a background in population ecology, and an interest in economics as it pertains to the conservation and management of large mammals.

Current controversies

Over-exploitation: causes and solutions

The exploitation of a renewable resource can have two outcomes, assuming that the resource stock is stable at its carrying capacity before exploitation begins and that exploitation continues at a constant rate. Either the stock can decline to a level at which recruitment equals offtake, at which point the population stabilizes at the new level, or the population can continue to decline, eventually to extinction. Which of these occurs depends on various factors. At equilibrium, the precise function for the relationship between hunting effort and population size will be biologically determined. The standard assumption made is that a population will compensate for a reduction in density by an increase in growth rate, usually represented using the logistic equation, where N is the population size, K is the carrying capacity and r is the intrinsic growth rate of the population:

$$\frac{dN}{dt}=rN\left(1-\frac{N}{K}\right) \tag{10.1}$$

In practice, the population growth rate per individual is unlikely to decline linearly with an increase in population size (Fowler 1984). An extreme example of nonlinearity is depensation, when population growth rate per individual increases with population size at very low population sizes. For example, when a population is very dispersed, finding a mate may be difficult at low densities.

The logistic equation, describing linear density-dependence, can be combined with a simple economic model to produce a model of the outcome of harvesting. The offtake by the harvesters is described by $h = qEN$, implying that offtake h is a function of the effort level E (number of boats, or hours spent hunting, for example) and of the stock size N. The constant q reflects the ease of catching a particular species (the 'catchability coefficient' in fisheries). The bioeconomic model for the equilibrium situation is shown below and in Fig. 10.2.

1 The population is at equilibrium (denoted by N^*) so that the population growth rate equals the offtake rate:

$$rN^*\left(1-\frac{N^*}{K}\right)-qEN^*=0 \tag{10.2}$$

2 The profits at equilibrium are zero, so that total revenues equal total costs:

$$\text{TR} = \text{TC}, \text{or } pqEN^* - cE = 0 \tag{10.3}$$

where p is the price per unit of offtake and c is the cost per unit of effort expended. The solution E_∞ shown in Fig. 10.2 can be obtained from Equations 10.2 and 10.3 by substitution.

Rather than assuming that the profits are zero at equilibrium, the other option is to assume that profits are maximized, so that $\Pi = \max(\text{TR} - \text{TC})$, which gives the solution E_0 shown in Fig. 10.2.

Economic factors are important in determining which of the two scenarios E_∞ or E_0 occur for a particular resource. As the x-axis of Fig. 10.2 is effort, the figure translates inversely into stock size—the higher the level of effort, the lower the stock size. Thus the situation at E_∞ is worse for the stock than E_0. It is the structure of the industry that affects whether profits are

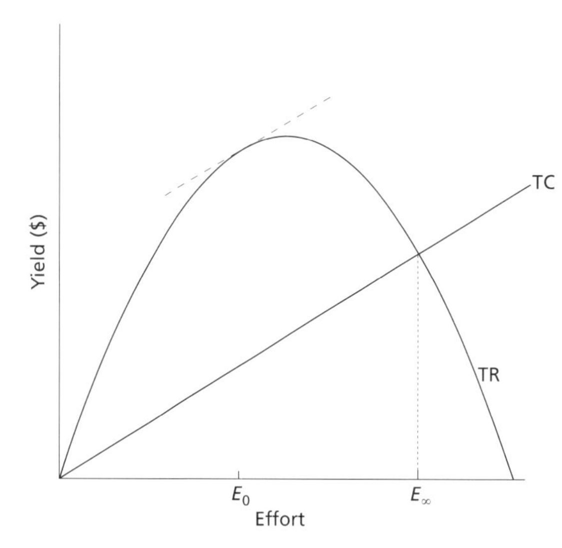

Fig. 10.2 The bioeconomic model of resource exploitation, plotted as the monetary yield against the effort expended by harvesters. TC is total costs, TR is total revenues. E_∞ is the equilibrium position where TC = TR (profit is zero), while E_0 is where profits (TR − TC) are maximized.

maximized or dissipated, and thus the structure of the industry has major ramifications for the equilibrium size of the exploited stock. The results will also depend on the assumptions made about costs and prices, and, of particular relevance to ecologists, the way in which the costs of exploiting the resource vary with the size of the stock, for example does the stock's aggregative behaviour change as density changes?

The two extremes of industry structure are perfect competition and monopoly. In a monopoly, there is a sole user of the resource, which is able to maximize its profits from the resource over time. A user with sole control of the resource will exploit the stock at the level E_0 in order to maximize profits. This is a static analysis, however. If time is brought into the analysis, there is a trade-off between offtake now and the stock left as capital for future harvests. The stock size at equilibrium will depend on the user's discount rate compared with the growth rate of the resource. As Clark (1990) shows, E_0 represents the situation with a zero discount rate, and in simple models the higher the discount rate, the nearer the equilibrium moves to the heavily exploited equilibrium of E_∞. One caveat applies for stocks with very low growth rates, lower than the rate of return in the general economy, when the optimal strategy for a sole user can be to liquidate the stock and invest the proceeds in some other asset with a higher rate of return. Clark (1973) demonstrated that this could be the case for the blue whale.

A resource being exploited under perfect competition is subject to harvesting by an unlimited number of individuals, who can enter or leave the industry at will. An individual user has no control over the total harvest rate of the stock, and so there is no incentive for an individual to practise restraint for the sake of stock conservation. Collusion between participants is impossible due to the free access to the industry. New entrants will continue to enter the industry until profits are reduced to zero, at which point the industry will be in equilibrium. This will occur at a stock size of E_{∞}, which is considerably lower than the monopolistic one. There is no consideration of stock conservation for the future in this situation; in other words the discount rate approaches infinity. This scenario is referred to as *open-access harvesting*.

In general, resource exploitation will occur within an industrial structure somewhere between these two extremes, although often a good approximation to one or other extreme can be made. For example, in the Luangwa Valley, Zambia, there were two types of ivory poacher operating in the 1980s: individuals hunting predominantly for meat, for whom perfect competition was an adequate assumption, and gangs sent by dealers, who had an effective monopoly over a particular area. In this case, the monopolistic structure led to 'optimal' population sizes that were very low, due to the high discount rates in the area and the low growth rates of elephants (Milner-Gulland & Leader-Williams 1992). One other industry structure that deserves a special mention is oligopoly, when only a few firms are present in the industry. This structure means that the decisions of each firm do have an effect on the market, depending on the decisions of the others. Game theoretical approaches are often used to model the equilibrium outcome in this case (e.g. Kennedy 1987).

The simple theoretical model of open-access harvesting has been at the heart of some of the bitterest arguments in the analysis of resource exploitation, with some anthropologists and economists lined up against what they see as the refusal of modellers to recognize the real-life complexities of systems of resource tenure and exploitation. The controversy started with Hardin's (1968) work in which he applied the ideas of Malthus to the problem of the overgrazing of common land; simply stated, when everyone has access to the area, there is no incentive for any one user to conserve the stocks of grass by limiting their livestock herd. This example, known as the 'tragedy of the commons' has been seen to have wide applicability in open access systems in general, and corresponds with the theoretical analysis of perfect competition developed above. Opponents of the concept, such as McCay & Acheson (1990), argue that the concept is deeply flawed in that it fails to recognize that traditional societies in particular tend to have developed systems of sustainable resource use, and that problems arise only when outside interference is imposed on these cultural norms. This argument has been useful in highlighting some important issues, but in most respects involves an attack on straw men. Part of the problem seems to have been an unfortunate choice of example in Hardin's original exposition—the common grazing land of a village is rarely actually an open-access resource. Another major factor has been the impingement of political ideology, in which the debate has implicitly become one about the privatization of resources versus community ownership. For the open-access system which he describes, Hardin's analysis is perfectly sound; the argument should be about the relative prevalence of open-access and other types of system in the real world.

One way in which the debate could be clarified would be the realization that the ownership of a resource is fundamentally an irrelevance to the outcome of exploitation. Privately owned, state, communally or genuinely unowned resources (rare as these last are) can all be subject to open-access exploitation, or to profit-maximization (effective monopoly). It is the control of the resource that is the key issue. Stewart (1985) puts this argument clearly for forest resources, and the example of the Luangwa Valley cited above demonstrates the same point—a state-owned national park was exploited as a monopoly by poaching gangs, the state-owned game management areas as an open-access resource by individual poachers. Thus the communally owned resources used as examples by McCay & Acheson (1990) are protected from open-access exploitation because they are communally controlled, not because of their ownership. Ownership and control do not always go together. One caveat to this argument, however, is that *de facto* sole control of a resource may not be adequate to promote a long-term conservation ethic—control must be perceived to be secure in the long term, otherwise the user will not have an incentive to conserve. This is where ownership and control do intersect—legal ownership can be seen as a guarantee that control will continue in perpetuity.

Game theory is another tool by which some of the decisions surrounding resource over-exploitation have been analysed, and it sheds a different light on the problem of the tragedy of the commons (Axelrod 1984). The classic game of the prisoner's dilemma has obvious parallels with the tragedy of the commons. The original game involved two prisoners, who were kept apart, and who had to decide whether to confess to a crime ('defect') or to keep quiet about it ('cooperate'). A typical payoff matrix is shown below, with A's payoffs for a given decision in bold, B's in italics:

		B	
		Cooperate	*Defect*
A	**Cooperate**	**+5**, *+5*	**–10**, *+10*
	Defect	**+10**, *–10*	**–5**, *–5*

Whatever the story woven around it, if the game is played once then the rational choice for both participants is to defect, with potential yields of +10 and –5, rather than to cooperate, with potential yields of +5 and –10 (see also Chapter 4). Clearly, in order to jointly maximize the payoff, the best option is for both to cooperate, while two defections is jointly the worst payoff. So individual rationality leads to the worst possible result occurring. Here we see the parallels with open-access harvesting—a monopolistic harvester is able to reach the jointly maximized optimum, while in open-access harvesting it is in the best interests of each individual harvester to defect and harvest as much as possible.

It has been argued that in fact the prisoner's dilemma will lead to cooperation, because each individual will realize that the other person is similar to themselves, and so will also realize that cooperation is best. Neither total self-interest nor similarity seem to describe social systems perfectly, but in situations where there are many players, so that no deals are possible (such as is the case in open-access harvesting), self-interest is likely to be the better model. If cooperation is to occur, there should not be too many players, and they should be able to communicate and to transfer payments between themselves (see the extension of this in the case of pair approximation, described in Chapter 4). Axelrod (1984) showed that the prisoner's dilemma can also result in cooperation if it is iterated over time, so that strategies can adapt to each other. As well as having implications for the evolution of cooperation in biological systems, these results can be used to produce a theory of the kinds of societies within which resource harvesting might evolve to sustainability and joint maximization, and those in which it is likely to remain open-access. Given these insights from game theory, it is no surprise that the majority of the examples of successful communal resource management cited by McCay & Acheson (1990) are small, closed societies in which the same people have interacted over a long period.

Game theory is not always useful for analysing natural resource management problems, particularly if there is only one player and the 'opponent' is the unpredictable environment. However, when people are interacting with each other, the technique can be useful. For example, Mesterton-Gibbons (1993) analyses the incentives for community members to steal communal water resources using game theory.

Sustainability

The debate on the institutional structures which promote long-term sustainable use of resources is just one part of the research that has been done on sustainability. This research is an important component of ecological economics at several levels. On the global level, there is the sustainability of human resource consumption and population increase—at what population density and consumption rate per capita can the human population be sustained indefinitely? On a level more applicable to most ecologists is the current trend in conservation towards advocating 'sustainable use' as a method for the conservation of threatened species and ecosystems. Is sustainable use actually an achievable and useful goal for conservation programmes, is it merely the least damaging of the many damaging ways in which humans affect endangered ecosystems, or is it a philosophy that puts all our protected ecosystems up for grabs by exploiters? These questions are discussed in depth in Milner-Gulland & Mace (1998).

One major problem with the whole sustainability debate is that of definitions. 'Sustainable' has an intuitive meaning to the lay person that allows policy-makers and researchers to get away with discussing it in a less-than-rigorous way. Thus claims can be made that a particular process is sustainable without any criteria for sustainability being fulfilled. When definition is attempted, it becomes apparent that an operational definition with practical meaning is not easy to achieve. There are various aspects to sustainability: economic, cultural, and ecological. A sustainable-use project has to be sustainable on all these levels to succeed—it must make a profit over the long term, it must be culturally appropriate to the users and it must not cause harm to the ecosystem in the long term. Evaluating ecological sustainability is relatively easy for a theoretical single-species harvesting operation—the population should reach a long-term equilibrium size at any given level of harvesting. But the ecological sustainability of a project that involves one or more species within an ecosystem is much less tangible. Another problem is that a time limit must be imposed for a project's sustainability to be evaluated. Sustainability in perpetuity cannot be guaranteed under the unpredictable conditions that prevail in the world. Finally, the ease with which sustainability appears to be achieved depends on the method used to compare costs and benefits at different points in time. This will be discussed further below.

One major disagreement between ecologists and economists concerns the feasibility of 'sustainable development' and 'sustainable growth', growth being defined as increases in standard economic indicators such as gross national product (GNP) and development as a continuing improvement in human well-being, not necessarily linked to conventional growth indicators (a conveniently vague concept!). This comes back to the different views on limits to growth in the two disciplines. The arguments were set out in various reports in the 1970s and 1980s (Meadows 1972; Simon & Kahn 1984; WCED 1987). The point of contact between the two camps can be made in the definitions of growth and development. The economists' argument is that sustainable growth and development are possible because economies will substitute away from scarce resources towards other resources, and because of the advances in technology which will occur. This argument is borne out by past experience—at the time of the 1973 oil crisis, consumers rapidly substituted away from the suddenly scarce oil into other fuels, although the crisis led to major repercussions for the world economy. Technology has certainly continued to advance since the Industrial Revolution. It is generally accepted in ecological economics that although sustainable growth may not be feasible, sustainable development is. However, in the end, there are fundamental differences in outlook and ecologists tend to remain deeply sceptical that previous experience will guarantee long-term sustainable development as a meaningful concept for the future. There is an analogy here with our previous success in the control of pest and disease agents' resistance to chemicals, which is seen as no guarantee for future success.

Time

The treatment of time is a key issue in ecological economics. The ecological effects of human use of the environment, both as a source and a sink for the economy, are often long-lasting and slow to appear. Harvested populations often require a dynamic treatment when modelling their population ecology, and the influence of economic decision-making is to impose human time preference on these dynamics. The standard way to treat time as a variable in economic decision-making is by the inclusion of a discount rate. Thus the bioeconomic model of harvesting discussed above can be generalized to include the optimization of monopolistic profits over time in the following way:

Maximize the present value of profits over time, where profits (Π) depend on stock size, hunting rate h and time, and the discount rate is δ.

$$\max \text{PV} = \int_{t=0}^{t=\infty} e^{-\delta t}\Pi(N,h,t)\,dt \tag{10.4}$$

given that $N_t \geq 0, h_t \geq 0$.

This is a problem in optimal control theory, with a solution that the steady-state population size N^* is given by:

$$f'\left(N_{\mu}^{*}\right) - \frac{c'(N^*)f(N^*)}{p - c(N^*)} = \delta \tag{10.5}$$

where $f(N)$ is the population growth rate, for example logistic growth. A detailed derivation of this result is given in Clark (1990) and Burghes & Graham (1986).

In general, assets are assumed to decline according to a negative exponential with time, so that the present value of a future payout of P at time t is represented as $\text{PV} = Pe^{-\delta t}$. The present value of a continuous stream of payouts into the infinite future is given by:

$$\text{PV} = \int_0^{\infty} Pe^{-\delta t}\,dt \tag{10.6}$$

which is equivalent to PV $= P/\delta$ (see Price 1993, for further discounting formulae).

This result is of importance to ecologists in that it shows that a stream of investments in perpetuity has a finite value of P/δ. In a short time, the value of the flows of benefits (or equally, costs) from an investment made now becomes negligible to the outcome of the calculation. Thus, projects with early benefits and deferred costs will tend to be favoured under this model of the change in asset value with time. At a discount rate of more than 5%, for example, any costs or benefits incurred after about 40 years will be largely irrelevant to the investment decision, and half the value of the investment has already accrued 7 years after the start of the project. This result should ring alarm bells for those accustomed to ecological time-scales of damage and recovery.

Any discount rate $\delta > 0$ implies that the future has less importance than the present. A discount rate of zero (as used in the calculation of E_0 above) implies that the future has equal weight with the present, as there is no trend in value with time (thus if spending £10 now would lead to a gain of £11 in 50 years time, it would be worth doing). A rate $\delta < 0$ intuitively means that the value of assets increases with time, giving the impractical outcome that we would constantly save and never spend. The base discount rate for the British economy has been in the region of 2–4% in recent times.

Discounting has three major uses in economics: as a descriptive model of human behaviour; for the prediction of human behaviour; and as an aid to investment appraisal (Price 1993). The first two uses are of lesser concern here—it is clear that humans do tend to show time preference for consumption in the present over the future, and although the negative exponential is an inadequate model of human behaviour, it is simple and is a relatively good predictor in most cases. Thus, for example, discounting can be used to compare the time at which poachers would find it worthwhile to kill a dehorned rhino (whose horn will grow back slowly) with the rotation period at which it would be most profitable for a rhino manager to dehorn, given that the manager is trying to maximize the returns from selling the horn (Milner-Gulland *et al.* 1992). Similarly, Hodson *et al.* (1995) discuss the present value of an area of tropical rainforest as a perpetual source of nontimber products compared with the present value of clearing it for agriculture.

It is the use of discounting in investment appraisal that has caused the major disagreements in ecological economics. Clearly, there has to be some method by which projects with different income and expenditure schedules can be compared, and with which one can judge whether a particular project is worth funding, compared to the performance of the general economy. Equally clearly, discounting as a method of carrying out this appraisal is flawed for projects in which deferred costs are large, such as many projects affecting the environment. For example, take a very simplistic analysis of a forest that is being harvested rapidly (after Price 1993). If harvesting continues, you can obtain an output worth £25 000 per year for the next 10 years, followed by no further yield. Alternatively, a 40-year moratorium on timber harvesting will lead to the recovery of the forest, followed by a sustained yield of £50 000 per year in perpetuity. Which option is economically more worthwhile?

Clearance: $PV = \frac{Y_1}{\delta}(1 - e^{-\delta T_1})$, where $Y_1 = 25000$ and $T_1 = 10$

Moratorium: $PV = \frac{Y_2}{\delta}(e^{-\delta T_2})$, where $Y_2 = 50000$ and $T_2 = 40$

The strategies are equally good where:

$$\frac{Y_1}{\delta}(1 - e^{-\delta T_1}) = \frac{Y_2}{\delta}(e^{-\delta T_2})$$

and solving for δ shows that harvesting to extinction is preferred at $\delta > 4.3\%$, which is a rather low discount rate.

The knee-jerk reaction to this problem is to suggest that a low discount rate should be set for environmentally sensitive projects, or for projects that have an important social function. This is not a useful response, firstly because the basic problem of assuming a negative exponential remains, and secondly because, as Pearce & Turner (1990) point out, high discount rates discourage investment. Investment tends on the whole to be bad for the environment—it uses resources and may lead to the development of countryside. Several authors have suggested various alterations to the discounting procedure, including adding a sustainability constraint (Pearce & Turner 1990) and using variable discount rates (Kula 1992). Price (1993) provides a comprehensive critique of discounting and the modifications suggested by others, and argues strongly for the use of methods which do not involve discounting, but deal explicitly with changes in value with time. The problem is that the negative exponential is a very simple and generally accepted model for change in value over time, and thus is subject to an enormous inertia—rather like the logistic equation in population ecology. And, also analogous to the logistic equation, the fact that people are constantly attempting to find ways to tweak it for greater realism demonstrates the deep underlying problems of the model. However, the constituency using the negative exponential is orders of magnitude larger and more entrenched than that using the logistic, so the challenge for those trying to get it abandoned is far greater.

One major issue that should be highlighted in the treatment of time, particularly with respect to the previous discussion on sustainability, is that of intergenerational equity. Using discounting as a convenient way to appraise our own investments is one matter; using it to decide on investments whose costs will be borne by future generations rather than ourselves is quite another (Norgaard & Howarth 1991). Lawrence H. Summers of the World Bank (quoted in Price 1993) puts the standard position on the subject very clearly, arguing that the compelling needs of those alive today and the uncertainty of the future costs of global warming justify using the same discount rate for present-day sanitation projects and projects protecting against global warming (the latter will, of course, receive a very low priority under discounting):

> Do I sacrifice to help those in the future or help the 1 billion extremely poor people who share the planet earth with me today? I hold no greater grief for the people who will die 100 years from now from global warming than for people who will die tomorrow from bad water.

The major assumption lying behind discounting projects with long-term consequences is that of the diminishing marginal utility of wealth over time—as the human race becomes richer, the unit value of wealth will decline, just as it does for an individual. It is assumed that future generations will be better off than us, and so value each unit of wealth less. This assumption, that mean affluence will increase with time, is borne out by past experience, but is by no means guaranteed for the future.

Valuation

The issue of the valuation of natural resources is a particularly thorny one for ecological economics. If ecological assets are properly valued within the economy, the argument goes, then they will be conserved and used sustainably. For example, a proposed harbour development could be quantified in terms of the ecological damage caused as well as economic benefit obtained. As was demonstrated above, the timing of costs and benefits is one of the things that impedes the proper weighting of ecological costs and benefits against economic ones. Another more fundamental issue is the scale of measurement upon which assets should be valued. In economics, the basic unit of measurement is utility, the amount of pleasure accruing to an individual from the consumption of a particular good. Money is generally used as a proxy for utility, on the assumption that the amount of money an individual is prepared to pay for a good is an adequate representation of the amount of satisfaction that will be obtained from its consumption. In fact, there is diminishing marginal utility to wealth—as for any other good, the satisfaction obtained per extra unit of money obtained declines as the amount you have increases.

Clearly, if ecological assets are going to be brought fully into the mainstream economy, there needs to be a single scale of measurement on which to compare ecological and economic values. Money is the obvious scale to use. Other scales have been tried, such as energy or materials balance (Perrings 1987), but in order to be acceptable as an economic scale of measurement, a scale must be able to measure individual preferences, and so utility. So the approach to the valuation of ecological assets has been to capture as much as possible of their monetary value to individuals, despite the fact that researchers' best efforts will not capture all of ecological value this way. Perhaps it would be better to start by asking whether the preferences of individuals are an appropriate basis for ecological valuation, if not, a single scale of measurement for ecological and economic value is not achievable. But even a single linear scale that captures ecological value alone seems unlikely. It is probably not appropriate in a policy-making setting to use the calculated monetary value of an ecological asset if the results of a cost–benefit analysis happen to favour environmental conservation, and to argue that the calculated value is an unquantifiable underestimate of the true value if the results happen to favour environmental destruction.

Given the above caveats, a lot of progress has been made recently in attempting to value the nonmarket costs and benefits inherent in ecological systems. Some of the values of ecosystems that are not clearly quantifiable in monetary terms will none-the-less be expressed to a degree by individuals through their use and appreciation of the natural world, their sense of responsibility towards its preservation and their willingness to accept state management of ecological assets for the public good. The problem is then one of detective work, attempting to get people to express these feelings in monetary terms, which can be very difficult. Often, a major reason why these values are not expressed in the market in the first place is that they are very hard to value monetarily, and so attempts to get people to reveal their preferences will be uphill struggles, for example how much would you be prepared to pay for seeing one more beautiful view a year? And for 10 more? Individuals' valuation of nonmarket benefits are usually divided into two types: use values and intrinsic (or existence) values. The former includes the value of the individual's direct use of the environment, the value to them of retaining their option for future use, and satisfaction at the use of the resource by others. If the natural resource is traded, there will also be a direct market value from the sale of the produce. Intrinsic value is much less tangible than use value, and represents individuals' feelings about the rights of natural systems to exist irrespective of the presence of humans, and their sense of justice and responsibility towards nature.

There are various other problems encountered in quantifying nonmarket costs and benefits. One is that ecological costs and benefits tend to be diffuse and indirect in both space and time, making them hard to trace. They are likely to accrue to different individuals to those getting the economic costs and benefits of a particular development. This problem is known as 'externalities'. If different people are experiencing the benefits to those that experience the costs, then the market has failed, because the incentives that the individuals experience cannot bring the market to the socially efficient

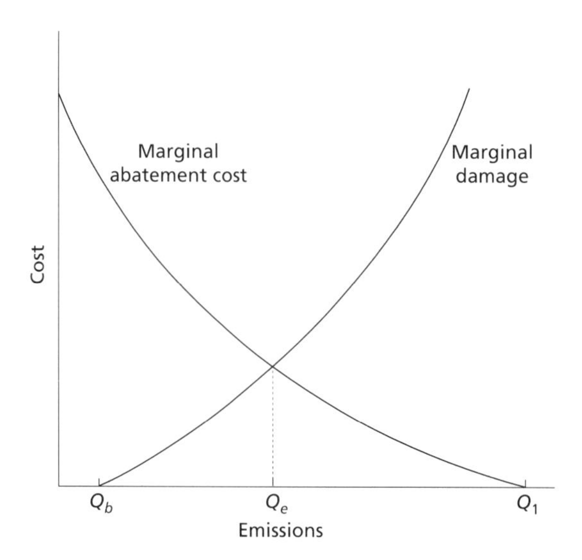

Fig. 10.3 The socially optimal level of resource use, in this case pollution emissions, plotted as marginal damage (MD; the extra damage incurred to the environment per extra unit of pollution) against marginal abatement cost (MAC; the cost to the firm of cleaning up each extra unit of pollution, assumed equivalent to the marginal net private benefits to the firm of polluting). Q_b represents the buffering capacity of the environment, so that there is no lasting damage at low levels of emissions. Q_1 is the level at which the firm wishes to produce, because all the damage costs are external. Q_e is the socially optimal level of pollution, where the costs of pollution balance the costs of not polluting; MD = MAC. If the firm pays no damage costs, then as far as they are concerned MD = 0, so their optimal level of pollution is MAC = MD = 0, which is where the MAC curve crosses the *x*-axis at Q_1.

optimum. Figure 10.3 illustrates the problem, which occurs both for external costs and external benefits. If there are external costs, for example if a factory is polluting a river causing costs to people wishing to fish downstream, the fact that the cost of the pollution is not included in the cost–benefit analysis of the factory means that the factory operates at Q_1, not at the socially efficient level Q_e Similarly, external benefits occur, particularly with public goods—goods that, once provided, are available for all to use for free, such as clean air and radio programmes. Public goods tend to be undervalued by individuals due to the free-rider problem—if I pay less for the good than I truly value it at, others will pay for it, and I will be able to use it anyway. Many ecosystem values are public goods, and many economic activities have ecological external costs. Often, the proposed solution is 'polluter pays': if the firm causing the pollution is required to clear it up, it will 'internalize the externality' and will produce output at the socially optimal level. This may work for a single factory polluting a river, but when the situation is transformed into a more realistic one, such as many poor farmers deforesting a mountainous region and causing flooding destroying crops downstream, 'polluter pays' becomes a completely impractical solution.

Add up the problems of time-scale, spatial diffusion, externalities and costs and benefits that are fundamentally difficult to value monetarily, and the situation becomes seemingly hopeless for the valuation of the non-market costs and benefits of ecological services. The assets that are most amenable to research under these circumstances are of two types: recreational facilities and goods that are consumed and whose values are represented indirectly in the market. Thus we have the valuation of national parks and the game within them in North America (Keith & Lyon 1985); the valuation of elephant viewing in Kenyan parks (Brown 1989); and the valuation of the use of tropical forests as providers of traditional medicines, compared to their value for logging (Peters *et al.* 1989). One of the most illuminating studies of this sort is that of Norton-Griffiths & Southey (1995), who calculated the value of Kenya's national park land for the production of tourist revenue and compared it with its potential value as farmland—in purely economic terms the present use of the land presents a substantial shortfall over what could be achieved from farming the land, begging the question of whether that shortfall is more or less than the nonmarket value of the national parks as providers of ecological services.

The Norton-Griffiths & Southey (1995) study involves the gathering of data from various indirect sources to calculate monetary values. Other methods of economic valuation are of two types: finding a surrogate market for the ecological good being investigated and carrying out an experiment to determine value directly. The surrogate market approach includes two common methods: the hedonic price method and the travel cost method. The former might estimate the value put on living in a city with low pollution levels by carrying out a multiple regression of house prices against the major factors that determine those prices, including pollution level. Clearly there are a number of confounding variables (pollution levels might be confounded with quality of public transport, for example), and this method is statistically difficult while being open to many biases. In its favour, it uses actual data in its calculations, unlike the experimental methods. The travel cost method is based on the fact that the time and money spent travelling to an environmental attraction such as a national park will be a measure of the value placed upon that attraction by the individual (having controlled for wealth, presence of alternative attractions and so on). Keith & Lyon (1985) and Donnelly & Nelson (1986) use travel cost methods to estimate the value of increasing or reducing the deer herd in a park by a single deer. Experimental methods of valuation create a hypothetical market for the good in question, often involving a questionnaire (contingent valuation), but in more rigorous cases involving a simulated market, in order to get people to reveal their preferences and to enable the researcher to calculate a utility function for the good. There are obvious problems with this method, including the fact that the respondent knows that the market is hypothetical, and has had no previous experience in valuing the good (which will be hard to value, being a nonmarket good). Various design issues are also important, including the bias introduced by the prior information given to the respondent by the experimenter. Despite these problems, the contingent valuation method has on the whole produced results similar to those obtained from surrogate market methods. Nonmarket valuation has been used to get quite large existence values for ecological assets, for example in 1983 the American people were estimated to put a value of $3.5 billion on ensuring the visibility of the Grand Canyon in the face of increasing air pollution (quoted in Field 1994).

National accounting is one area in which valuation of ecological assets has important and far-reaching consequences for governmental policy. The standard methods of calculation of GNP and other measures of the economic well-being of a country do not take the ecological assets of a country into account, even if those assets are relatively easy to value monetarily, for example the value of exported timber from a country would count in the national accounts, but the value of the standing forest as a capital asset for the country would not. Similarly, a major oil spill would show up as a benefit in the national accounts, because the ecological damage caused would be ignored, but the clear-up costs would be included as spending within the country—a positive contribution to GNP. Clearly, even if the true ecological values of the ecosystem services concerned were impossible to include, such glaring omissions should be rectified, and moves are now being made in this direction (DESIPA 1993).

Recent advances

Economic aspects of modelling

The fields that I have discussed above are controversial, and thus they are all areas where research is being undertaken and advances made. But advances are also being made in other less high profile areas. I highlight one such area here, which is particularly interesting, partly because it is one area in which bioeconomic modelling can be directly applied to practical problems of policy, and partly because it is an example of the need for an interdisciplinary approach that encompasses not just biology and economics, but other subjects—in this case law and sociology. This is the modelling of law enforcement as a component of resource management.

One of the advances that has been made in the modelling of the economics of natural resource management involves the recognition that there are costs involved in the implementation of policy that have to be taken into account if one is to reach the socially optimal level of resource use. The neoclassical view has been that to identify the socially optimal level, one needs to specify the marginal damage function (or marginal external cost) from the resource use and the marginal abatement cost to the exploiter (assumed equivalent to the marginal net private benefit of pollution to the polluter). The intersection of these two functions is the socially optimal level of resource use, where the cost to society of the environmental damage caused by resource use is equivalent to the benefit provided by that resource use (Q_e in Fig. 10.3). Who actually bears the costs and who receives the benefits is irrelevant in theory, though important in practice. Policy measures such as taxes, transferable quotas or standards can then be implemented in order to achieve Q_e. Which economic measure you choose to use depends on factors such as the measure's efficiency, cost-effectiveness, enforceability, equity, moral message and the incentives for innovation that it provides (Field 1994).

The criterion of enforceability has started to be investigated more thoroughly now, and it has been found to be of great importance for natural resource management. Graphically illustrated, it is clear that the greater the cost of enforcing a particular measure, the higher the socially optimal level of pollution will be (Fig. 10.4). If the socially optimal level of polluton is Q_e under costless enforcement, then under costly enforcement, the level will increase to Q_a. If resource use is to be controlled by a standard, then the

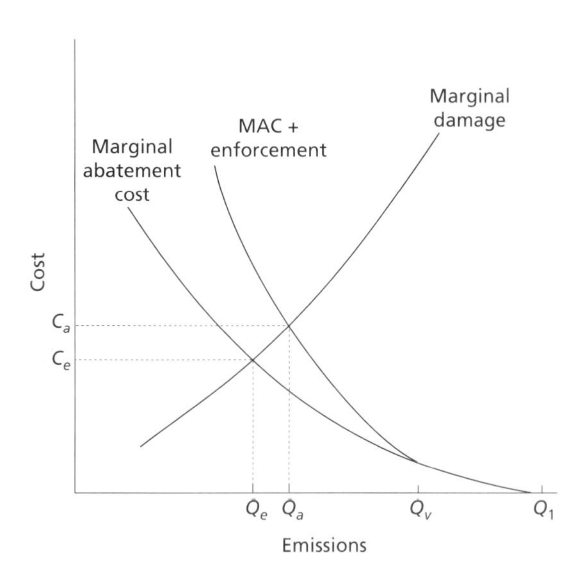

Fig. 10.4 The effect of enforcement costs on the socially optimal rate of resource use (in this case pollution emissions, as in Fig. 10.3). Q_v is the level of abatement that will occur voluntarily when laws are introduced; it is not adequate to reach Q_e. The cost of enforcement to the state (CE) must be added to the firm's marginal abatement cost so that at equilibrium, MD = MAC + CE. Because enforcement is costly (and progressively more costly for each extra unit of abatement), the socially optimal level of pollution increases to Q_a and the level of environmental damage MD rises.

stricter that standard is, the more costly it will be to enforce it, because there will be a stronger incentive to breach the standard. Thus a laxer standard properly enforced may lead to a greater overall level of pollution reduction for a given enforcement budget.

Sutinen and colleagues realized the importance of the observation that enforcement is generally imperfect, and have advanced the mathematical treatment of its consequences on resource management, working chiefly on fisheries (Sutinen & Anderson 1985; Anderson 1989; Mazany *et al.* 1989). They started by assuming that the exploiter of a resource is acting as an economically rational being, and weighs up the costs and benefits from breaching regulations from a purely economic perspective. This 'economic theory of crime' was first propounded by Becker (1968). Clearly the assumption is simplistic in the light of social norms, risk aversion and the differences between the true risk of capture and people's perception of the risk, but as a first approximation it is adequate. For the exploiter of a resource, the penalties for breaching regulations can be expressed as $E(P)$, the expectation of a penalty, which has two components, p_c, the probability of receiving a fine, and F, the fine incurred if caught. In fact, p_c is composed of several probabilities: of being detected, captured once detected and convicted once captured, all of which would have to be estimated separately to obtain the composite probability p_c. The expected penalty $E(P)$ can simply be included in the resource user's profit maximization, in the following way:

$$\max \Pi = \int_{t=0}^{t=\infty} e^{-\delta t}[pqEN - cE - E(P)]\mathrm{d}t \qquad (10.7)$$

This can be solved to give a similar result to Equation 10.5 (Milner-Gulland & Leader-Williams 1992).

If $E(P)$ is too low, then it is economically worthwhile for the exploiter not to comply with the rules—the value of the extra resource gained from breaking the rules exceeds the expected cost of the penalty. To raise $E(P)$, either the fine or the probability of capture can be increased. It seems at first glance that to increase compliance, one could increase one or the other component equally well. This in turn implies that the best strategy is to increase the fine imposed as much as possible, since imposing a large fine is costless to the state, while increasing detection rates is costly. However, the maximum fine imposable is in fact bounded, since if fines are too high, they will not be implemented by the judiciary. So should one aim at the maximum fine acceptable to the judiciary, and tailor p_c to produce a value for $E(P)$ that is just large enough to deter law-breakers? Some work has been carried out on the attitude of natural resource users to law enforcement measures (e.g. Sutinen & Gauvin 1989), but one can also turn to work on theft and burglary in the USA to find an answer. In these studies, the probability of being caught and convicted has been found to be a strong deterrent to crime (Cook

1977). Opinion is divided as to whether the severity of the penalty imposed has any deterrent effect at all, but studies agree that the penalty level is less of a deterrent than the detection rate (Ehrlich 1973; Avio & Clark 1978). This result contradicts the intuitively optimal strategy for the state, of a large fine and low probability of capture.

This research in the sociology of crime could have far-reaching policy consequences if it were applied widely to the deterrence of environmental crime. It is particularly appropriate for application to wildlife poaching (e.g. Leader-Williams & Milner-Gulland 1993). Although framed in terms of fines, the results apply with equal force to prison sentences, the other typical penalty for wildlife poaching. An offender's a priori perception of the severity of a prison sentence will depend on the rate of time preference and time horizon—how much the present is valued over the future and how far the individual looks into the future. If someone is indifferent between the value of time in the present and the future, an expected prison sentence of 0.2 years can be expressed equally well as a 10% chance of 2 years in prison or a 20% chance of 1 year in prison. However, if the value of the future declines as it gets more distant, the second year in prison is valued less than the first year, so 1 year with a probability of 0.2 is the worse option. In countries with a high degree of uncertainty about the future, such as many developing countries, rates of time preference will tend to be high, and so concentrating on detection rates will be a much more effective method of reducing poaching on a limited budget (particularly because prison sentences are very costly to the state as well as to the offender, unlike fines). However, the force of this observation is only apparent if the prisons budget is drawn from the same source as the antipoaching budget—which is rarely the case.

This last point leads on to a consideration of transactions costs, which are also important in the successful enforcement of laws on natural resource exploitation. This is now being increasingly recognized, particularly in the prosecution of large corporations in Western countries. Transactions costs are the costs of making an economic transaction, such as obtaining information and ensuring that the transaction is correctly carried out. The costs of legal proceedings are one significant area of transactions costs for environmental resources. If large law suits are required to determine the socially efficient levels of pollution, in order to gauge the compensation that a polluting company must pay for example, then the costs of this process are likely to be very large, and may even exceed the costs of the original damage imposed. Transactions costs are important for law enforcement on a smaller scale as well, and they interact with the problems of the institutional structures within which environmental agencies act. For example, if a wildlife authority in an African national park wished to prosecute a poacher, the costs involved in transporting the offender to court would be large, particularly if there was a shortage of vehicles and guards—the number of animals poached while the park officials were in court could be significant. Added to this, the courts are completely separate from the wildlife authorities, and are unlikely to set the same priorities on wildlife offences. Attempts to impose large fines will be thwarted as magistrates compare poaching with crimes they consider to be of similar magnitude (theft of personal property, for example) and reduce the penalties accordingly. For example, in Zambia, the government was concerned about ivory and rhino horn trafficking and the loss of elephants and rhinos, and in 1982 imposed mandatory 5–15 year prison sentences for ivory and horn poachers. The magistrates did impose more prison sentences after this ruling, but not all offenders received them and most were only of a few months; by 1985, the longest sentence imposed was 3 years, and Zambia's elephant and rhino populations had virtually disappeared (Leader-Williams *et al.* 1990). So the standard reaction to over-exploitation of resources—imposing stiffer penalties—can be demonstrated to be an ineffective way to tackle the problem on several levels.

Ecological aspects of modelling

It is still the case that the ecological viewpoint is under-represented in ecological economics. The field where most of the advances have been made is the one that has always been the preserve of ecologists: the management of exploited species. There has been a paradigm shift in the modelling of the exploitation of harvested species over the last few decades, which is most clearly exemplified by the history of the management plans of the International Whaling Commission, up to the recent publication of its Revised Management Procedure (Kirkwood 1996).

The International Whaling Commission has employed scientists to perform stock assessments and recommend hunting quotas for whales since its inception in 1946. The stock assessments of the 1970s were based on the maximum sustainable yield (MSY) concept—the aim of management being to produce yields at the maximum level that could be continued indefinitely. This is the maximum of the total revenue curve in Fig. 10.2. This aim has gone out of favour, for both biological and economic reasons. On the economic side, the analysis surrounding Fig. 10.2 made clear the irrelevance of MSY to economic profit-maximization, principally because it ignores costs (a point well made by Clark 1990). On the biological side, the objection is the knife-edge nature of the MSY—if harvesting were slightly higher than the MSY level, then the population would begin to decline, to extinction if the situation were not rectified. A higher harvesting level might arise due to a miscalculation of the stock size (particularly easy to do for whales), or to the stochasticity of natural systems, since there will always be variability in the system, rather than the deterministic parameter values assumed by the simple models from which MSY is calculated. Also, MSY models assume that there is no problem in reversing the effects of over-exploitation by reducing the harvest, i.e. that there is no problem with hysteresis in the system.

Although MSY is still valid as a constraint on hunting rate, the emphasis has shifted towards the explicit treatment of stochasticity in the modelling of natural resource harvesting. Walters and others (Holling 1978; Walters 1986) have championed adaptive management as a tool for resource managers, whereby a feedback loop is established between management actions and their outcomes, through monitoring. Others have explored the inclusion of stochasticity into harvesting decisions using Bayesian methods, which produce results that are easier to interpret than traditional methods (Clark & Kirkwood 1986). The Revised Management Procedure is particularly innovative in that it considers several sources of bias and uncertainty independently, and incorporates a *precautionary approach* (Kirkwood 1996). The precautionary approach is gaining stature fast as the basis upon which decisions with uncertain effects should be made—the benefit of the doubt should be given to ecosystem services whose loss would be irreversible (see also Chapter 11).

In many ways, as would be expected, ecological developments in ecological economics have paralleled similar developments in mainstream ecology. Thus the treatment of stochasticity has gained in prominence. Optimization models involving stochastic dynamic programming are being used for human decision-making in similar ways to their use in behavioural ecology. This is an example of a technique that is well established in neoclassical economics (Puterman 1978), was applied in the 1970s to natural resources (e.g. Reed 1974) then moved to theoretical ecology (Houston & McNamara 1988; Mangel & Clark 1988) and thence back into ecological economics, where it is realizing its potential as a tool in the analysis of optimal resource management. For example, Mace & Houston (1989) used stochastic dynamic programming to analyse the optimal herd composition for a pastoralist attempting to avoid destitution. Similarly, ecological advances in individual-based and spatial modelling are now well established in ecology (see Chapters 1 and 3), and so progress is likely to be made in their application to resource management problems (e.g. Clayton *et al.* 1997).

As ecology has developed a concern for the whole ecosystem to be represented in its models (Chapter 11), so progress has been made in ecological economics away from the harvesting decisions of a single exploiter of a single species, and towards the complexities of several exploiters and several species being harvested. Similar models and difficulties exist as for the equivalent ecological models of predator–prey relationships. One result from the simplest generalization of the one-species model is that if two or more species are being harvested together, then the species with the lower population growth rate and/or higher catchability may be driven to extinction on the back of the harvesting of the other species, despite the fact that if it were being exploited alone, harvesting would stop before extinction because of the escalating costs of hunting (Clark 1990). This has been suggested as a mechanism for the continued decline of the blue whale arising from the harvesting of other whale species; the rhino in the Luangwa Valley, Zambia, due to elephant hunting; and the babirusa in Sulawesi, due to the hunting of the Sulawesi wild pig (Clayton *et al.* 1997).

Of particular interest are models using game theory as a tool for understanding the decisions of competing harvesters. Game theory has followed a similar route to ecological economics as stochastic dynamic programming—from applied mathematical economics to theoretical ecology and back to ecological economics, though again with some early application to ecological economics problems. An example of the kind of modelling that can be done is Kaitala & Pohjola (1988), on the dynamics of competing Japanese and Australian tuna fishing industries. However, as well as specific applications such as this one, there is more general power in game theory as a tool for the understanding of resource management problems, as was discussed above.

Potential future areas for advances

Costanza *et al.* (1991) have set out a comprehensive agenda for the future of ecological economics, including the major research questions that they feel require addressing. They identify five major areas of interest to ecological economists: sustainability; valuation of natural resources and natural capital; ecological economic system accounting; ecological economic modelling at all scales; developing innovative instruments for environmental management. This is the consensus of a conference of ecological economists (with the emphasis on economists), and includes the major issues discussed above. Here, I offer a supplement to their list, with a bias towards the areas which I feel are presently neglected by ecologists and are desperately in need of some ecological input.

Sustainability, in particular the ecological component

If there is to be general consensus that sustainable use is a positive goal, then the ecological definition of sustainability needs some theoretical work. It may be relatively simple to predict sustainable levels of harvesting for single species, but the sustainable levels of use of more complex, real world systems have not been adequately studied. This is especially pertinent in the light of the continuing popularity of the sustainable use of *protected areas* as a conservation tool (e.g. Western & Henry 1995). If it becomes essential for protected areas to pay their way, then the sustainable levels of use, be it subsistence hunting, tourism or commercial harvesting, must be gauged. This should be done in the light of the precautionary principle and the requirement of a protected area truly to protect the species and ecosystems within it. Are there levels of use low enough to allow the structure of the protected community to remain basically intact, even if the relative importance of certain species within the community is altered? What level of ecological community disruption is acceptable within a protected area?

Environmental economics: the whole subdiscipline

This is an area in which substantial economic progress has been made, but with little input from ecologists. For example, a lot of mainstream ecological work is done in the field of pollution, its effects and control, but there seems to be little interaction between the theoretical and policy-based work of the economists and the research of the ecologists. Ecologists need to make their voices heard if policies for pollution control are to be both economically and ecologically sound. For example, one policy instrument popular with ecological economists is the transferable emissions permit, with the cost of a permit to emit pollution zoned by proximity to human settlements, the idea being that the further from human settlements the factory is, the cheaper it should be to pollute.

The linkage between environmental economics and resource economics (Fig. 10.1)

This has been researched in a few cases, for example the effects of acid rain on the production of the forests and fisheries of Central and Northern Europe (McCormick 1989), and clearly demonstrated in others, such as the effects of the pollution of the Aral Sea area on human health (Reznichenko 1992). The demonstration of other clear linkages in which pollution has compromised the ability of the environment to produce inputs for the economy will strengthen the link closing the circle on the neoclassical open view of the economy.

Flows of energy and matter within ecosystems

Much ecological research has been devoted to the study of energy and nutrient cycling in space and time (e.g. Baird *et al.* 1990). This might well be helpful knowledge in the research area attempting to express bioeconomic values in physical terms, rather than in terms of money or individual utility.

Valuation of ecosystem services

Although the valuation of nonmarket benefits has drawbacks, as discussed above, useful research could be done if ecologists got more involved, particularly in two areas. The first is the importance of ecosystem services for the protection of economic activity, for example the importance of forested watersheds in the generation of hydroelectric power (Garcia 1984) or of coral reefs as sea walls for vulnerable coastal settlements (Wells & Price 1992). Costanza *et al.* (1997) have estimated that the average annual value of the world's ecosystem services are $33 trillion, compared with a global national product total of $18 trillion per year. The second area of interest is the inclusion of ecological assets as natural capital in a country's national accounts, so that a more realistic picture of a country's bioeconomic health is obtained.

In short, there is considerable scope for ecologists to get involved in ecological economics. This needs to be fulfilled if the true value and complexity of the environment is to be recognized in economic analyses. Perhaps one of the problems of the discipline of ecological economics is that it is still predominantly a branch of economics, divorced (often acrimoniously) from the ecologist's equivalent applied discipline of conservation biology. A recent example of this is the debate over the African elephant (compare Poole & Thomsen's 1989 analysis with that of Barbier *et al.* 1990). Perhaps we ecologists should make more effort to bring the two disciplines together, so that ecological knowledge gets a higher priority, and the discipline changes from ecological economics into economic ecology!

Acknowledgements

I am very grateful to Jim Cannon, Andrew Price, Lynn Clayton and Ruth Mace for helpful comments and suggestions on the manuscript.

References

Anderson, L. (1989) Enforcement issues in selecting fisheries management policy. *Marine Resource Economics*, **6**, 261–277.

Avio, K.L. & Clark, C.S. (1978) The supply of property offences in Ontario: evidence on the deterrent effect of punishment. *Canadian Journal of Economics*, **10**, 1–19.

Axelrod, R. (1984) *The Evolution of Co-operation*. Basic Books, New York.

Baird, D., McGlade, J.M. & Ulanowicz, R.E. (1990) The comparative ecology of six marine ecosystems. *Philosophical Transactions of the Royal Society of London*, **B333**, 15–29.

Barbier, E., Burgess, J.C., Swanson, T.M. & Pearce, D.W. (1990) *Elephants, Economics and Ivory*. Earthscan, London.

Becker, G.S. (1968) Crime and punishment: an economic approach. *Journal of Political Economy*, **76**, 168–217.

Brown, G. (1989) The viewing value of elephants. In: *The Ivory Trade and the Future of the African Elephant*. Ivory Trade Review Group, Queen Elizabeth House, Oxford.

Burghes, D. & Graham, A. (1986) *Introduction to Control Theory including Optimal Control.* Ellis Horwood, Chichester.

Clark, C.W. (1973) Profit maximisation and the extinction of animal species. *Journal of Political Economy*, **81**, 950–961.

Clark, C.W. (1976, 1990) *Mathematical Bioeconomics: The Optimal Management of Renewable Resources.* Wiley Interscience, New York.

Clark, C.W. & Kirkwood, G.P. (1986) On uncertain renewable resource stocks: optimal harvest policies and the value of stock surveys. *Journal of Environmental Economics and Management*, **13**, 235–244.

Clayton, L., Keeling, M.J. & Milner-Gulland, E.J. (1997) Bringing home the bacon: a spatial model of wild pig hunting in Sulawesi, Indonesia. *Ecological Applications*, **7**, 642–652.

Cook, P.J. (1977) Punishment and crime. *Law and Contemporary Problems*, **5**, 164–204.

Costanza, R., Daly, H.E. & Bartholomew, J.A. (1991) Goals, agenda and policy recommendations for ecological economics. In: *Ecological Economics* (ed. R. Costanza). .Columbia University Press, New York.

Costanza, R., d'Arge, R., de Groot, R. *et al.* (1997) The value of the world's ecosystems services and natural capital. *Nature*, **387**, 253–260.

Dasgupta, P.S. & Heal, G.M. (1979) *Economic Theory and Exhaustible Resources.* Cambridge University Press, Cambridge.

Department for Economic and Social Information and Policy Analysis (DESIPA) (1993) *Integrated Environmental and Economic Accounting.* Studies in Methods, Series F, No. 61. United Nations, New York.

Donelly, D. & Nelson, L. (1986) *Net Economic Value of Deer Hunting in Idaho.* USDA Forest Service Resource Bulletin RM-13.

Ehrlich, I. (1973) Participation in illegitimate activities: a theoretical and empirical investigation. *Journal of Political Economy*, **81**, 521–565.

Field, B.C. (1994) *Environmental Economics.* McGraw-Hill, Singapore.

Fowler, C.W. (1984) Density dependence in cetacean populations. *Report of the International Whaling Commission*, Special Issue **6**, 373–379.

Garcia, J.R. (1984) Waterfalls, hydropower, and water for industry: contributions from Canaima National Park, Venezuela. In: *National Parks, Conservation and Development: The Role of Protected Areas in Sustaining Society* (eds J.A. McNeely & K.R. Miller). Smithsonian Institution Press, Washington DC.

Hardin, G. (1968) The tragedy of the commons. *Science*, **162**, 1243–1247.

Hodson, T.J., Englander, F. & O'Keefe, H. (1995) Rainforest preservation, markets and medicinal plants: issues of property rights and present value. *Conservation Biology*, **9**, 1319–1321.

Holling, C.S. (1978) *Adaptive Environmental Assessment and Management.* John Wiley, New York.

Houston, A.L. & McNamara, J.M. (1988) A framework for the functional analysis of behaviour. *Behavioral and Brain Sciences*, **11**, 117–163.

Kaitala, V. & Pohjola, M. (1988) Optimal recovery of a shared resource stock: a differential game model with efficient memory equilibria. *Natural Resource Modeling*, **3**, 91–119.

Keith, J.E. & Lyon, K.S. (1985) Valuing wildlife management: a Utah deer herd. *Western Journal of Agricultural Economics*, **10**, 216–222.

Kennedy, J.O.S. (1987) A computable game theoretic approach to modelling competitive fishing. *Marine Resource Economics*, **4**, 1–14.

Kirkwood, G.P. (1996) The Revised Management Procedure of the International Whaling Commission. *Fisheries Management Global Trends Conference, Seattle, USA.*

Kula, E. (1992) *Economics of Natural Resources and the Environment.* Chapman & Hall, London.

Leader-Williams, N., Albon, S.D. & Berry, P.S.M. (1990) Illegal exploitation of black rhinoceros and elephant populations: patterns of decline, law enforcement and patrol effort in Luangwa Valley, Zambia. *Journal of Applied Ecology*, **27**, 1055–1087.
McCay, B.J. & Acheson, J.M. (1990) *The Question of the Commons.* University of Arizona Press, Tucson, AZ.
McCormick, J. (1989) *Acid Earth: The Global Threat of Acid Pollution*. Earthscan, London.
Mace, R. & Houston, A. (1989) Pastoralist strategies for survival in unpredictable environments. *Agricultural Systems*, **30**, 1–19.
Mangel, M. & Clark, C.W. (1988) *Dynamic Modeling in Behavioral Ecology*. Princeton University Press, Princeton.
Mazany, R.L., Charles, A.T. & Cross, M.L. (1989) Fisheries Regulation and the Incentives to overfish. Paper at Canadian Economic Association Meeting, 2–4 June 1989, Laval University, Quebec.
Meadows, D.H. (1972) *The Limits to Growth, a Report for the Club of Rome's Project on the Predicament of Mankind*. Earth Island, London.
Mesterton-Gibbons, M. (1993) Game-theoretic resource modeling. *Natural Resource Modeling*, **7**, 93–147.
Milner-Gulland, E.J. & Leader-Williams, N. (1992) A model of incentives for the illegal exploitation of black rhinos and elephants. *Journal of Applied Ecology*, **29**, 388–401.
Milner-Gulland, E.J. & Mace, R.H. (1998) *Conservation of Biological Resources.* Blackwell Science, Oxford.
Milner-Gulland, E.J., Beddington, J.R. & Leader-Williams, N. (1992) Dehorning African rhinos: a model of optimal frequency and profitability. *Proceedings of the Royal Society of London*, **B249**, 83–87.
Norgaard, R.B. & Howarth, R.B. (1991) Sustainability and discounting the future. In: *Ecological Economics* (ed. R. Costanza). Columbia University Press, New York.
Norton-Griffiths, M. & Southey, C. (1995) The opportunity costs of biodiversity conservation in Kenya. *Ecological Economics*, **12**, 125–139.
Pearce, D. & Turner R.K. (1990) *Economics of Natural Resources and the Environment*. Johns Hopkins, Baltimore, USA.
Perrings, C. (1987) *Economy and Environment*. Cambridge University Press, Cambridge.
Peters, C.M., Gentry, A.H. & Mendelsohn, R.O. (1989) Valuation of an Amazonian rain forest. *Nature*, **339**, 655–656.
Poole, J.H. & Thomsen, J.B. (1989) Elephants are not beetles. *Oryx*, **23**, 189–198.
Price, C. (1993) *Time, Discounting and Value*. Blackwell, Oxford.
Puterman, M.L. (1978) *Dynamic Programming and its Applications*. Academic Press, NY.
Reed, W.J. (1974) A stochastic model for the economic management of a renewable animal resource. *Mathematical Biosciences*, **22**, 313–337.
Reznichenko, G. (1992) *The Aral Sea Tragedy*. Novosti, Moscow.
Simon, J.L. & Khan, H. (1985) *The Resourceful Earth*. Blackwell, Oxford.
Stewart, P.J. (1985) The dubious case for state control. *Ceres*, **18**, 14–19.
Sutinen, J. & Anderson, P. (1985) The economics of fisheries law enforcement. *Land Economics*, **61**, 387–397.
Sutinen, J.G. & Gauvin, J.R. (1989) An econometric study of regulatory enforcement and compliance in the commercial inshore lobster fishery of Massachusetts. In: *Rights Based Fishing* (eds P.A. Neher, R. Arnason & N. Mollet). Kluwer Academic Publishers, Lancaster.
Walters, C. (1986) *Adaptive Management of Renewable Resources*. Macmillan, London.
Wells, S.M. & Price, A.R.G. (1992) *Coral Reefs — Valuable but Vulnerable.* WWF, Gland, Switzerland.
Western, D.W. & Henry, W. (1995) *Natural Connections: Perspectives in Community-based Conservation*. Open University Press, Milton Keynes.
World Commission on Environment and Development (WCED) (1987) *Our Common Future*. Oxford University Press, Oxford.

11: Ecosystem analysis and the governance of natural resources

J.M. McGlade

Introduction

Despite the growing hegemony of the environment, many of the earth's living resources are in decline or under serious threat of collapse. Such changes are occurring everywhere as a result of human population growth, alterations to the biosphere, via long-term trends in the climate, and increased levels of exploitation.

Given that it is impractical to control the dynamics of most natural ecosystems, the current emphasis in many resource management systems on determinism, stability and optimization is clearly inappropriate. Instead, resource governance requires a new theoretical basis upon which to create management tools that will mitigate against the effects of change and cope with increasing levels of conflict over use demands.

In previous chapters, new concepts and theoretical approaches have been presented to address the impacts of stochasticity, biological variation and external forcing functions on resource exploitation and ecosystem functioning. These have included different types of spatial models, individual-based models, pairwise correlations, trophic interactions, life-history dynamics and ecological economics. In this chapter I explore how these and other approaches can be integrated over a range of physical, biological and socioeconomic scales to respond to the problems of resource governance that now face us.

System description is still one of the least resolved aspects of ecology. There are numerous approaches available, each using different criteria and operating over different space and time scales. They include biogeography, habitat delimitation, population and metapopulation dynamics, tropho-dynamics, etc. Once an approach and numeraire (e.g. numbers of species, units of carbon flow between trophic levels, etc.) has been adopted, the boundaries of an ecosystem need to be determined and the groupings within it established. But the problem is that ecosystems, by their very nature, are continuously changing, and the elements within them shifting between groups (e.g. age classes). Fuzzy logic (Zadeh 1965) provides a tool to address this problem and later in the chapter I explore its use and suggest why it is potentially a more appropriate approach for ecologists than simple Bayesian methods or other statistical or empirical approaches of pattern matching and classification.

The development of any model depends upon its purpose, so the degree of generality and the suitability of approach needs to be judged accordingly. One useful approach is to make a primary subdivision into *implicit* and *explicit* models. For example, in implicit patch models, space is represented by subdivision of ecological systems into different regions. In explicit models, fixed spatial locations are ascribed to individuals or populations; these models can then be further divided into continuous and discrete models. When space is treated as a continuum in explicit continuous models, a partial differential equation is used; when space is treated discretely, *cellular automata* and *coupled map lattices* are used for discrete and continuous state variables respectively (McGlade 1993).

Resource scientists have attempted to approximate many of the implicit ecological processes in simple mathematical models or numerical simulations with sometimes catastrophic results (McGlade 1989; Cook *et al.* 1997; Healey 1997; Hutchings *et al.* 1997a,b). The most common output required from such models are predictions about the effect of interventions on a particular ecological resource; e.g. water, fish stock, agricultural land, etc. Most existing resource-use models are designed to simulate or in a crude way anticipate the future—by implication these models are supposed to include all the key interactions and exchanges of materials as well as those processes directing the forward evolution of an ecosystem. All future states are thus contained within the dynamical description of the current system. This is rarely the case (and even if it were, the inner dimensions of such a model would contain so much working detail that it would be of little practical use).

Another reason why such chronic failures occur is that in many instances highly relevant knowledge held within the local community is ignored. In the last section of this chapter I examine how expert systems can be used to exploit the knowledge and insights that exist, in order to predict the outcome of change in systems that are inherently 'unmodellable' or uncertain. Such systems are highly appropriate for ecosystem management because they allow managers to make extensive use of the specialized and sometimes highly localized and nontextual information held by individuals and communities.

This last point is ultimately of great interest, because in the context of uncertainty about the future of the biosphere, a variety of policy responses can arise depending on which constituency is consulted. Different groups have different collective mindsets and attitudes to decision-making and risk. For example, in an earlier paper on human behaviour in the fishing industry two classes of fishermen were found, *stochasts* (high risk-takers) and *carte-*

sians (low risk-takers) (Allen & McGlade 1986, 1987). Stochasts were inevitably high-liners, who could work successfully, even with highly uncertain or imperfect information, to hunt out new schools of fish, harvest them quickly and then move on. Cartesians on the other hand would take their time responding to coded calls, waiting for a number of vessels to confirm their catches before moving to the area, by which time the catch rates would already be relatively low. Separately, these behaviours were only marginally efficient; together they became a deadly combination, allowing fleets to act like super-predators. Policies aimed at individual vessels were therefore inappropriate for controlling such pack behaviour. The perception of risk-taking is however another matter, as in interviews everyone wanted to be thought of as a stochast!

From a policy and institutional perspective, the difference between perceived and actual human responses to uncertainty about resources is thus very important. For example, Ausubel *et al.* (1990) and Bretherton (1994) characterize four types of response to predictions about climate change: entrepreneurs who see change as inevitable and consider it likely to stimulate creativity and cash flow; pessimists who consider change as disastrous; waverers who respond to uncertainty by denial; and evaluators who see the wisdom of recognizing limits to growth. The influence of these groups to a large extent follows the pedigree of the scientific evidence and theories. And from this stems different forms of governance. In the final section I also examine the emergence of governance in relation to the sustainable use of natural resources and discuss the role of theoretical ecology in determining the mindset of society and hence the future of our biosphere.

Ecosystem description: components, internal relationships and fuzzy sets

Ecological components

Apart from rate-determining processes, ecosystem models contain a number of elements or building blocks. A number of new analytical techniques (molecular probes, acoustic imaging, earth observation, compact airborne sensor imaging) have helped considerably in the definition of ecosystem components; thus in addition to the range of taxonomic levels and classes commonly used (e.g. strains, subspecies, species, populations and individuals), other elements now include trophic groups, communities, assemblages, keystone species, metapopulations and habitats.

Whilst most resource management models are based on populations, there are a number of instances, such as in the case of some endangered species, plants, bacteria or viruses, colonies, ramets, etc., where an individual-based approach has been used to support conservation and management decisions (see Chapters 1, 3 & 4 and DeAngelis & Gross 1992; Holt *et al.* 1995; Turner *et al.* 1995). For species such as the northern spotted owl, California spotted owl, the desert tortoise and Kirtland's warbler, individual-based models have had a tremendous influence on the practical management of the whole ecosystem. However, by providing immense detail and 'realism', they have also led to an over-estimation of their reliability. For example, Bart (1995) points to a number of instances where individual-based models have failed to predict the densities of animals, despite continuous sampling and data collection. Field workers and conservationists thus need to establish guidelines concerning the types of analyses required before accepting any model output as a basis for management decisions. In the case of the northern spotted owl recovery team, guidelines now include secondary predictions and outputs (e.g. the pathways followed by dispersing juveniles, distribution of ages at first breeding, locations of individuals across a landscape and distribution of suitable habitat within territories). Similar guidelines have been suggested in the management of the red grouse (see Chapter 1).

Techniques to characterize classes of individuals, sub-populations and populations have expanded from simple morphological descriptions to include behavioural and molecular characteristics (Begon *et al.* 1996). The results from these studies have shown that whilst the concept of a population was important in the past, *metapopulations* are potentially of greater relevance to resource management. A metapopulation is a set of populations that can be effectively separate, weakly coupled or globally interacting through strongly coupled patches (Frank *et al.* 1994; Hanski *et al.* 1995). Metapopulation models are used to describe the balance between local and global dynamics (Hanski *et al.* 1996; Sutcliffe *et al.* 1997). In recent years, metapopulation approaches have been used to address critical issues such as recruitment in marine populations, loss of endangered species, habitat conservation and the spread of diseases in the wild (e.g. Grimm *et al.* 1994; Grenfell *et al.* 1995; Grenfell & Harwood 1997).

The next level of aggregation is the *community* and/or *assemblage* (Magurran 1991). Definitions of the term community include 'a group of populations of plant and animals in a given place' (Krebs 1985); 'an assemblage of species populations which occur in space and time' (Begon *et al.* 1990); and 'an organized body of individuals in a specified location' (Southwood 1988). Underwood (1986) sees a community as tightly structured, consisting of many types of organisms at different trophic levels, whereas an assemblage is a neutral term referring to a collection of organisms at a particular site at a particular time. Others see that assemblages are made up of populations and species of the same phyla. Bianchi (1992) uses the term assemblage to indicate 'an association of co-existing species with similar environmental tolerance, possibly trophic relationships, but not totally interdependent'. Tyler *et al.* (1982) however adopt the criterion that an assemblage is based solely on its geographic distribution. Whichever is used, it is important to recognize that many contemporary assemblages are geologically recent (often post-Pleistocene), but their properties might be anachronistic, having been shaped by interactions that may no longer exist because of extinctions (Janzen & Martin 1982).

The paucity of long-term ecological data sets means that many community analyses rely on statistical pattern matching between location and/or environmental factors, or the presence and absence of species (see Kooijman 1977; ter Braak 1986). For example, Hawkins (1988, 1990) found that having distinguished six host feeding-niches, global parasitoid communities could be broadly characterized as either interactive (i.e. structured by interspecific competition) or noninteractive (i.e. not structured by competition) from an ANOVA of logarithmically transformed data of parasitoid richness, environmental and biogeographical variables.

Empirical studies have also given rise to the concepts of *keystone species* and *trophic cascades*. Keystone species play a significant role in the functioning of an ecosystem (see Paine's original experiments on the intertidal zone, 1966; 1974); if lost, the species composition of the community itself may drastically change or collapse, e.g. through cascading changes in trophic dynamics (Carpenter *et al.* 1985). A number of keystone species have been described, including the army ant (*Eciton burchelli*) in neotropical rainforests. Many birds, lizards and invertebrates associate with this ant and more than 50 species of ant-birds are obligate trail followers (Willis 1967; Ray & Andrews 1980). The foraging activities of this ant also create a patchwork of areas in different stages of ecological succession (Franks 1982; Franks & Bossert 1983; Otis *et al.* 1986).

Another type of aggregation found in ecological modelling is the *trophic group*. The concept of a trophic level made up of groups of organisms eating resources from a similar level in the energy cycle arose out of Elton's (1927) pyramid of numbers and Lindeman's (1942) work on decomposers. In a trophic framework, the ecosystem is described via energy flows between levels or groups, the internal behaviour of which can be independently modelled. There are difficulties and ambiguities when assigning species to particular trophic groups or levels when they consume organisms and materials from more than one trophic level, e.g. omnivorous fishes. Ulanowicz (1983) uses proportions to assign a species to different trophic positions, but in this way a trophic level or group is no longer a discrete entity. Nevertheless, models of trophic groups can lead to useful insights about ecosystems as a whole. For example, recent papers by Pauly and co-workers have shown the long-term significance of fishing on shifting the trophodynamics further down foodchains (Pauly & Christensen 1995; Pauly *et al.* 1998).

Community versus functional relationships

An ecosystem is a spatio-temporal component of the biosphere, determined by past and present environmental forcing functions and interactions amongst the biota. Within the boundaries of an ecosystem we can expect to see homogeneous and/or characteristic patterns in the dynamics, structure and evolution of its biological components (e.g. Stommel 1963; Steele 1978; Powell 1988; Hogg *et al.* 1989; May 1989). Ideally, as Rosen (1977) has stressed, ecologists should have no preconceived notions about the relative importance of subunits, rather identification should emerge from an analysis of the system behaviour, so that any function-preserving fractionation remains compatible with the original system dynamics.

Traditionally this has not been the case. Biotic elements have always played a critical role in establishing the boundaries of different habitats, which are then treated synonymously with ecosystems. The underlying assumption is that ecosystems are networks of interacting populations subject to natural selection, predation, competition and population growth and that abiotic elements are external influences. This population–community approach, which was built on the work of Lotka (1956) and Volterra (1931), has been well described in previous texts by May (1973, 1989) and Pimm (1982).

The alternative, functional view of ecosystems was developed by Lindeman (1942) and particularly E.P. Odum (1969) in their work on trophodynamics. The main emphasis was to understand energy and material flows, rather than the role of any one specific species or population, in order to explain ecosystem structure (see review by DeAngelis 1992). At its extreme, such a functional approach leads to analyses of total flows in and out of whole landscape units (e.g. Bormann & Likens 1979).

Tansley (1935) recognized the importance of both abiotic and biotic components in defining ecosystems, a view echoed by Hutchinson (1978) in his work on the niche, which he saw as the environmental context in which the community underwent its dynamics. There are now numerous classification schemes for terrestrial and marine ecosystems, where the boundaries are determined by both environmental and biological criteria (e.g. Bailey 1989; Longhurst 1995).

With the increase in ecological modelling and simulation there has been a trend towards combining community and functional approaches to examine fundamental problems relating to resource management. For example, in the determination of migration schedules in bird populations (e.g. Ens *et al.* 1994); using interactions amongst species within food webs, to examine the influence of energy flows and hence resource biomass production (e.g. see Chapter 5); modelling the ideal free distribution to describe the distribution of organisms in relation to different predators and food sources (e.g. Fretwell 1972); using the levels of recycling to explain redundancies in communities (DeAngelis *et al.* 1989); and describing the dynamics of savannah vegetation to ascertain the relative importance of elephants and fire (Dublin *et al.* 1990).

Another combined approach is mosaic theory. This asserts that an ecosystem never reaches equilibrium but rather remains in flux—a mosaic landscape consists of mosaic stones or patches of communities which cycle continually through a set of states, with adjacent patches cycling synchronously. This concept has been applied widely to forests and marine systems (see below and Whittaker & Levin 1977; Remmert 1985, 1991; Reise 1991; Wissel 1992; Hendry & McGlade 1995). (see also Chapter 1, Fig. 1.1.)

In marine systems a number of combined approaches have been used to define ecosystem boundaries; these include biological assemblages in relation to environmental forcing (e.g. Mahon 1985; do Carmo Gomes 1993; Koranteng 1998); marine communities under different flow rates of nutrients (e.g. Ulanowicz 1986; Baird *et al.* 1991); trophic mass-balance assessments of different communities (Pauly & Christensen 1995; Walters *et al.* 1997); biological and physical oceanography in relation to trophically dependent populations (e.g. MacCall 1990) and for large marine ecosystems (AAAS 1986, 1989, 1990, 1991, 1993; Sherman *et al.* 1996); bio-optical and physiological properties in relation to vertical biogeochemistry (e.g. Sathyendranath *et al.* 1991); and climate-ocean forcing on productivity patterns (e.g. Longhurst 1995). What is surprising is the relatively high degree of consistency in the definition of marine boundaries arrived at via different approaches.

Fuzzy sets

In our everyday life we continually provide descriptions of entities, events, processes and issues, using imprecise, linguistic phrases which are understood clearly, e.g. pollution levels are **getting worse**, fish are **becoming scarcer**. One way to utilize such qualitative, linguistic or imprecise information is to adopt a *fuzzy logic* approach to system characterization, using *fuzzy sets* (Zadeh 1965; McGlade & Novello-Hogarth 1997). Fuzzy set theory follows the principles of conventional set theory with one major exception: in conventional set theory elements are divided into two categories, i.e. those that belong to a set and those that do not. The conventional, non-fuzzy or crisp set thus maintains a clear difference between elements which are members and those which are not. In fuzzy set theory, the linguistic variables are context-dependent variables whose values are thus words or sentences, for example **oil spill size** (*small, medium, large*); **age** (*young, middle age, old*), etc. The range of possible values is known as the universe of discourse. Elements within the universe of discourse are assigned a grade of membership between 0 and 1, although in some cases the membership functions can be single values, or *singletons*.

Initially, all input variables are converted into fuzzy variables using membership functions—a process known as *fuzzification*. The shape of the membership function (e.g. a simple vector, S-function, triangular, trapezoid) is optimized through successive observations and may differ depending on the application and the need to capture different levels of uncertainty. For example, in an area where oil spills ranging in size from 10 to 100 km^2 have been observed, the fuzzy set 'about 50 km^2' can operate over the entire range of 0–100 km^2. The membership value decreases progressively from 1 to 0 as the distance from the set point (50 km^2) increases; thus at the 25 km^2 position, the membership is 0.5. In conventional set theory, this point would have been assigned a membership value of 0.

Fuzzy theory thus allows lesser points to be recognised within the universe of discourse which may signify other key attributes, e.g. heterogeneity in growth due to nutrient status. It also allows uneven observation of ecosystem components to be taken into account. For example, given sufficient time and access to a school of fish it would be possible to assign it to a particular crisp set of school sizes, e.g. 100, 200, 300, etc. However, it is more likely that even with repeated observations, the observer would only be able to provide an estimate of the size of the school in the field; in this case a fuzzy set can be created, e.g. where '100' = 50–149, '200' = 150–249, '300' = 250–349, etc. so that a fish school estimated to contain 120 fish is assigned to the fuzzy set 100, a school of 210 assigned to fuzzy set 200 and so on (see Fig. 11.1a). The measure of the observational uncertainty in assigning a particular school to a particular crisp set is thus taken into account. This situation is akin to one where it is difficult to assign sets in the first instance, e.g. as with trophic groups where membership can change due to migration or omnivory. In this case the trophic groups are placed in a fuzzy set from the start, and this can then also include observational uncertainties.

Uncertainty in the basic definition of a set or the observation of it can be further captured through the spread, shape and overlap with adjacent fuzzy sets through the manipulative operations of union, intersection and fuzzy relationships. The union operation, when applied to two fuzzy sets both of the same universe of discourse, is equivalent to a connective OR: the operation of union is indicated by the use of '+' sign instead of the conventional $\cup$ sign. For example, in the fuzzy sets describing **oil spill size**, sets linguistically named *small* and *medium* can be defined (see Fig. 11.1b). Applying the principle of union to the sets *small* and *medium* creates a *small* OR *medium* set (see Fig. 11.1c).

In a similar manner, the operation of intersection when applied to two fuzzy sets of the same universe of discourse is equivalent to a connective AND; the operation of intersection is indicated by the sign $\cap$. By the application of the operation of intersection to the fuzzy sets, *small* and *medium*, describing oil spill size, a new *small* AND *medium* set is created (Fig. 11.1d).

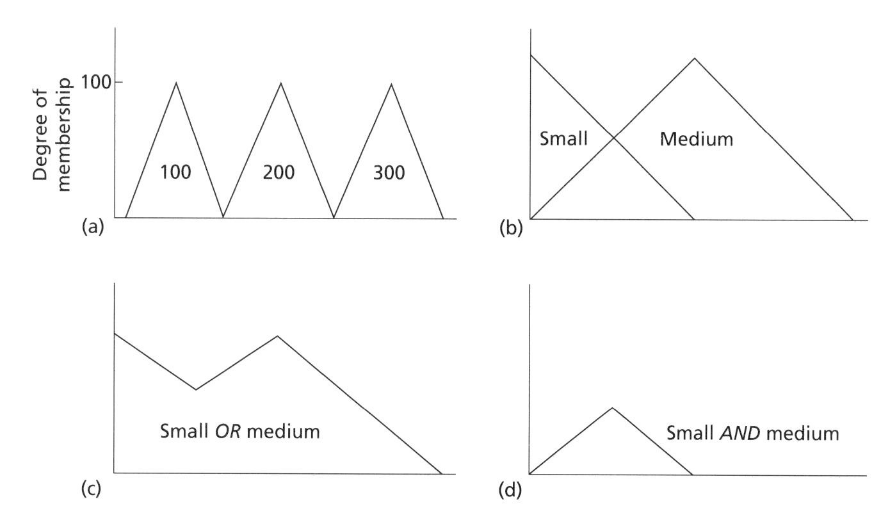

Fig. 11.1 (a) The fuzzy sets of fish school sizes '100', '200', '300'; (b) the fuzzy sets *small* and *medium*; (c) union of fuzzy sets *small* OR *medium*; (d) intersection of fuzzy sets *small* AND *medium*.

Hedges are used to emphasize (e.g. to reflect the phrase *very*, as in *very large*) or de-emphasize (e.g. to reflect *somewhat*) the fuzzy shape of the set. Finally, to extract a crisp out value for practical needs, *defuzzification* is undertaken, e.g. using the centroid or the maximum value of the set.

In this way even relatively poorly studied ecosystems can have their constituent parts described for further analysis. Later on in this chapter I examine a further use of fuzzy logic in the creation of expert systems for resource management.

Ecosystem analysis and resource management

As shown above, descriptive analyses and empirical data collection have largely dominated ecology, involving intensive experimentation and field work. It is not surprising then that theoretical approaches to ecology, often seen as lacking extensive support from observational data, have been considered either controversial or irrelevant. But in the last two decades, advances in many areas of mathematics, especially nonlinear dynamical systems and the study of complexity (see, for example, texts by Drazin 1992; Langton *et al.* 1992; Kauffman 1993; Mosekilde & Mouritsen 1995; Tong 1995), have shown that the way in which data are collected and their interpretation are not only highly dependent on the underlying assumptions of the investigator, but also on the form of the model used. It is even more important then that models are not only used in data interpretation but also in a priori definition of empirical investigations. In this section I look at advances in ecosystem analysis in relation to the management and governance of natural resources.

Ecosystem analysis and emergent properties

Nearly a decade ago, Cherrett (1989) identified 10 major concepts that were considered to have contributed to our understanding of the natural world: these included fluxes of matter and energy, food web dynamics, the ecological niche, diversity and stability, predator–prey and host–pathogen interactions, population regulation, competition, life-history strategies and optimization. Each continues to play a part in helping to understand more about ecosystem functioning, but as it is generally impossible to undertake full-scale ecosystem experiments it is often difficult to discriminate between them in any meaningful way. As a result, there continue to be conflicts over the relative importance of any one approach or the properties that should be studied in order to manage ecosystems and resources effectively.

Ecology does not have a set of general laws, so it is virtually impossible to provide robust predictions about individual organisms, populations or whole ecosystems. Some large-scale approximations do exist (e.g. individuals give rise to individuals) but the question remains as to whether they occur more broadly and produce emergent properties or whether ecosystems are always unpredictable thereby making it invalid to look for deeper universal laws (Stewart 1998). If as Cohen and Stewart (1994) conclude 'complexity is one of the least conserved properties in the universe' then by studying ecosystems through their component parts, will metacharacteristics or properties emerge that can then be more easily monitored or managed?

Using 'toy models', Bak, Tang and others (Bak *et al.* 1988; Flyvbjerg *et al.* 1995) have demonstrated that large dynamical systems show a tendency to evolve or self-organize into a critical, non-equilibrium state characterized by bursts or avalanches of dynamical activity of all sizes. The idea is very appealing in its possible explanation of intermittency of biological evolution. In the so-called 'NKC-models' for coevolving species evolving with periods of stasis interrupted by coevolutionary avalanches, Kauffman (1993) has argued that the ecology as a whole is the most fit organizational structure (even though some external forcing of the system had to be applied to achieve this!).

Other emergent properties include *emergy*, one of the thermodynamic conjectures of H.T. Odum (1983, 1988), whereby the maximum power principle is proposed as a generic feature of evolution in ecosystems. However, there are formidable obstacles to obtaining an operational definition of a general, aggregated available-work concept, which is a prerequisite for the systems approach and concepts of *emergy* used by Odum, and these strongly mitigate against its widespread use for the management of resources (Månsson & McGlade 1993).

Yet despite the fact that a detailed mechanistic understanding is fundamentally harder to achieve for biological systems compared to physics and chemistry (e.g. see Schoener 1986), there are still a number of reasons for encouraging research into general models that expose emergent properties. For example, general models can be simple and comprehensible and can be used to investigate those factors causing systems to differ in their behaviours and thereby eliminate implausible or impossible mechanisms. Such models frequently produce counter-intuitive results, e.g. the inability of super-viruses to invade certain spatially extended systems (Rand 1994), and the subsequent experimental investigations can lead to novel insights into how ecosystems function, e.g. see the role of parasites in driving oscillations in Anderson & May's (1991) significant survey of infectious diseases.

Indeed, as more attention has been paid to the role of underlying processes in defining ecosystems, the effect of changes in the rates of processes on ecosystem resilience, stability and structure have become more apparent. Simon (1962), in his seminal work on complexity, came to this conclusion for social systems. His ideas were subsequently taken up within many academic disciplines including ecology (see the work on hierarchy theory Allen & Starr 1982, O'Neill *et al.* 1986). The main tenet of this approach is that different rates in the system cause different behaviours that can be grouped into classes. Slower rates occur at higher organizational levels; thus changes in species through succession represents the integration of shorter-term changes. In this way even highly complex systems can be decomposed into discrete entities on the basis of response times. Although there have been few examples where this approach has been directly applied to ecosystem function or resource management, it has yielded a wide range of insights into how interventions at one level might lead to counter-intuitive results at another.

By contrast, the inclusion of too much detail into numerical simulations often obscures understanding and limits the use of any results to a narrow range of systems. Thus highly parameterized models (as are often found in ecology) are neither powerful nor reliable predictors, because the numerical predictions can be precisely fitted to almost any data (Levin 1992). Holling (1964) also makes the point that because many numerical simulations involve estimation or condensing of the parameter space, they tend to make little reference to underlying mechanisms. Thus the value of quantitative predictions for ecosystem prediction and resource management from very complicated models should generally be questioned.

The fundamental works of May (1973, 1981), Holling (1973) and Maynard-Smith (1974) have highlighted the importance of nonlinear processes in determining the functioning of ecosystems and the subsequent impacts on community structure. Their continued insights into evolution and dynamics have led to a gradual shift away from the idea that ecosystems are fundamentally stable and in equilibrium, to one where instability, nonequilibrium and even chaos can be considered the norm. Moreover they introduced key ideas in game theory and spatial modelling, which have developed rapidly over the past two decades. There are now many useful techniques that can be applied to the management and conservation of natural resources, many of which are described in this volume; other examples include:

1 identification of evolutionary stable attractors and invasion exponents in resource–predator–prey with local chaotic dynamics (e.g. Rand *et al.* 1994);
2 identification and measurement of the characteristic or underlying natural length scales of ecosystems (Keeling *et al.* 1997);
3 evolution to criticality in ecosystems (Bak *et al.* 1988);
4 determination of the fractal and multifractal structure of communities (e.g. Burrough 1981; Morse *et al.* 1985; Hastings & Sugihara 1993; Virkkala 1993; Hendry 1995);
5 correlation models to examine the effects of sites or individuals whose states are changing at rates determined by the states of their neighbours (e.g. Matsuda *et al.* 1992; Satō *et al.* 1994);
6 advances in stochastic spatial models (e.g. Durrett & Levin 1994); and
7 development of Markov chain Monte Carlo methods, e.g. for the spread of plant disease epidemics (Gibson 1997).

Ecosystem analysis in relation to resource management models

Although ecologists use a wide variety of analytical approaches to describe and interpret the patterns and processes that they observe (e.g. statistical mechanics, thermodynamics, linear algebra, nonlinear dynamics, geometry, neural networks, number theory, time series analysis and multivariate and spatial statistics), resource managers have generally relied on a far smaller subset of approaches. These include statistical correlations, ode's and pde's, plus a few examples of coupled map lattice models and cellular automata in certain horticulture and forestry problems (Hendry & McGlade 1995; Hendry *et al.* 1996). The delay amongst conservation and resource managers in moving from largely deterministic, nonspatial models to the types of models and analyses described in this book and listed above is partly due to the analytical demands of using these models but also due to the lack of appropriate governance structures to implement them (McGlade & Price 1993).

The majority of current resource management models aim to optimize the harvest of certain economically important species by commercial means, with virtually no reference to the other parts of the ecosystem. However, over the past three decades, there has been a gradual, but sometimes not-so-gradual, collapse of many commercially important populations of fish and mammals (e.g. Cook *et al.* 1997; Milner-Gulland & Mace 1998). Despite sometimes catastrophic results, little reference has been made to the role that models played in their demise. The reasons as highlighted in the Canadian case of the cod fishery were a mixture of technical and human errors, combined with a lack of understanding about the level of uncertainty in the system and how uncertainties were being propagated through the system by the very models that were generating them in the first place (Hutchings *et al.* 1997a,b).

From the analysis of the spatial dynamics of a mixed demersal fishery (Allen & McGlade 1986), it could be seen that many of the signals characteristic of the typical boom–bust cycle of a fish stock being exploited by a strong predator (i.e. a fishing fleet directed at a target species) were simply not picked up in the standard fisheries assessment models. The situation was further complicated by the spatial heterogeneity in the fleet caused by perceived success rates in finding new areas of high fish concentrations. Worse still, time delays in the management procedures meant that advice was often based on evidence that was up to 2 years out of date. Thus in any one year the industry could be under pressure to cut back fishing effort at a time when fish were abundant, thereby creating over-the-side wastage, or hunting for more fish when in fact the resource was collapsing (McGlade 1989).

Generally speaking, few resource management systems stray far beyond logistic or Leslie matrix analyses of target species. Thus metapopulation models, which incorporate predator–prey dynamics, are hardly used, even though it is well known that in many instances the effects of an intervention often affect the predator more than the prey. For example predator populations can be more severely suppressed by pesticides than pest populations (Waage *et al.* 1985). From theoretical studies it is also known that predator aggregation can play an important role in metapopulation dynamics (Godfray & Pacala 1992; Murdoch *et al.* 1992; Nisbet *et al.* 1992), and that migration can reduce the fluctuations in metapopulation density through an increase in asynchrony (Allen *et al.* 1993; Jansen 1994).

Other approaches that are also overlooked in many management systems include community analyses and trophic models. The development of trophic models for resource management by Polovina was primarily in response to a lack of age-structured data for many tropical aquatic populations (Polovina 1984; Christensen & Pauly 1992, 1993). Recent analyses using these techniques have generated disturbing results regarding the impacts of overfishing on the world's marine fish communities (Pauly *et al.* 1998). The analysis shows a gradual transition in landings from long-lived, high trophic level, piscivorous demersal fish towards short-lived, low trophic level invertebrates and planktivorous pelagic fish. Pauly (1998) concludes that without marine protected areas and changes to management the oceans might end up as a marine junkyard dominated by plankton.

Ecosystem level analyses are also important for the conservation of biodiversity, where the collapse of one resource clearly has an impact on many others. Mangel & Tier (1994) have shown that if such catastrophes as indicated above are ignored, errors in estimates of the expected extinction times of many orders will occur. Also, Andrén (1994), and Bascompte Solé (1996) have shown that extrinsic variations in habitat quality, either natural or arising from partial habitat destruction, are very important in determining the success of practical conservation measures (Pulliam 1988).

Costanza and co-workers (Costanza *et al.* 1997) have used the basic ecosystem approach of estimating flows of energy to derive estimates of the value of the planet's ecosystem services (given as $16–54 trillion per year): the comparison with global gross national product (GNP) ($18 trillion per year) underlines the point they wished to make, i.e. the extent to which the biosphere is undervalued in relation to human activity.

Expert systems and rule-based knowledge in resource management

Expert systems: brief background

As events around the world have shown, resource managers often think they know what an intervention in an ecosystem will do, but rarely do they know why or even how it is doing it! This problem arises because the managers are either acting locally and do not realize that they can affect and be affected by global ecosystem processes or that by acting at the ecosystem level they have not anticipated local effects on certain resources. This latter situation often arises because essential information about local conditions, which is not necessarily present in a formal context, e.g. traditions, folklore and other community-based knowledge, has not been adequately captured.

The issue then is how best to bring together information about the key processes and elements of the local and global systems, quantify or highlight the linkages between them, and identify the nature of the uncertainties in the system, so as to be able to assess the impacts of interventions on specific resources.

Intelligent soft-computing expert systems present one avenue to achieving several elements of this. These systems are now employed in many areas such as control in consumer electronics (e.g. auto-focus cameras), decision support in medicine (e.g. computer-assisted management of childbirth, automatic interpretation of brain activity, intensive care monitoring), business and finance decision support (e.g. credit rating, stock market analysis and forecasting, direct marketing) and natural resource management (e.g. development of coastal zone management schemes). An important goal of such systems is to simulate one or more forms of natural intelligence, e.g. learning, knowledge and skills, expert behaviour, adaptive and evolutionary strategies, in light of information and data on key processes.

Three key intelligent systems techniques, in conjunction with approaches such as fuzzy logic, are potentially useful for resource management: neural networks (e.g. Recknagel *et al.* 1997), expert systems and genetic algorithms systems. Fuzzy logic is used heavily in expert systems to handle uncertainty and imprecision in knowledge and data, and in neural networks to learn about hidden patterns within data sets in order to generate membership functions.

An expert system is a computer-based system that contains specialist knowledge in a given area and is able to use the stored knowledge to solve problems or make inferences and deductions that would normally require human expertise. In simple terms, to build an expert system, the necessary knowledge is obtained from one or more experts and encoded into a computer system. A typical expert system contains a knowledge base, which holds the expert knowledge, an inference engine, which decides how the knowledge in the knowledge base should be used, and a user interface, for communication with the user.

Rule-based knowledge

In the majority of expert systems the representation of knowledge is based on logic, using syllogisms and sets, and represented in the form of rules. Rules represent the long-term knowledge of the system. Inferences are made between rules and new information to derive new knowledge, using a variety of methods depending on the pedigree or quality of the research in the area; examples include:

1 *Deduction*: logical reasoning in which conclusions must follow from their premises.

2 *Induction*: inference from the specific case to the general.

3 *Intuition*: no proven theory. The answer just appears, possibly unconsciously recognizing an underlying pattern. Expert systems do not implement this type of inference yet, although a neural net will always give the best guess for a solution.

4 *Heuristics*: rules of thumb based on experience.

5 *Generate and test*: trial and error. Often used with planning for efficiency.

6 *Abduction*: reasoning back from a true conclusion to the premises that may have caused the conclusion.

7 *Default*: in the absence of specific knowledge, assume general or common knowledge by default.

8 *Autoepistemic*: self-knowledge. This is always context-sensitive.
9 *Nonmonotonic*: previous knowledge may be incorrect when new evidence is obtained.
10 *Analogy*: inferring a conclusion based on the similarities to another situation.

Although not explicitly identified, *commonsense knowledge* may be a combination of any of these methods. It is a type of reasoning that is very difficult for computers to mimic, and it is for this reason that fuzzy rule-based logic is generally used for commonsense reasoning in expert systems to support decision-making based on imprecise data (see below and examples in McGlade & Novello-Hogarth 1997; Roberts 1997; Silvert 1997; Wu *et al.* 1997; Zhu *et al.* 1997). Induction is used to infer new rules and rediscover known rules. The *metaknowledge* about known rules is usually of two types: *metarules* that tell how the rules are to be applied and a *rule model type* that determines whether the new rule is in an appropriate form to be entered into the knowledge base. The general format of a rule is:

IF ⟨*condition1*⟩ AND ⟨*condition2*⟩ (11.1)
THEN ⟨*action*⟩

IF ⟨*condition1*⟩ OR ⟨*condition2*⟩ (11.2)
THEN ⟨*action*⟩

In crisp expert systems, the variables for the rules (1) and (2) are either true or false. In fuzzy logic, as in conventional logic, the functions AND, OR and NOT are used to combine the fuzzy variables in the premise (the IF part of a rule). Thus fuzzy rules can refer to crisp or fuzzy sets, in the same way that experts do, to develop an heuristic rule set. An example of a simple fuzzy rule set would be:

IF catch rate for cod is ⟨*very high*⟩ AND price in port is ⟨*medium*⟩
THEN the occurrence of illegal landings is ⟨*high*⟩

i.e.

IF X is A AND Y is C
THEN Z is B

where A, B and C are fuzzy sets which are known in advance, and X, Y and Z are crisp or fuzzy measurements or the outcome of another fuzzy rule in the inference net. Given X and Y, a *fuzzy inference* process is used to deduce the extent of the *belief* in the conclusion of the rule or the degree to which a rule will activate.

One of the most common forms of fuzzy inference is the *max-min*. If there are multiple premises, then the induced fuzzy sets depend on whether the premise is joined by an AND or an OR function. Thus where an AND occurs, the inference is that both premises must be valid for the rule to be activated; hence the output is true at least to the degree that the least valid one is fulfilled, i.e. the minimum of the two membership degrees. The same is true for the OR, where only one or the other needs to be valid for the output to be valid; hence the output is true at least to the degree that the most valid one is fulfilled, i.e. the maximum of the two membership degrees.

Reasoning under uncertainty

Although much has been written about managing resources under conditions of uncertainty, it is important to clarify what is meant by the term. Uncertainty is not merely the spread of data around some arbitrary mean, known with confidence; it is a systemic form of error that can swamp an otherwise easily calculated random counterpart. As the complexity of a system increases, simple measures of inexactness such as error bars and estimates of variance cannot fully describe the uncertainty implicit in any analysis, and instead it must be clarified and dealt with explicitly. So for example, the errors associated with data points represent the *spread*, i.e. the tolerance or random error in a calculated measurement. *Confidence limits* refer more to risk; for example, in an analysis of future scenarios resulting from different policies, confidence limits are reflected in estimates when they are qualified as optimistic, neutral or pessimistic.

Reasoning under uncertainty is a crucial issue in resource management, mainly because of the heterogeneity of the information to be utilized and the pedigree of the field. When a field is largely based on uneducated guesses, and only definitions exist, logical reasoning is very difficult. As a field progresses, it moves to statistical procedures, computer models, theoretical models and finally to established theories, and the data underpinning this move from educated guesses, to historical/field data, and on to experimental data. Generally, it is true to say that natural resource management is still based on statistical procedures and historical data.

Different types of errors can also contribute to uncertainty: there is *ambiguity*, *incompleteness*, *incorrectness*, *false positive* (i.e. accepting a hypothesis when it is not true), *false negative* (i.e. rejecting a hypothesis when it is true), *imprecision* (i.e. how well the truth is known), *accuracy*, *unreliability* (if the measuring equipment supplying facts is unreliable the data are erratic), *random*, *systematic* (introduced via some bias), *invalid induction* and *invalid deduction.*

Unfortunately even experts are not immune to making mistakes, especially under conditions of high uncertainty! Different theories of uncertainty attempt to resolve some or all of these errors to provide the most reliable inference. For example, in contrast to the other errors in the list, the last two are errors of reasoning. This can be a major problem in knowledge acquisition when the experts' knowledge must be quantified in rules, because then inconsistencies, inaccuracy and all other possible errors of uncertainty may show up.

The major quantitative way of dealing with uncertainty has traditionally been through probability theory. *Classical probability*, which was first proposed by Pascal and Fermat in 1654, also called a priori probability, deals with ideal systems. In such systems all numbers occur equally, so an experiment or trial involving tossing a die a number of times will generate a number of equally possible *events*. When repeated trials give the same result the system is said to be deterministic.

An event is a subset of the sample space. A certain event is assigned probability 1 and an impossible event assigned probability 0. Mutually exclusive events have no sample point in common, e.g. an animal cannot be both gestating and not gestating at the same time. The corollary means that the probability of an event occurring plus it not occurring is equal to 1. These axioms, devised by Kolmogorov, refer to the *objective theory of probability*.

In contrast to the a priori approach, *experimental probability* or a posteriori probability, defines the probability of an event as the limit of a frequency distribution. The idea is to measure the frequency that an event occurs for a large number of trials and from this induce the experimental probability. This is the most common approach used in ecosystem or resource management systems, e.g. flood defence where the probability of an n year event horizon is used as the criterion for flood warnings to be issued.

Next there is *subjective probability*. This deals with events that are not reproducible and have no historical basis on which to extrapolate, such as drilling an oil well or building a wetland at a new site. However, a subjective probability by an expert is better than no estimate at all and is usually very accurate (or the expert would not be an expert for long). A subjective probability is actually a belief or opinion expressed as a probability rather than one based on axioms or empirical measurement. In the real world events tend to compound each other; so it is important to be able to separate those events that really are pairwise or stochastically independent. For the remainder, the effect on calculating the mutual independence of events is to enormously increase the number of equations that need to be satisfied. In this case an *additive law* is applied. Beliefs and opinions of an expert play an important role in expert systems, even though they cannot be explicitly described mathematically and the relative frequency method is impossible to apply.

Finally there is *conditional probability*, which describes the fact that events which are not mutually exclusive influence one another. Knowing that one event has occurred may cause us to revise the probability that another event will occur. In this instance the generalized multiplicative law is used. The conditional probability states the probability of event A given that event B occurred.

Inverse probability is the probability of an earlier event given that the later one has occurred. This type of probability often occurs where symptoms appear and the problem is to find the most likely cause. The solution to this problem is given by *Bayes' theorem*. Bayes' theorem is commonly used for decision-making and is also used in an existing expert system PROSPECTOR (Duda & Reboh 1984) to decide on favourable sites for mineral exploration. (PROSPECTOR achieved a great deal of fame as the first expert system to discover a valuable molybdenum deposit worth $100 million.)

Prospecting is an obvious activity where it is appropriate to use Bayesian approaches for decision-making under uncertainty. Initially, the prospector must decide what the chances are of finding a particular resource. If there is no evidence either for or against it being present, the prospector may assign the subjective prior probabilities for or against as both 0.5. Without evidence, an assignment of probabilities that are equally weighted between possible outcomes is said to be made *in desperation*. This term does not mean that the prospector is (necessarily) in desperate need; rather it is the term for unbiased prior assignment. If there is a survey there may be some information to help make a decision. However surveys are rarely 100% accurate. The test may be false positive or false negative. Using the conditional and prior probabilities an initial probability tree can be constructed, and from this a set of expected payoffs can be calculated at each event node, so that the overall best action can be determined. The decision tree is an example of hypothetical reasoning or 'what if' type situations. By exploring alternate paths of action, the paths that do not lead to optimal payoffs can be pruned.

Reasoning about events that depend on time is called *temporal reasoning* and it is something that humans do very well. However, it is difficult to formalize temporal events so that a computer can make temporal inferences. Yet it is just such systems that could be most useful in determining policies and planning options. Different temporal logics have been developed, based on different axioms (Turner 1984). The different theories are based on the answers to such questions as whether time is continuous or discrete; or is there only one past but many futures? If the temporal reasoning is best dealt with using probabilities, e.g. for weather, storms and epidemics, a *Markov chain process* should be used. For resource management it is also important to consider the different temporal scales over which changes are likely to occur, and the possibility of history playing a key role in determining the future possible states. More recently, Monte Carlo Markov chain (MC^2) approaches have been developed, using samplers such as Gibbs, to model complex biological data and construct the Markov chain in such a way that reversibility is achieved, i.e. its limiting distribution is the required posterior (Spiegelhalter *et al.* 1995).

So far we have only dealt with probabilities as measures of repeatable events. However in many instances the events or contexts are unique. In such a case the probability should be interpreted as the *degree of belief*, and the conditional probability is then referred to as the *likelihood*. It is important in determining the likelihood of some event occurring that the degree of uncertainty in each piece of evidence is understood.

The final aspect is the use of an *inference net*. For real-world problems the number of inferences required to support a hypothesis or to reach a conclusion are often very large. In addition many inferences are made under uncertainty of the evidence and rules themselves. An inference net is thus a good architecture for expert systems that rely on taxonomy or knowledge (see, for example, McGlade & Novello-Hogarth 1997). The basic idea is to encode the experts' knowledge into the system. The data for each model are organized as an inference net, where nodes can represent evidence to support other nodes that represent the ideas or hypotheses of other experts. Each model inside the system can be encoded as a network of connections or relations between evidence and hypotheses. Thus, an inference net is a type of semantic net. Observable facts, such as the type of meteorological conditions obtained from field observations, comprise the evidence to support the intermediate hypotheses; groups of intermediate hypotheses are used to support the *top-level idea*. Because experts often find it difficult to specify posterior probabilities or likelihood ratios, a certainty factor can also be applied.

Most expert systems can create partitioned semantic nets to enable the user to group portions of the net or rule-base into meaningful units. Partitioned semantic nets were developed by Hendrix (1979) to allow the power of predicate calculus such as quantification, implication, negation, disjunction and conjunction. The important point however is that most inference nets have a *static knowledge structure*; i.e. the nodes and connections between them are fixed in order to retain the relationships between nodes in the knowledge structure. More recent expert systems have been designed to allow a *dynamic knowledge structure* either with no fixed connections between rules or via changes in the probabilities associated with each node.

An expert system is thus not a pure probabilistic system because it can use fuzzy logic and certainty factors for combining evidence. The weighted combinations are also examples of *plausible relations*. The term 'plausible' means that there is some evidence for belief. Other terms also have a clear meaning, for example: *impossible* (evidence definitely known against); *possible* (not definitely disproved); *plausible* (some evidence exists); *probable* (some evidence for); *certain* (evidence definitely known supporting).

Probability theory can be considered as a way of capturing and reproducing uncertainty. However as much of what happens in resource management is based on human belief rather than the classic frequency interpretation of probability, an expert system is more appropriate because it allows *inexact reasoning*, i.e. where the antecedent, the conclusion and even the meaning of the rule itself is uncertain to some extent. Generally a rule with more *specific* patterns generally has a higher priority. There is also *recency of facts*; given differences in the rule-base through time, it is possible to undertake hindcasts and forecasts to examine the effects that different decisions using the system might have made at any one time. The ordering within a rule also makes a difference.

Certainty factors can also be assigned to rules to address the problem where one piece of evidence might give one value, whereas five give another. The expert system should not then fire on the basis of only one piece of evidence. Certainty factors are a special case of the Dempster–Shafer theory which deals with inexact reasoning (Gordon & Shortliffe 1985) but are only used in certain areas for the following reason. A fundamental difference between the Dempster–Shafer theory and probability theory is the treatment of *ignorance*. Ignorance is a measure of the gaps in our knowledge, which may be due to undersampling, or simply reflect the maturity of the subject. Probability theory distributes an equal amount of probability even in ignorance. The Dempster–Shafer theory does not force belief to be assigned to ignorance or refutation of a hypothesis, but only to those subsets of the environment to which the user wishes to assign belief.

Assessing resource management policies and interventions

The traditional approach to assessing the effectiveness of resource management policies and interventions is through long-term monitoring of target species, environmental variables and indices; these historical time-series are then used as qualifiers to determine the current state of the system. However, because monitoring whole ecosystems is very costly, few long-term ecological time-series exist, and so models are routinely used instead to produce hindcasts against which to evaluate performance. However, there is growing evidence that many of the processes which are of interest to resource managers are not only changing more rapidly than previously thought, but also interacting in such a way as to create nonlinear feedbacks amongst themselves and thereby subverting the original intention of particular policies. The fact that many processes are either not studied to any large extent, require more data than are available, or lie beyond current technologies is thus clearly of some concern.

To be able to properly assess the impacts and likelihood of success of different resource management interventions and policies it is therefore important to be able to derive an interaction matrix between the different parts of an ecosystem over a range of time and space scales. For example, logging in an upland area could lead to significant siltation in an intertidal zone in the short term, which could in turn lead to a long-term decline in artisanal fisheries as a result of the loss of coral reef cover; the decline in local fishing could in turn lead to more fishing in adjacent areas, a displaced migrant population, and increased farming pressure on the adjacent land to make up the lost revenues from fishing. Thus the effect of logging on distant marine and terrestrial biodiversity levels, for example, could be significant. But the actual size of the impact depends on a variety of inputs including biophysical features, such as terrain, surface vegetation, inclination and surface run-off, precipitation and time of year, as well as institutional structures, such as the land-ownership, presence of a system of environmental protection laws, enforcement capabilities, investment banks, etc., as these could have a moderating effect on the values in the interaction matrix.

Given the problems of managing resources in such complex institutional and natural environments, resource managers and policy-makers first require analytical tools to assess the efficacy of their own interventions in light of those processes, activities and issues that are likely to have a significant impact on the targets within their own sphere of responsibility but which lie outside their own control (see, for example, McGlade & Novello-Hogarth 1997). By using local knowledge, statutory observations and research, it is possible to build up a local expert system of models and rules and thereby generate sets of interaction matrices based on different inputs and current and future scenarios for different parts of the system. The matrices can then be assessed using analytical tools such as singular value decomposition, designed to extract patterns from multivariate, qualitative data. The scaled outputs indicate which components of the system are having or have had the most impact on a particular policy or resource (e.g. see McGlade & Price 1993). The results can also be used to assess the basis on which policy changes and plans need to be developed and prioritized. As different parts of society are likely to have different perceptions of what the issues, problems and needs are, a similar exercise can be repeated by various peer groups using different rules and the results compared. In this way, it is possible to build up a far more integrated 'expert' approach to resource management which explicitly contains knowledge about human responses to environmental risk and uncertainty.

The governance of natural resources

Why look at governance? Throughout the chapters in this book there is an implicit assumption that it is a good thing for us to know more about the internal workings of ecosystems. Human activities, especially population growth and resource extraction, are clearly having a vast impact on the functioning of ecosystems and the biosphere. In this last section, I examine some of the reasons why it is not only important for theoretical ecologists to grapple with issues such as ecosystem dynamics, epidemiology or earlier diagnoses of resource over-exploitation, but also to address the issue of how to achieve the longer-term solution of improved forms of governance.

Types of governance

The term 'governance' has a number of meanings: it can be the activity or process of governing, a condition of ordered rule, those people charged with the duty of governing or the manner/method/system by which a particular society is governed. The current use of governance does not treat it as a synonym of government, but rather signifies a change in the meaning of government. Rhodes (1996) has reported at least six separate uses for the term.

1 The minimal state: reducing the extent and form of public intervention and the use of markets and quasi-markets to deliver public services.

2 Corporate governance: referring to the system by which organizations are directed and controlled. There are usually three principles associated with successful governance: openness or disclosure of information, integrity or straightforward dealing and accountability.

3 New public management: has two meanings, managerialism and new institutional economics. Managerialism involves hands-on professional management, explicit standards, measures of performance, managing by results, value for money and closeness to the user. New institutional economics implies incentive structures, disaggregating bureaucracies and greater competition through contracting-out and quasi-markets. This type of management is relevant to natural resources because *steering* is central and synonymous to governance. It promotes competition between providers; empowers citizens by pushing control out of the bureaucracy; measures performance by outcomes; is driven by goals and a mission, not rules and regulations; anticipates problems; decentralizes authority–creating participatory management; prefers market mechanisms; and catalyses all sectors into action to solve community problems.

4 Good governance: there is a worldwide trend towards good governance. This is seen as the exercise of political power to manage a nation's affairs, and is achieved via encouragement of competition and markets, privatization of public enterprise, reform of the civil service and greater use of nongovernmental organizations. Good governance involves an efficient public service, an independent judicial system and legal framework to enforce contracts, an accountable administration of public funds, an independent public auditor, respect for law and human rights and a pluralistic institutional structure.

5 Socio-cybernetic system: the pattern or structure that emerges in a sociopolitical system as an outcome of the interacting intervention efforts of all persons involved. In this sense, central government is no longer supreme, the system is increasingly differentiated, i.e. a polycentric state with self- and co-regulation, public–private partnerships, cooperative management and joint entrepreneurial ventures. It highlights the limits to governments.

6 Self-organizing networks: involving the transformation of a system of local government into a system of local governance, with complex sets and networks of organizations drawn from the public and private sector. The key to understanding the importance of this type of governance comes from the observation that integrated networks resist government steering, they develop their own policies and mould their environment. This leads to interdependence between organizations (governance is broader than government), continuing interactions between network members caused by the need to exchange resources and negotiate shared resources, game-like interactions, a significant degree of autonomy and a *hollowing out of the state*.

In managing natural resources it is possible to choose between governing structures such as markets, hierarchies and networks. None of these structures are intrinsically good or bad for allocating resources authoritatively or for exercising control and coordination, and the choice is not inevitably a matter of ideological conviction, rather practicality. However, given a world where governance is increasingly operative without government, where lines of authority are increasingly more informal than formal, where legitimacy is increasingly marked by ambiguity, society is increasingly capable of holding its own by knowing when, where and how to engage in collective action. For resource management this is a critical issue, because as has been witnessed in many areas, the social constraints on over-exploitation of resources are rapidly diminishing (Berkes 1989; McGlade 1995).

Governance in the face of uncertainty

In recent years there have been a number of instances where the incompatibilities and interplay between policy, politics and science have had serious consequences in terms of natural resources and human health. They include the emergence of evidence on bovine spongiform encephalopathy (BSE) in cattle in the UK and the significant effect it had on EU agricultural policies, the potential link between badgers and bovine tuberculosis; the impacts of industrial fishing on human consumption species in the North Sea (Robertson *et al.* 1996) and the collapse of the northern cod (Healy 1997). It is therefore absolutely critical that in future the extent of understanding about the relevant parts of the ecosystem, plus the level of uncertainty, should be made explicit.

The boundaries of ignorance are however difficult to map but relate strongly to the pedigree or state of the art of a particular field from which the quantity derives. In the case of relativity, there was a progression from an embryonic field in 1905 through to the 1950s when experimental results had largely corroborated the theory. Natural resource management on the other hand relies on data that are often qualitative and heterogeneous, and well-structured and accepted theories are conspicuous by their absence. For example, two key events shaped the environmental debate: the Villach conference in 1985 on human-induced climate change and the Brundtland report on sustainable development (World Commission on Environment and Development 1987). Despite the fact that they have led to the International Panel on Climate Change, Agenda 21, the 1992 Earth Summit in Rio de Janeiro and the 1997 Kyoto Conference on Climate Change there is still no clear consensus in the collective mindset of policy-makers.

The extent of any consensus reflects the social strength of the paradigms against which the information is cast. Thinking that we can make exact predictions about natural resources under highly complex circumstances is still likely to be premature, and potentially dangerous as it often leads those involved in decision-making towards a misdirected sense of concreteness in overall policy judgement. Worse still is the fact that the credibility of science is also at risk. Managing uncertainty in the context of responsibility cannot be side-stepped, neither can it be eliminated.

Unfortunately, many of today's institutions have been developed to undertake planning and policy development from a standpoint of determinacy rather than complexity. This is largely because the problems created by indeterminacy in policy-related research are known to increase as a result of societal interactions. Scientific advice is judged by the public on its performance in relation to sensitive issues such as human health hazards, the over-exploitation of natural resources, dumping of hazardous wastes, the dangers of oil spills, and environmental pollution, etc. All involve much uncertainty, as well as inescapable social and ethical aspects, so simplicity and precision in predictions or even setting safe limits are not always feasible. Yet policy-makers tend to expect straightforward information to use as input into their own decision-making process.

The problems become manifest at several levels, the simplest one being the representation of uncertainty in only qualitative terms. Any scientific advisor knows that a prediction such as a 'one in a million' chance of a serious accident or health incident should be hedged with statements of many kinds so as to caution any user as to the reliability of the numerical assertions. But if these were all expressed, policies would become tedious and incomprehensible, yet if omitted then the same policies could convey a level of certainty unwarranted by the facts.

In addition to low-frequency hazards, there are also problems relating to higher probability events such as the failure of computing networks, diffused hazards such as the long-term usage of chemicals or possible large-scale environmental perturbations such as global warming. The dilemma is that any definite advice is liable to go wrong: a prediction of danger will appear alarmist if nothing happens in the short term, whilst reassurance can be condemned if it retrospectively turns out to be wrong. Thus the credibility of research, based on the supposed certainty of its conclusions, is endangered by giving any scientific advice on inherently uncertain issues. But if the researcher prudently refuses to accept vague or even qualitative expert opinions as a basis for quantitative assessments, and declines to provide definitive advice when asked, then research itself is regarded as obstructionist, not performing its public functions and its legitimacy is called into question. It is not surprising then that most natural resource policy and planning institutions have been unable or unwilling to respond in a locally adaptive way, especially when in most cases the burden of proof in instances of resource collapse lies with the managers themselves.

Looking more closely at governing structures, it is clear that good governance effectively means that a balance has been achieved between governing needs (problem situations or the impact of new opportunities) and governing capacities (creating patterns of solutions or developing new strategies) (Kooiman 1993). But the governing needs and capacities or *common pool* resources (i.e. those which are difficult to bound or divide) are very different from *noncommon pool* resources (i.e. those bounded and to some extent private). At an international level, an external perspective generally exists, i.e. there is sufficient understanding of the various price and property mechanisms that goals, such as maximum sustainable yields, can be determined externally and will operate through *negative reciprocity* or self-interest. At a local level many local communities operate solely with an internal perspective; they reject objective efficiency and presume that any market failure can be determined independently of the existence or magnitude of an external cost. Indeed they avert many of the problems by operating through *generalized reciprocity* arrangements relying on mutual help and solidarity.

Thus in the face of uncertainty about natural resources and ecosystem dynamics, the level and type of participation in the governance of resources is critical. In reality, there is a continuum from total state control, liaison with industry, consultation by industry, representation, co-management, community-based management and individual control. Co-management, communal control and community-based management are all bottom-up, rather than top-down, and participatory forms of governance, and the type of knowledge that is used to inform the management of resources is quite different from that used in top-down or state-controlled management. It is widely recogized that it has been the absence of communal control that has caused the collapse of many resources; but to support any move towards more local governance it will be important to provide expert systems based on clear ecosystem theoretic approaches. Without such a transition, the form of governance controlling natural resources will rapidly move away from regulation and conservation, based on the formal authority of the state or economic controls based on the market, to one of communicative governance based on the force of the argument and political rhetoric.

Hollowing out of the state is already occurring as a result of transitions in society towards self-organizing networks at one end of the scale and globalization at the other. The danger is that without a sustained effort on the part of theoretical ecologists, the impacts of human activities on the biosphere will not be made clear to society. The challenge is therefore to ensure that the new advances in theoretical ecology discussed in this book are used to further elucidate the dynamics of ecosystems, and thereby help to inform the decision-making processes set up to protect and conserve the biosphere.

Acknowledgements

I would like to thank Jan Kooiman and John Marshall for helpful discussions and insights, and Tielale Viegas and the European Commission for their continued support.

References

Allen, J.C., Schaffer, W.M. & Rosko, D. (1993) Chaos reduces species extinction by amplifying local population noise. *Nature* **364**, 229–232.

Allen, P.M. & McGlade, J.M. (1986) Dynamics of discovery and exploitation: the case for the Scotian Shelf groundfish fisheries. *Canadian Journal of Fisheries and Aquatic Sciences*, **43**, 1187–1200.

Allen, T.F.H. & Starr, T.B. (1982) *Hierarchy: Perspectives for Ecological Complexity*. University of Chicago Press, Chicago.

American Association for the Advancement of Science (AAAS) (1986) *Variability and management of large marine ecosystems*. Westview Press, Boulder, CO.

American Association for the Advancement of Science (AAAS) (1989) *Biomass yields and geography of large marine ecosystems*. Westview Press, Boulder, CO.

American Association for the Advancement of Science (AAAS) (1990) *Large Marine Ecosystems: Patterns, Processes and Yields*. AAAS Press, Washington, DC.

American Association for the Advancement of Science (AAAS) (1991) *Food Chains, Yields, Models and Management of Large Marine Ecosystems*. Westview Press, Boulder, CO.

American Association for the Advancement of Science (AAAS) (1993) *Large Marine Ecosystems: Stress, Mitigation and Sustainability*. AAAS Press, Washington, DC.

Anderson, R.M. & May, R.M. (1991) *Infectious Diseases of Humans: Dynamics and Control*. Oxford University Press, Oxford.

Andrén, H. (1994) Effects of habitat fragmentation on birds and mammals in landscapes with different proportions of suitable habitat: a review. *Oikos*, **71**, 355–366.

Ausubel, J., Arrhenius, E., Benedick R.E. *et al.* (1990) Social and institutional barriers to reducing CO_2 emissions. In: *Limiting Greenhouse Effects; Controlling Carbon Dioxide Emissions* (ed. G.I. Pearmann). John Wiley, Chichester.

Bailey, R.G. (1989) *Ecoregions of the Continents*. US Department of Agriculture, Forest Service, Washington, DC.
Baird, D., McGlade, J.M. & Ulanowicz, R.E. (1991) The comparative ecology of six marine ecosystems. *Philosophical Transactions of the Royal Society of London*, **B333**, 15–29.
Bak, P., Tang, C. & Wiessenfeld, K. (1988) Self-organized criticality: an explanation of 1/f noise. *Physics Review Letters*, **59**, 381–384.
Bart, J. (1995) Acceptance criteria for using individual-based models to make management decisions. *Ecological Applications*, **5**, 411–420.
Bascompte, J. & Solé, R.V. (1996) Habitat fragmentation and extinction thresholds in spatially explicit models. *Journal of Animal Ecology*, **65**, 465–471.
Begon, M., Harper, J.L. & Townsend, C.R. (1990) *Individuals, Populations and Communities*. Blackwell Scientific Publications, Oxford.
Begon, M., Mortimer, M. & Thompson, D.J. (1996) *Population Ecology: A Unified Study of Animals and Plants*, 3rd edn. Blackwell Science, Oxford.
Berkes, F. (ed.) (1989) *Common property resources*. Belhaven Press, London.
Bianchi, G. (1992) Study of the demersal assemblages of the continental shelf and upper Angolan slope. *Marine Ecology Progress Series*, **81**, 101–120.
Bormann, F.H. & Likens, G.E. (1979) *Pattern and Process in a Forested Ecosystems*. Springer-Verlag, New York.
Bretherton, F. (1994) Perspectives on policy. *Ambio*, **23**, 96–97.
Burrough, P.A. (1981) Fractal dimensions of landscapes and other environmental data. *Nature*, **294**, 240–242.
Carpenter, S.R., Kitchell, J.F. & Hodgson, J.R. (1985) Cascading trophic interactions and lake productivity. *BioScience*, **35**, 634–639.
Cherrett, J.M. (ed.) (1989) *Ecological concepts*. Blackwell Scientific Publications, Oxford.
Christensen, V. & Pauly, D. (1992) ECOPATH II—a software for balancing steady-state ecosystem models and calculating network characteristics. *Ecological Modelling*, **61**, 169–185.
Christensen, V. & Pauly, D. (eds) (1993) (eds.) Trophic models of aquatic ecosystems. *ICLARM Conference Proceedings*, **26**, Manila, 390pp.
Cohen, J. & Stewart, I. (1994) *The Collapse of Chaos*. Viking, New York.
Colinvaux, P.A. (1983) *Introduction to Ecology*. John Wiley, New York.
Cook, R.M., Sinclair, A. & Stefansson, G. (1997) Potential collapse of North Sea cod stocks. *Nature*, **385**, 521–522.
Costanza, R., d'Arge, R., de Groot, R. *et al.* (1997) The value of the world's ecosystem services and natural capital. *Nature*, **387**, 253–260.
Darzin, P.G. (1992) *Nonlinear Systems*. Cambridge University Press, Cambridge.
DeAngelis, D.L. (1992) *Dynamics of nutrient cycling and food webs*. Chapman & Hall, London.
DeAngelis, D.L. & Gross, L.J. (eds) (1992) *Individual-Based Models and Approaches in Ecology*. Populations, Communities and Ecosystems, Chapman & Hall, New York.
DeAngelis, D.L., Bartell, S.M. & Brenkert, A.L. (1989) Effects of nutrient recycling and food-chain length on resilience. *Nature*, **134**, 778–805.
do Carmo Gomes, M. (1993) Prediction under uncertainty. Fish assemblages and food webs on the Grand Banks of Newfoundland. In: *Social and Economic Studies*, **51**. Institute of Social and Economic Research, Memorial University of Newfoundland, St. John's.
Dublin, H.T., Sinclair, A.R.E. & McGlade, J. (1990) Elephants and fire as causes of multiple stable states in the Serengeti-Mara woodlands. *Journal of Animal Ecololgy*, **59**, 1147–1164.
Duda, R.O. & Reboh, R. (1984) AI and decision making: the PROSPECTOR experience. In: *Artificial Intelligence for Business* (ed. W. Reitman). Ablex Publicity Corporation, TX.
Durrett, R. & Levin, S.A. (1994) Stochastic spatial models — a user's guide to ecological applications. *Philosophical Transactions of the Royal Society of London*, **B343**, 329–350.
Elton, C.S. (1927) *Animal Ecology*. Sedgewick & Jackson, London.

Ens, B.J., Piersma, T. & Tinbergen, J.M. (1994) Towards predictive models of bird migration schedules: theoretical and empirical bottlenecks. *Netherlands Institute for Sea Research (NIOZ) Report*, **Vol. 5**.

Flyvberg, H., Bak, P., Jensen, M.H. & Sneppen, K. (1995) A self-organized critical model for evolution. In: *Modelling the Dynamics of Biological Systems. Nonlinear Phenomena and Pattern Formation* (eds E. Mosekilde & O.G. Mouritsen). Springer, Berlin.

Frank, K., Dreschler, M. & Wissel, C. (1994) Überleben in fragmentierten Lebensräume—Stochastiche Modelle zu Metapopulationen. *Zeitschrift fur Ökologie uber Naturschutz*, **3**, 167–178.

Franks, N.R. (1982) Ecology and population regulation in the army any *Eciton burchelli*. In: *The Ecology of a Tropical Forest: Seasonal Rhythms and Long-Term Changes* (eds E.G. Leigh, A.S. Rand & D.M. Windsor). Smithsonian Institution Press, Washington, DC.

Franks, N.R. & Bossert, W.H. (1983) The influence of swarm raiding army ants on the patchiness and diversity of a tropical leaf litter ant community. In: *The Tropical Rain Forest: Ecology and Management* (eds E.L. Sutton, A.C. Chadwick & T.C. Whitmore). Blackwell, Oxford.

Fretwell, S. (1972) *Populations in a Seasonal Environment.* Princeton University Press, Princeton, NJ.

Gibson, G.J. (1977) Markov chain Monte Carlo methods for fitting spatio-temporal stochastic models in plant epidemics. *Applied Statistics*, **46**, 251–233.

Godfray, H.C.J. & Pacala, S.M. (1992) Aggregation and the population dynamics of parasitoids and predators. *American Naturalist*, **140**, 30–40.

Gordon, J. & Shortliffe, E.H. (1985) The Dempster-Shafer theory of evidence. In: *Rule-based expert systems* (eds B. Buchanan & E. Shortliffe). Addison-Wesley, Reading.

Grenfell, B.T., Bolker, B.M. & Kleczkowski, A. (1995) Seasonality and extinction in chaotic metapopulations. *Proceedings of the Royal Society of London*, **B259**, 97–103.

Grenfell, B.T. & Harwood, J. (1997) Metapopulation dynamics of infectious diseases. *Trends in Evolution and Ecology*, **12**, 395–399.

Grimm, V., Stelter, C., Reich, M. & Wissel, C. (1994) Ein Modell zur Metapopulations-dynamik von *Bruodema tuberculata* (Saltatoria, Acrididae). *Zeitschrift für Ökologie uber Naturschutz*, **3**, 189–195.

Hanski, I., Poyry, J., Pakkala, T. & Kussaari, M. (1995) Multiple equilibria in metapopulation dynamics. *Nature*, **353**, 618–621.

Hanski, I., Moilanen, A., Gyllenberg, M. (1996) Minimum viable metapopulation size. *American Naturalist*, **147**, 527–541.

Hastings, A. & Sugihara, G. (1993) *Fractals: A User's Guide for the Natural Sciences.* Oxford University Press, Oxford.

Hawkins, B.A. (1988) Species diversity in the third and fourth trophic levels: patterns and mechanisms. *Journal of Animal Ecology*, **57**, 137–162.

Hawkins, B.A. (1990) Global patterns of parasitoid assemblage size. *Journal of Animal Ecology*, **59**, 57–72.

Healey, M.C. (1997) Comment: the interplay of policy, politics and science. *Canadian Journal of Fisheries and Aquatic Sciences*, **54**, 1427–1429.

Hendrix, G. (1979) Encoding knowledge in partitioned networks. In: *Associative Networks* (ed. N. Findler). Academic Press, Oxford.

Hendry, R.J. (1995) *Spatial modelling in plant ecology*. PhD thesis, University of Warwick, Warwick.

Hendry, R.J. & McGlade, J.M. (1995) The role of memory in ecological systems. *Proceedings of the Royal Society of London*, **B259**, 1153–1159.

Hendry, R.J., McGlade, J.M. & Weiner, J. (1996) A coupled map lattice model of the growth of plant monocultures. *Ecological Modelling*, **84**, 81–90.

Hogg, T., Huberman, B.A. & McGlade, J.M. (1989) The stability of ecosystems. *Proceedings of the Royal Society of London*, **B237**, 43–51.

Holling, C.S. (1964) The analysis of complex population processes. *Canadian Entomologist*, **96**, 335–347.
Holling, C.S. (1973) Resilience and stability in ecological systems. *Annual Review of Ecology and Systematics*, **4**, 1–23.
Holt, R.D., Pacala, S.W., Smith, T.W. & Liu, J. (1995) Linking contemporary vegetation models with spatially explicit animal population models. *Ecological Applications*, **5**, 20–27.
Hutchings, J.A., Walters, C. & Haedrich, R.L. (1997a) Is scientific inquiry incompatible with government information control? *Canadian Journal of Fisheries and Aquatic Sciences*, **54**, 1198–1210.
Hutchings, J.A., Walters, C. & Haedrich, R.L. (1997b) Reply: Scientific inquiry and fish stock assessment in the Canadian Department of Fisheries and Oceans and Reply: The interplay of policy, politics, and science. *Canadian Journal of Fisheries and Aquatic Sciences*, **54**, 1430–1431.
Hutchinson, G.E. (1978) *An Introduction to Population Ecology*. Yale University Press, New Haven, CT.
Jansen, V.A.A. (1994) *Theoretical aspects of metapopulation dynamics*. PhD thesis, University of Leiden.
Janzen, D.H. & Martin, P.S. (1982) Neotropical anachronisms: the fruits the gomphortheres ate. *Science*, **215**, 19–27.
Kauffman, S. (1993) *The Origins of Order*. Oxford University Press, New York.
Keeling, M., Mezi', I., Hendry, R., McGlade, J. & Rand, D. (1997) Measuring characteristic length scales in spatially extended systems. *Philosophical Transactions of the Royal Society of London*, **B352**, 1589–1601.
Koranteng, K. (1998) *The impacts of environmental forcing on the dynamics of demersal fishery resources of Ghana*. PhD thesis, University of Warwick.
Kooijman, S.A.L.M. (1977) Species abundance with optimum relations to environmental factors. *Annual Review of Systematics and Evolution*, **6**, 123–138.
Kooiman, J. (1993) *Modern Governance. New Government–Society Interactions*. Sage Publications, London.
Krebs, C.J. (1985) *Ecology, The Experimental Analysis of Distribution and Abundance*. Harper & Row, New York.
Langton, C.G., Farmer, J.D., Rasmussen, S. & Taylor, C. (eds) (1992) Artificial Life II. *Proceedings Volume in the Santa Fe Institute Studies in the Sciences of Complexity*, **10**. Addison-Wesley, Reading.
Levin, S. (1992) The problem of pattern and scale in ecology. *Ecology*, **73**, 1943–1967.
Lindeman, R.L. (1942) The trophic dynamic aspects of ecology. *Ecology*, **23**, 399–418.
Longhurst, A. (1995) Seasonal cycles of pelagic production and consumption. *Progress in Oceanography*, **36**, 77–167.
Lotka, A.J. (1956) *Elements of Mathematical Biology*. Dover Publications, New York.
McCall, A. (1990) *Dynamic Geography of Marine Fish Populations*. Washington Sea Grant Programme, University of Washington Press, Seattle, WA.
McGlade, J.M. (1989) Integrated fisheries management models: understanding the limits to marine resource exploitation. *American Fisheries Society Symposium*, **6**, 139–165.
McGlade, J.M. (1993) Alternative ecologies. *New Scientist* (Suppl.), **137**, 14–16.
McGlade, J.M. (1995) Integrating social and economic factors into fisheries management in the European Union—an issue of governance. *Environmental Management: Review of 1994 and Future Trends*, **2**, 77–80.
McGlade, J.M. & Novello-Hogarth, A. (1997) Sustainable management of coastal resources: SimCoast™ a fuzzy-logic rule-based expert system. *Hydro International*, **1**, 6–9.
McGlade, J.M. & Price, A.R.G. (1993) Multi-disciplinary modelling: an overview and practical implications for the governance of the Gulf region. *Marine Pollution Bulletin*, **27**, 361–377.
Magurran, A.E. (1991) *Ecological Diversity and its Measurement*. Chapman & Hall, London.

Mahon, R. (ed.) (1985) Groundfish assemblages on the Scotian Shelf. In: Towards the inclusion of fishery interactions in management advice. *Canadian Technical Report of Fisheries and Aquatic Science*, **1347**, 153–162.

Mangel, M. & Tier, C. (1994) Four facts every conservation biologist should know about persistence. *Ecology*, **75**, 607–614.

Månsson, B.Å. & McGlade, J.M. (1993) Ecology, thermodynamics and H.T. Odum's conjectures. *Oecologia*, **93**, 582–596.

Matsuda, H., Ogita, N., Sasaki, A. & Satō, K. (1992) Statistical mechanics of populations—the lattice Lotka–Volterra model. *Progress in Theoretical Physics*, **88**, 1035–1049.

May, R.M. (1973) *Stability and Complexity in Model Ecosystems.* Princeton University Press, Princeton, NJ.

May, R.M. (1991) *Theoretical Ecology: Principles and Applications.* (2nd ed.), Sinauer Associates, Massachusetts.

May, R.M. (1989) Levels of organisation in ecology. In: *Ecological Concepts. 29th Symposium of the British Ecological Society* (ed. J.M. Cherrett). Blackwell Scientific Publications, Oxford.

Maynard Smith, J. (1974) *Models in Ecology*. Cambridge University Press, Cambridge.

Milner-Gulland, E.J. & Mace, R. (1998) *Biological Conservation and Sustainable Use*. Blackwell Science, Oxford.

Morse, D.A., Lawton, J.H., Dodson, M.M. & Williamson, M.H. (1985) Fractal dimension of vegetation and the distribution of arthropod body lengths. *Nature*, **314**, 731–733.

Mosekilde, E. & Mouritsen, O.G. (1995) (eds) Modelling the dynamics of biological systems. Nonlinear phenomena and pattern formation. *Springer Series in Synergetics*, **65**. Springer, Berlin.

Murdoch, W.W., Briggs, C.J., Nisbet, R.M., Gurney, W.S.C. & Stewart-Oaten, A. (1992) Aggregation and stability in metapopulation models. *American Naturalist*, **140**, 41–58.

Nisbet, R.M., Briggs, C.J., Gurney, W.S.C., Murdoch, W.W. & Stewart-Oaten, A. (1992) Two-patch metapopulation dynamics. In: *Patch Dynamics in Terrestrial, Freshwater and Marine Ecosystems* (eds S.A. Levin, J.H. Steele & T. Powell).

Odum, E.P. (1969) The strategy of ecosystem development. *Science*, **164**, 262–270.

Odum, H.T. (1983) *Systems Ecology*. John Wiley, New York.

Odum, H.T. (1988) Self-organization, transformity and information. *Science*, **242**, 1132–1139.

O'Neill, R.V., DeAngelis, D.L., Waide, J.B. & Allen, T.F.H. (1986) A hierachical concept of ecosystems. Princeton University Press, Princeton.

Otis, G.W., Santana, C.E., Crawford, D.L. & Higgins, M.L. (1986) The effect of foraging army ants on leaf-litter arthropods. *Biotropica*, **18**, 56–61.

Paine, R.T. (1966) Food web complexity and species diversity. *American Naturalist*, **100**, 65–75.

Paine, R.T. (1974) Intertidal community structure: experimental studies on the relationship between a dominant competitor and its principal predator. *Oecologia*, **15**, 93–120.

Pauly, D. (1998) Research news. *Science*, **279**, 809.

Pauly, D. & Christensen, V. (1995) Primary production required to sustain global fisheries. *Nature*, **374**, 255–257.

Pauly, D., Christensen, V., Dalsgaard, J., Froses, R. & Torres, F. Jr. (1998) Fishing down marine food webs. *Science*, **279**, 860–863.

Pimm, S. (1982) *Food Webs*. Chapman & Hall, London.

Polovina, J.J. (1984) Model of coral reef ecosystem I. The ECOPATH model and its application to French Frigate Shoals. *Coral Reefs*, **3**, 1–11.

Powell, T.M. (1988) Physical and biological scales of variability in lakes, estuaries and the coastal ocean. In: *Perspectives in Ecological Theory* (eds J. Roughgarden, R.M. May & S.A. Levin). Princeton University Press, Princeton.

Pulliam, H.R. (1988) Sources, sinks and population regulation. *American Naturalist*, **132**, 652–661.

Rand, D.A. (1994) Measuring and characterising spatial patterns, dynamics and chaos in spatially extended dynamical systems and ecologies. *Philosophical Transactions of the Royal Society of London*, **A348**, 497–514.

Rand, D.A., Wilson, H. & McGlade, J.M. (1994) Dynamics and evolution: evolutionarily stable attractors, invasion exponents and phenotype dynamics. *Philosophical Transactions of the Royal Society of London*, **B343**, 261–283.

Ray, T.S. & Andrews, C.C. (1980) Antbutterflies: butterflies that follow army ants to feed on antbird droppings. *Science*, **210**, 1147–1148.

Recknagel, F., French, M., Harkonen, P. & Yabunaka, K.-I. (1997) Artificial neural network approach for modelling and prediction of algal blooms. *Ecological Modelling*, **96**, 11–28.

Reise, K. (1991) Mosaic cycles in the marine benthos. In: *The mosaic-cycle concept of ecosystems* (ed. H. Remmert). Springer, New York.

Remmert, H. (1985) Was geschieht in Klimax-Stadium? Ökologisches Gleichgewicht durch Mosaik aus desynchronen Zyklen. *Naturwissenschaften*, **72**, 505–512.

Remmert, H. (ed.) (1991) The mosaic-cycle concept of ecosystems. Springer, New York.

Rhodes, R.A.W. (1996) The new governance: governing without government. *Political Studies*, **44**, 652–665.

Roberts, D.W. (1997) Modelling forest dynamics with vital attributes and fuzzy systems theory. *Ecological Modelling*, **90**, 161–173.

Robertson, J., McGlade, J.M. & Leaver, I. (1996) *Ecological effects of the North Sea industrial fishing industry on the availability of human consumption species*. Unilever Commissioned Report, Univation, Aberdeen.

Rosen, R. (1977) Observation and biological systems. *Bulletin of Mathematical Biology*, **39**, 663–678.

Sathyendranath, S., Platt, T., Horne, E.P.W. *et al.* (1991) Estimation of new production in the ocean by compound remote sensing. *Nature*, **353**, 129–133.

Satō, K., Matsuda, H. & Sasaki, A. (1994) Pathogen invasion and host extinction in lattice structured populations. *Journal of Mathematical Biology*, **32**, 251–268.

Schoener, T.W. (1986) Mechanistic approaches to community ecology: a new reductionism? *American Zoologist*, **26**, 81–106.

Sherman, K., Jaworski, N.A. & Smayda, T.J. (1996) *The Northeast Shelf Ecosystem: Assessment, Sustainability and Management*. Blackwell Science, Cambridge, MA.

Silvert, W. (1997) Ecological impact classification with fuzzy sets. *Ecological Modelling*, **96**, 1–10.

Simon, H. (1962) The architecture of complexity. *Proceedings of the American Philosophical Society*, **106**, 467–482.

Southwood, T.R.E. (1988) The concept and nature of the community. In: *Organisation of Communities: Past and Present*. Blackwell Scientific Publications, Oxford.

Spiegelhalter, D.J., Thomas, A., Best, N. & Gilks, W.R. (1995) *BUGS: Bayesian Inference Using Gibbs Sampling. Manual of Version 0.5*. Medical Research Council Biostatistics Unit, Cambridge.

Steele, J.H. (1978) *Spatial Pattern in Plankton Communities*. Plenum Press, New York.

Stewart, I. (1998) *Life's Other Secret. The New Mathematics of the Living World*. John Wiley, London.

Stommel, H. (1963) Varieties of oceanographic experience. *Science*, **139**, 572–576.

Sutcliffe, O.L., Thomas, C.D., Yates, T.J. & Greatorex-Davies, J.N. (1997) Correlated extinctions, colonizations and population fluctuations in a highly connected ringlet butterfly metapopulation. *Oecologia*, **109**, 235–241.

Tansley, A.G. (1935) The use and abuse of vegetational concepts and terms. *Ecology*, **16**, 284–307.

ter Brak, C.J.F. (1986) Canonical correspondence analysis: a new eigenvector technique for multivariate direct gradient analysis. *Ecology*, **67**, 1167–1179.

Tong, H. (ed.) (1995) *Chaos and forecasting*. World Scientific, London.

Turner, M.G., Aarthaud, G.J., Engstrom, R.T. *et al.* (1995) Usefulness of spatially explicit population models in land management. *Ecological Applications*, **5**, 12–16.

Turner, R. (1984) *Logics for artificial intelligence*. Ellis Horwood Ltd, Chichester.

Tyler, A.V., Gabriel, W.L. & Overholtz, W.J. (1982) Adaptive management based on the structure of fish assemblages of northern continental shelves. In: *Multispecies Approaches to Fisheries Management Advice* (ed. M.C. Mercer). *Canadian Special Publication on Fisheries and Aquatic Science*, **59**.

Ulanowicz, R.E. (1986) *Growth and Development: Ecosystem Phenomenology*. Springer-Verlag, New York.

Underwood, A.J.T. (1986) What is a community? In: *Patterns and Processes in the History of Life* (eds D.M. Raup & D. Jablonski). *Dahlem Workshop (1985)*. Springer-Verlag, Berlin.

Virkkala, R. (1993) Ranges of Northern Forest passerines: a fractal analysis. *Oikos*, **67**, 218–226.

Volterra, V. (1931) *Theorie Mathematique de la Luute Pour la Vie*. Gauthier-Villars, Paris.

Waage, J.K., Hassell, M.P. & Godfray, H.C.J. (1985) The dynamics of pest–parasitoid–insecticide interactions. *Journal of Applied Ecology*, **22**, 825–838.

Walters, C., Christensen, V. & Pauly, D. (1997) Structuring dynamic models of exploited ecosystems from trophic mass-balance assessments. *Reviews in Fish Biology and Fisheries*, **7**, 139–172.

Whittaker, R.H. & Levin, S.A. (1977) The role of mosaic phenomena in natural communities. *Theoretical Population Biology*, **12**, 117–139.

Willis, E.O. (1967) The behaviour of bicolored antbirds. *University of California Publications in Zoology*, **79**, 1–127.

Wissel, C. (1992) Modelling the mosaic cycle of a middle European beech forest. *Ecological Modelling*, **63**, 29–43.

Wu, H.-W., Li, B.-L., Stoker, R. & Li, Y. (1997) A semi-arid ecosystem simulation model with probabilistic and fuzzy parameters. *Ecological Modelling*, **90**, 147–160.

Zadeh, L.A. (1965) Fuzzy sets. *Information and Control*, **8**, 338–353.

Zhu, A.-X., Band, L.E., Duttin, B. & Nimlos, T.J. (1997) Automated soil inference under fuzzy logic. *Ecological Modelling*, **90**, 123–145.

Zimmermann, A.-T. (1987) *Fuzzy sets, decision making and expert systems*. Kluwer Academic, Boston.

Index

Please note: page numbers in *italic* refer to figures; those in **bold** to tables.

abundance curves, and species diversity 276–8
acquired immune deficiency syndrome (AIDS) 24
  epidemic models 45
  incubation periods 46
  transmission 48
activators, and inhibitors 70
AD see Always Defect (AD)
AE *see* artificial ecologies (AE)
age-structured pair models *see* ASP models
aggregation
  definition 68
  and predation 69
  spatial models 68
AIDS *see* acquired immune deficiency syndrome (AIDS)
altruism
  evolution of 102, 132–9
  and migration 137–8
  types of 133
  see also unconditional altruism
Always Defect (AD), strategies 138–9
Amoeba proteus **147**
Arcadia (model), applications 16
army ants, as keystone species 313
artificial ecologies (AE) 4, 86–7
  host–parasitoid models 89–90
  parasites *87*
ASP models 125
  stochastic 126, *127*
assemblages, definitions 312–13
assembly dynamics *144*
  models 158
  use of term 147
asymptotic stability 148–9
  global 149

Ball and O'Neill equations 55, 57
Bayes' theorem, applications 327
beech forests
  long-term effects of 11
  memory models 11–12
  mosaic cycles 10–11
behaviour, stabilizing, and refuges 69
behaviour change, and population dynamics 47–8
behavioural studies, and individual-based models 13–15
Belousov–Zhabotinsky reaction 65
Bernoulli trial statistics 108–9, 110, 111, 120, 129, 130–1, 136, 139
Beverton and Holt model 200–1
bias, reduction 250–5
biodiversity, conservation 322
biogeographical provinces
  definition 260–1
  see also island biogeography
biological growth mechanisms, Verhulst–Pearl assumptions 33
biological studies, historical background 23
biological systems
  correlations in 101
  models 5
    research issues 319
biology, and mathematics, developments 23–4
biomass
  flows 194, 224–6
    buffering 225
    translocation 225
  herbivore 184
  resilience 173, *174*
  resistance 173, *174*
birds
  behavioural studies 14
  range size studies 263
birth rates
  density-dependent 55
  and Grenfell and Roberts process equation 50
birth–death processes 24–30
  crowding effects 32–3
  deterministic approaches 25–6
  immigration factors 29–30
  simulations 27–9
    stochastic logistic realizations 35, *36*
  stochastic analyses 26–7, *29*
  see also population dynamics
birth–death–migration processes
  N-colony 53–6
  two-colony 52–3
Blepharisma japonicum **147**
body size
  and gestation period 235, *236*

body size (*Cont.*)
and mortality rates 235–7
and survival 240–1
body weight, life history variations 233–4
brain weight, life history variations 235
breeding density 208
Brownian motion, in spatial processes 55

Cakile edentula, source–sink dynamics 269
canonicity, assumption of 277
capture–recapture methodology 253–5, 258
closed-model theory 254
open-model theory 254
Pollock's robust method 254–5
caricature models 87–9
cellular automata 11, 310
analyses 12
forest fire studies 15
host–parasitoid models 89, 90–1
limitations 100–1
spatial distributions *85*
see also probabilistic cellular automata (PCA)
Chaoborus, food web studies 179–82
Chao's estimators 258, 259
determination 253
chaos theory, developments 1
Charnov's model 232–3, 238–41
criticisms 241, 246
clumped networks
and disease transmission 116–18
and oscillations 118
structures 109–10
CML *see* coupled map lattices (CML)
coexistence, criteria 148–53
Colpidium striatum **147**
commons, overgrazing 287–8
commonsense knowledge 324
communities
definition 312
invasion 161–4
succession effects 158–60
community analyses 313
limited use of 321
community assembly
algorithms 155–7
dynamics 157–64
endstates 164–7
establishment conditions 154–5
and invasion resistance 161–4
models, Markov processes 146–7, 167
sequential 153–7
succession 159–60
theoretical perspectives 143–71
three-species systems *158*
transients 155–6
community dynamics *144*
community ecology
field of study 143
state variables 146–7
theoretical perspectives 143–71
competition
absolute asymmetry 7
absolute symmetry 7
multispecies interactions 36–7
plants 6–7
relative asymmetry 7
relative symmetry 7
complexity, studies 317, 319
computer programs, food web simulations 178–9
concentration, Simpson's index of 251–3
confidence limits 325
configuration, use of term 103
consumer goods, valuation 297
consumer–resource stacks 210–12
consumer–resource systems
continuous-time models 194, 210–26, 227
definition 194
discrete-time models 194, 200–10, 227
fixed resources 200–1
population and evolutionary dynamics 194–231
research trends 226–8
contact process, infection dynamics 112
cooperation, evolution of 102, 132–9
correlation equations
applications 101–3, 140
approximations 108–11
clumped network structures 109–10
derivation 103–11
and events 104–6
master equation 102
derivation 104–6
notation 103–4
for spatial ecologies 100–42
use of term 101
correlations
in biological systems 101
and infection dynamics 139–40
local, spatial models 68
triple 110
coupled map lattices (CML) 4, 77–80, 310
applications 77
characteristics **74**
grid size factors 89
neighbourhood structure 77, *78*
plant studies 6–7, *8*, *9*
crime, economic theory of 300
crisp sets, vs. fuzzy sets 315–16
critical transmissibilities 115, 116–22
lower 119, 120
upper 118, 120–1
cuckoo doves, regional pools 145

Daphnia
food web studies 184–6
stressor studies 17
Darwin, Charles Robert (1809–82), *On the Origin of Species* (1859) 133, 172
death rates
density-dependent 55
and Grenfell and Roberts process equation 50
see also mortality rates
decision-making
behavioural studies 14
and uncertainty, Bayesian approaches 327

deer
  antler sizes 234
  herds, valuation 297
defuzzification, processes 317
Dempster–Shafer theory, and probability theory compared 329
density dependence, abrupt, evolution 201–3
deterministic approaches
  birth–death processes 25–6
  and stochastic processes 23, 24–5, 58–61
    compromise 33–4
development, sustainable 290–1
die-back, forests 15–16
differential equations, stochastic 101
discount rates
  determination 291
  interpretation 292
discounting
  applications 292
  assumptions 294
  in investment appraisals 292–3
disease transmission 83–4
  and clumped networks 116–18
  SIR models 102, 116–18
  and vaccination 114–15
  waves 65, 66
  see also infection dynamics
diseases
  cross-infection 24
  establishment 113–14
  infectious, modelling 38–45
  seasonal factors 47
  spatial processes 50–2
disturbance, and species diversity 267–8
diversity
  effects
    on geographical range 261–2
    on populations 261–2
  and equability 255
  point 256–8
  and steady states 265–6, 274
  and time 259–60
  see also biodiversity; species diversity
Drosophila melanogaster, growth studies 33
drought, resistance to 187–8
dynamic systems, nonlinear 317

Eciton burchelli, as keystone species 313
ecological assets
  and gross national product 298
  research 297
  valuation 294–5
    and national accounting 298
ecological dynamics *144*
ecological economics 282–308
  developments 298–304
  future research 304–6
  historical background 283–4
  issues 284–98
  nature of the field 282
  and neoclassical economics 283, *284*
  and time 291–4
  see also environmental economics; resource economics
ecological modelling, developments 1
ecological processes, and epidemiological processes 44
ecology
  individual-based models 1–22
  and life history variations 245
  spatial heterogeneities 65
  see also community ecology
ecosystem analysis
  and emergent properties 318–20
  and fuzzy set theory 309–10, 315–17
  and governance of natural resources 309–41
  and resource management 317–22
    models 320–2
ecosystems
  bottom–up vs. top–down control 186
  changes
    permanent 173, 182–7
    simulations 176, *177*, *178*
    studies 188–90
    theories 187–8
    to components 186–7
    transient 173, 174–82
  classification schemes 314
  components 311–13, 314
  concepts 318
  dynamics 320
    matter and energy flows 172–93
  energy flows 305
  externalities 295–7
  global, valuation 322
  large-scale patterns 183–4
  marine, boundary definitions 315
  mosaic theory 315
  nonmarket costs and benefits 295–8
  process rates 319
  relationships, community vs. functional 314–15
  services, valuation 306
embedding theory, applications 91
emergy, concept of 318–19
emperor geese, population dynamics 17
endstates
  numbers of 166–7
  types of 164–6
energy flows
  between trophic groups 313
  dynamics 172–93
  early studies 314
  ecosystems 305
environmental crime, deterrence 301
environmental economics
  nature of the field 282–3
  policies 305
  and resource economics 305
environmental impacts
  human 291
  individual-based models 16–17
  and investment 293
epidemics
  incubation/digestive periods 46
  persistence 42
  probabilities 40–2

epidemics (*Cont.*)
rabies 60–1
recurrent
simulations 43–5
stochastic effects 42–5
simulations 41–2
spatial cross-infection 55–6
stochastic effects 37–42
see also infection dynamics
epidemiological processes, and ecological processes 44
equability, and diversity 255
equilibrium probabilities, in population dynamics 31–2
errors, and uncertainty 325–6
ESS *see* evolutionary stable strategies (ESS)
Euplotes patella **147**
events
and correlation equations 104–6
definition 105
edge 106
site 106
evolution
of altruism and cooperation 102, 132–9
of virulence 118
evolutionary dynamics
of consumer–resource systems 194–231
invasion analysis 196–8
parameter evolution 199–200
evolutionary stable strategies (ESS)
and abrupt density dependence 201–3
in population models 195–210
solutions
augmented evolutionary dynamics approach 195, 199–200
invasion approach 195, 196–8
expert systems
applications 322–3
fuzzy set theory in 323
goals 323
and local knowledge 310
in resource management 322–30
externalities, in ecosystems 295–7
extinction
effects of refuges on 69
harvesting to 293, 303–4
and population dynamics 31–2
species 173
issues 187–90
extinction rates
and metapopulations 263–4
and population size 262
and range size 262–4

fish
population dynamics 14
school sizes, via fuzzy set theory 316, *317*
Fisher, Sir Ronald Aylmer (1890–1962) 276
fisheries management 227
age-structured population models 207–8
models, limitations 321
overfishing impacts 321–2
fishermen, stochasts vs. cartesians 310–11
Fisher's α 255, 258
determination 250–1
fitness generating functions 196
flour beetles, spatial process studies 54
fluctuation analyses 12
food webs
consumer–resource interactions 227–8
discrete logistic models 195
individual-based models 16
patterns 172–4
simulations, computer programs 178–9
and species richness 16
storage in 224–6, 228
studies
frog ponds 184–6
lakes 179–82
forest dynamics
individual-based models 7–9, 15–16
simulations 89
SORTIE studies 7–9
forest fires, ecological impacts 15
forest mosaic cycles, individual-based models 9–12
forests
die-back 15–16
harvesting 292–3
FORET (simulator) 89
FORTRAN, pseudo-random number generators 28
foxes
behavioural studies 15
rabies, epidemic models 60–1
fuzzification, processes 316
fuzzy inference 324
fuzzy set theory
and ecosystem analysis 309–10, 315–17
in expert systems 323
in rule-based knowledge 324
fuzzy sets
vs. crisp sets 315–16
operators 316–17

gall wasp communities, on oaks 144, *145*
game theory
evolutionary 127–8
and harvesting 304
and resource over-exploitation 288–9
gap phase dynamics, and tree size 16
genes, evolution, and population size 264–5
genetic models, spatial patterns 70, *71*
genostasis, concept of 264
geographical barriers, types of 265
geographical range, diversity effects 261–2
gestation period
and body size 235, *236*
over age of weaning *236*
Gilbert and Sullivan, *The Mikado* 173
global warming, future costs 294
GNP *see* gross national product (GNP)
Gotelli's general model 263
governance
of natural resources 330–5
and ecosystem analysis 309–41
types of 331–2
and uncertainty 332–5

Grand Canyon (US), valuation studies 298
grazing
  resistance to 188
  studies 188–90
Great Mountain Forest (US) 8
green wood hoopoe, behavioural studies 14
Grenfell and Roberts process equation 49–50
  simulations 50
  stochastic realization *51*
gross national product (GNP), and ecological assets 298
growth
  rates, slow 240
  sustainable 290–1

habitats
  boundaries 314
  elevational 271–2
  patch models 3–4
  spatial distributions within 64
Hamilton's condition 137
Hanski distributions, and species diversity 278
hares, behavioural studies 15
harvesting
  effects on population dynamics 205–6, 208, 223–4
  forests 292–3
  and game theory 304
  generalized model 291–3
  open-access 287–8, 289
  outcomes 285
  resources 286
  to extinction 293, 303–4
  whales, policies 302–3
Hawaii, rainforests 15
hawk–dove game
  irregular networks 132
  payoffs **130**
  regular networks 129–31
  simulations *130*, *131*
herbivore biomass (HB), factors affecting 184
herbivores, and plants 173, 182, 183–6
heteroclinic cycles 165–6
heterogeneities
  aggregation factors 68
  causes of 66–8
  self-induced 67
  spatial models 65–8
  and species diversity 256
  temporal, and species diversity 269–70
  underlying 67
HIV *see* human immunodeficiency virus (HIV)
Holling type I response *214*
Holling type II function 215
Holling–Tanner process 37, *38*, 43, 48, 53
Holling's disc equation 212, 214
homogeneous models *see* mean-field models
host–parasite systems
  evolution to criticality 102, 118–23, 140
  simulations *121*, 122
  stochastic simulations *117*
host–parasitoid models 70, 72
  artificial ecologies 86–7, 89–90
  cellular automata 89, 90–1
  dynamics *88*
  embedding theory 91
  evolutionary 206–7
  patch models *76*
  probabilistic cellular automata 83–6
  spatial 79–80
  see also Nicholson–Bailey host–parasitoid equations
human immunodeficiency virus (HIV)
  infection 46
  transmission dynamics 45
Huxley, Julian Sorell (1887–1975), body size theories 233–4

i-state configuration models 3
i-state distribution models 3
IBMs *see* individual-based models (IBMs)
ignorance, and expert systems 329
immigration, and birth–death processes 29–30
individual-based models (IBMs)
  advantages 2
  applications 13–17
  criteria 18
  criticisms 5
  definition 1
  development 1–2
    historical background 2–5
  disadvantages 2
  in ecology 1–22
  evaluation 17–18
  examples 5–13
  and mean-field models compared 2
  reliability issues 312
  in resource management 311–12
  and rule evolution 18
  scale factors 72
  on World Wide Web 5
individuals, *i-state* 3, 14
infection dynamics
  contact process 112
  and correlations 139–40
  measles 102, 123–7
  modelling 111–27
  see also disease transmission; epidemics
inference nets
  applications 328
  dynamic knowledge structure 328
  static knowledge structure 328
inhibitors, and activators 70
interacting particle systems (IPS) 4, 82–7
  characteristics **74**
  use of term 83
  see also artificial ecologies (AE); probabilistic cellular automata (PCA)
intergenerational equity, issues 293–4
International Whaling Commission, management procedures 302–3
invasion analysis
  chaotic systems 198

invasion analysis (*Cont.*)
constant and periodic systems 196–8
invasion exponents *see* Lyapunov exponents
invasion resistance, and community assembly 161–4
investment, and environmental impacts 293
investment appraisals, discounting in 292–3
IPS *see* interacting particle systems (IPS)
Isham–Whittle normal approximation 47
island biogeography
definition 273–4
equilibrium theory 144
and species diversity 266–7
islands, species–area curves 271
ivory poaching 287, 288, 302

Kermack and McKendrick nonspatial threshold theorem 54
keystone species, concept of 313
kin altruism 133
kin-tolerance, *Lagopus lagopus scoticus* 13
knives (geographical barriers) 265
Kozlowski and Weiner's model 233, 241–6
simulations 244

Lagopus lagopus scoticus
kin-tolerant behaviour 13
population dynamics 12–13
resource management 312
territorial behaviour, individual-based models 12–13
lakes, phosphorus flows and stocks 179–82
largemouth bass, polychlorinated biphenyl studies 17
lattice neighbourhood, plants *6*
law enforcement
limitations 300
and resource management 298–302
and resource use, socially optimal levels of *299*
and transaction costs 301–2
Law of Mass Action 67
life history variations
body weight 233–4
phylogenetically controlled tests 234–8
Charnov's model 232–3, 238–41
and ecology 245
integrated models 237–8
Kozlowski and Weiner's model 233, 241–6
mammals 232–48
likelihood statistics 328
log-normal distributions
simulations 277
and species diversity 276–7
logistic processes 32–3
Lotka–Volterra dynamics **157**
Lotka–Volterra equations 37–8, 42, 47, 48, 53, 211, 218
Lotka–Volterra models 77, 178, 182–3, 195
continuous-time 227
Lotka–Volterra orbits *151*
Lotka–Volterra predator–prey equations 45
Lotka–Volterra systems 70, 86, 150, 152, 156, 161, 175, 314

Luangwa Valley (Zambia) 287, 288
Lyapunov exponents 198
determination 154–5
evaluation 12
Lyapunov functions 150–2

macroparasite–host models *see* host–parasitoid models
Malthus, Thomas Robert (1766–1834) 287
mammals, life history variations 232–48
marginal abatement cost, vs. marginal damage *296, 299*
marine ecosystems, boundary definitions 315
Markov chains
applications 327
Monte Carlo 327–8
Markov processes
in community assembly models 146–7, 167
mosaic cycle studies 11
mass effects, species diversity 268–9
mathematics, and biology, developments 23–4
matter flows, dynamics 172–93
maximum sustainable yield (MSY)
models, limitations 302–3
whales 302–3
mean-field models 119
assumptions 66–7
definition 3
and individual-based models compared 2
limitations 100
and pair approximations 5
measles
epidemics 43, 123
infection dynamics 102, 123–7
occurrence 110
patch models 77
memory
models, beech forests 11–12
and radiation, in mosaic cycles 11–12
metabolic rate, life history variations 235
metaknowledge, types of 324
metapopulation models *see* patch models
metapopulations
and extinction rates 263–4
and resource management 312
migration
and altruism 137–8
determinants 52
rates, density-dependent 55
and species diversity 268–9
see also birth–death–migration processes
Minnesota (US), grazing studies 188–90
moats (geographical barriers) 265
modelling
ecological aspects 302–4
economic aspects 298–302
models
caricature 87–9
comparative analyses 4
genetic 70, *71*
goals 3
limitations 321
moment–closure 92

models (*Cont.*)
  resource-use, limitations 310
  time-lag 46
  tinker-toy 174–9, 182–3, 318
  trophic 321–2
  see also ASP models; host–parasitoid models; individual-based models (IBMs); Lotka–Volterra models; mean-field models; patch models; population models; predator–prey models; RAS models; SIR models; spatial models
moment–closure models 92
money, as measure of utility 294
monocultures, individual-based models 6–7
Monte Carlo Markov chains 327–8
Moore neighbourhood, eight-cell 80, 90, 91
morphogens
  interactions, reaction–diffusion equations 82
  Turing ring processes 56–8
mortality rates
  adults 238–40, 243
  and body size 235–7
  see also death rates
mosaic cycles
  beech forests 10–11
  concept of 9–10
  forests 9–12
  stability 12
mosaic theory, ecosystems 315
multispecies processes 36–50
  see also single-species processes
multitrophic interactions, modelling 210–12
mutation rates, beneficial, and population size 264
myxomatosis, coevolutionary effects 47

N-colony processes 53–6
nanopopulations, recovery 80
national accounting, and ecological asset valuation 298
national parks, valuation 297
natural resources
  conservation 320
  governance of 330–5
    and ecosystem analysis 309–41
  management, techniques 320
  modelling, stochastic processes 303
  valuation 294–8
negative feedback, and species diversity 261–6
nematodes, population dynamics 49–50
neoclassical economics, and ecological economics 283, *284*
nest site selection, and population dynamics 14
networks
  irregular 107
  regular 107
  see also clumped networks
Nicholson–Bailey host–parasitoid equations 73–4, 75, 78, 84, 86–7, 91
  coupled map lattice version 79–80
Nicholson–Bailey map 77
nitrogen, effects on plant diversity 188–9
NKC-models 318
nonlinear averaging, concept of 269–70
nonmarket benefits
  ecosystems 295–8
  measurement problems 295–7
  types of 295
  valuation 306
nonmarket costs
  ecosystems 295–8
  measurement problems 295–7
northern spotted owl, resource management 312
nutrient cycles, models 179
nutrients, food web studies *185*

oaks, gall wasp communities 144, *145*
over-exploitation, issues 284–9
overfishing, impacts 321–2
overgrazing, commons 287–8

p-state 3
pair approximations
  applications 5, 92–3, 101–3
  development 4–5
  and mean-field models 5
  for spatial ecologies 100–42
  and spatial games 102, 128–9
  use of term 101, 111
pair closure 105, 107–11
Paramecium caudatum **147**
parameters
  noisy 47
  spatial models 70–2
  time-dependent 46–7
parasites
  artificial ecologies *87*
  communities, characterization 313
  densities *85*
  effects on resources 206–7
  population dynamics 48, 49–50
  spatial patterns *79*, 80
partial differential equations (PDEs) 80–2
  applications 101
    biological 82
  characteristics **74**
  theoretical criteria 81–2
  see also reaction–diffusion equations
patch models 74–7
  applications 312
  characteristics **74**
  development 3–4
  explicit 310
  extensions 76–7
  host–parasitoid systems *76*
  implicit 310
  limited use of 321
  research trends 76–7
  see also coupled map lattices (CML)
PCA *see* probabilistic cellular automata (PCA)
PCBs *see* polychlorinated biphenyls (PCBs)

PDEs *see* partial differential equations (PDEs)
per-capita growth rate function 196
Periodengeber 234
periphyton, food web studies *185*
permanence
  definition 149–50
  limitations 152–3
  test for 150–2
permanent endcycles 164–5
permanent endpoints 164
persistence, and community assembly 147
Phoeniculus purpureus, behavioural studies 14
phosphorus, in lakes 179–82
phytoplankton
  ecosystem change effects 174–9
  food web studies *185*
plankton, spatial models 73
plant communities, individual-based models 15–16
plant productivity (PP), factors affecting 184
plants
  competition 6–7
  coupled map lattice studies 6–7, *8*, *9*
  and herbivores 173, 182, 183–6
  individual-based models 6–7
  lattice neighbourhood *6*
  mean mass 6–7
  neighbourhood effects 6–7
plausible relations, use of term 328
poaching
  horn 302
  ivory 287, 288, 301–2
point diversity 256–8
Poisson statistics 108, 109, 132
pollutants, environmental, individual-based models 17
pollution, socially optimal levels *296*, *299*
polychlorinated biphenyls (PCBs), largemouth bass studies 17
ponds, food web studies 184–6
population density, and sustainability 289
population dynamics *144*
  and behaviour change 47–8
  of consumer–resource systems 194–231
  developments 45–8
  equilibrium probabilities 31–2
  evolutionary trade-offs 203–4
  and extinction 31–2
  factors affecting 146
  fish 14
  and fixed resources 200–1
  harvesting effects 205–6, 208, 223–4
  individual-based models 16–17
  Lagopus lagopus scoticus 12–13
  parasites 48, 49–50
  trophic interactions 206–7
  see also birth–death processes
population growth
  exponential and logistic 218–20
  models 25–7
population models
  continuous 194, 210–26
  discrete 194, 200–10
  evolutionary stable strategies analysis 195–200
  spatial, limitations 100–1
  stochastic effects 23–63
population size
  and beneficial mutation rates 264
  and extinction rates 262
  and speciation rate 264–5
populations
  age-structured 207–9
  cycling *88*
  demographic factors 47
  diversity effects 261–2
  interacting, spatial models of 64–99
  p-state 3
  semelparous 209–10
    resource competition 202–3
  subgroups, heterogeneous 47
  viscous
    cooperation in 138–9
    unconditional altruism in 135–8
  see also metapopulations
PP *see* plant productivity (PP)
predation
  and aggregation 69
  effects on resources 206
  models 37
  strategies, spiders 14–15
predation rates, nonlinear 37
predator–prey models 70, 72
  endstates 167
  and resource management 321
prey density 81, *82*
primates, reproductive rates 245
prisoner's dilemma (game) 101, 128, 138–9, 140, 288–9
  payoffs **138**
probabilistic cellular automata (PCA) 4, 83–6, 119
  definition 83
  host–parasitoid models 83–6
probabilities
  classical 326
  conditional 326–7
  degrees of belief 328
  epidemics 40–2
  equilibrium, in population dynamics 31–2
  experimental 326
  inverse 327
  subjective 326
  and uncertainty 326
probability theory
  and Dempster–Shafer theory compared 329
  objective 326
prospecting, expert systems 327
PROSPECTOR (expert system) 327
protected areas, sustainability 304–5
pseudo-random number generators 28

QBASIC, and stochastic processes 28

rabies, foxes, epidemic models 60–1

radiation, and memory, in mosaic cycles 11–12
rainforests, die-back 15–16
Rana, food web studies 184–6
RANDU generator 28
range size
  and extinction rates 262–4
  and fragmentation 263
  and speciation rate 265
rarefaction, determination 252–3
RAS models
  measles studies 123, 125–7
  stochastic *127*
ratio dependence, spatially homogeneous pure 213–16
reaction–diffusion equations 80–1
  applications 82
reaction–diffusion systems 3, 57
realistic age-structured models *see* RAS models
reasoning
  inexact 329
  temporal 327
  and uncertainty 325–9
reciprocal altruism 133, 134
recreational facilities, valuation 297
red grouse *see Lagopus lagopus scoticus*
refuges
  effects on extinction 69
  and stabilizing behaviour 69
relative nonlinearity, determination 270
reproduction, time-lag models 46
reproductive rates, primates 245
resilience, definition 173
resistance
  definition 173
  and species richness *189*
resource dependence, spatially homogeneous pure 213–16
resource economics
  and environmental economics 305
  nature of the field 282–3
resource exploitation
  bioeconomic model of *286*
  penalty expectations 300–1
resource management
  approaches 314–15
    issues 320–2
  and ecosystem analysis 317–22
  expert systems in 322–30
  individual-based models 311–12
  intelligent systems techniques 323
  interventions
    assessment 329–30
    issues 322–3
  and law enforcement 298–302
  models
    and ecosystem analysis 320–2
    types of 311–12
  policies, assessment 329–30
  precautionary approaches 303
  rule-based knowledge in 323–5
  techniques 320
  trophic models 321–2
resource use
  socially optimal levels *296*
    and law enforcement *299*
resource-use models, limitations 310
resource–consumer systems *see* consumer–resource systems
resources
  density vs. extraction efficiency 222–4
  distribution, heterogeneous 213–16
  extraction 212–16
    general expression for 216–18
    parameters **224**
  harvesting 286
  over-exploitation, and game theory 288–9
  per capita 200
  renewable
    exploitation 284–9
    vs. nonrenewable 282
  search efficiency vs. extraction rate 220–2
  sustainability 289–91, 304–5
  see also natural resources
return time 177
rhinoceros
  dehorning, discounting factors 292
  horn poaching 302
Ricker model 201
risk-taking behaviour, perceptions 311
rule-based knowledge
  inferences, methods 323–4
  in resource management 323–5
  rule formats 324
rules, evolution 18

salmon
  population dynamics 209–10
  smolt production studies 17
sample size
  bias 258
  species diversity measurement 250
scavenging processes 37
SCOPE 164
Scotland
  grouse studies 12–13
  vegetation transitions 159, *160*
SeaLab, applications 14
seed dispersal, models 5
seedlings, SORTIE studies 8–9
SEIR equations
  measles studies 123–7
  simulations *125*
semantic nets, partitioned 328
Serengeti National Park (Tanzania), grazing studies 188, *189*
Shannon–Weiner index 255
sheep, population studies 33
Simpson's index of concentration 255
  determination 251–3
simulations
  accurate 87–9
  of birth–death processes 27–9
    stochastic logistic realizations 35, *36*
  development 24
  ecosystem changes 176, *177*, *178*
  of epidemics 41–2

simulations (*Cont.*)
recurrent 43–5
food webs, computer programs 178–9
forest dynamics 89
Grenfell and Roberts process equation 50
hawk–dove game *130*, *131*
host–parasite systems *121*, 122
Kozlowski and Weiner's model 244
log-normal distributions 277
numerical, limitations 319
SEIR equations *125*
of single-species processes 35–6
spatial ecologies 100
stochastic, host–parasite systems *117*
weather predictions 88
single-species processes 30–6
equilibrium probabilities 31–2
local stochastic approximations 33–5
logistic 32–3
simulations 35–6
see also multispecies processes
sink species, use of term 268–9
SIR equations 112–14, 118
SIR models
and disease transmission 102, 116–18
mean-field vs. pair approximations 114
sites, state of 103
SOLVER, applications 40
SORTIE (simulator)
forest dynamics studies 7–9, 89
underlying theories 8
source species, use of term 269
source–sink dynamics 269
SPARs *see* species–area curves (SPARs)
spatial distributions, cellular automata *85*
spatial ecologies
correlation equations for 100–42
pair approximations for 100–42
phenomena 100
simulations 100
spatial games 127–32
and pair approximations 102, 128–9
rules 128
spatial models
aggregation 68
alternatives to 92–3
characteristics 68–73
data analysis 91–2
definition 64
heterogeneity 65–8
historical background 65
of interacting populations 64–99
interaction ranges 90–1
length scales 89–90
local correlations 68
parameters 70–2
research trends 87–93
scale factors 72–3
types of 73–87
see also coupled map lattices (CML); interacting particle systems (IPS); partial differential equations (PDEs); patch models
spatial patterns
generation 69–70
genetic models 70, *71*
occurrence 65–6
parasites *79*, 80
spatial processes 50–8
N-colony 53–6
Turing 56–8
two-colony 52–3
speciation rate
and population size 264–5
and range size 265
species
coexistence, criteria 148–53
economically important, management 321
extinction 173
issues 187–90
keystone 313
species distributions, individual-based models 16
species diversity 249–81
and abundance curves 276–8
components 256–60
and disturbance 267–8
dynamic processes 260–70
small-scale 267–9
enumeration 250
estimators 250–5
evolution 272
and Hanski distributions 278
and island biogeography 266–7
local, and productivity 275–6
and log-normal distributions 276–7
mass effects 268–9
measurement 249–60
and sample size 250
and migration 268–9
and negative feedback 261–6
temporal effects *275*
and temporal heterogeneity 269–70
theories, comparative 270–8
turnover rate 267
species pools
assembly sequences **159**
regional 144–5
species richness
and food webs 16
and resistance *189*
species–area curves (SPARs) 257–8, 259, 270–4
compilation 271–3
scales 273–4
spiders
behavioural studies 14–15
predation strategies 14–15
Splus, applications 40
spread (statistics) 325
stability index, definition 32
state variables, in community ecology 146–7
steady states, and diversity 265–6, 274
Stipa capensis, source–sink dynamics 269
stochastic analyses, birth–death processes 26–7, *29*
stochastic approximations, local 33–5
stochastic closure 105

stochastic effects
  in patch models 4
  in population models 23–63
stochastic processes
  and deterministic approaches 23, 24–5, 58–61
    compromise 33–4
  in natural resource modelling 303
stochastic threshold theorem 41
storage effects 270
subadditivity, determination 270
superindividuals, concept of 16–17
surrogate market approach
  hedonic price method 297
  travel cost method 297
survival
  and body size 240–1
  probabilities 208
susceptible, exposed, infectious and resistant individuals *see* SEIR equations
susceptible, infectious and resistant individuals *see* SIR equations
sustainability
  issues 289–90
  resources 289–91
    future research 304–5
  use of term 290
symbiosis, relationships 37

territorial behaviour, *Lagopus lagopus scoticus*, individual-based models 12–13
Tetrahymena pyriformis **147**
TFT see Tit For Tat (TFT)
threshold theorem 39
time
  and diversity 259–60
  and ecological economics 291–4
time-lag models 46
tinker-toy models 174–9, 182–3, 318
Tit For Tat (TFT), strategies 138–9
top-level ideas 328
transients, in community assembly 155–6
transmissibilities
  evolution of 121–2
  use of term 119
  see also critical transmissibilities
tree growth, SORTIE studies 7–9
tree size, and gap phase dynamics 16
Tribolium confusum, spatial process studies 54
trophic cascades, concept of 313
trophic groups, concept of 313
trophic models, in resource management 321–2
trophodynamics, early studies 314
Turing, Alan Mathison (1912–54), spatial pattern research 65
Turing ring processes 56–8
  limitations 59–60
  stochastic realizations 58, *59*
Turing structures
  generation 69–70
  occurrence 65, 81
turnover rate, species diversity 267
two-colony processes 52–3

uncertainty
  and decision-making, Bayesian approaches 327
  and errors 325–6
  and governance 332–5
  and models 321
  and probabilities 326
  and reasoning 325–9
  resources, perceptions 311
  use of term 325
unconditional altruism 133, 134
  in viscous populations 135–8
universe of discourse, use of term 315
utility, measurement of 294

vaccination, and disease invasion 114–15
valuation
  contingent 298
  global ecosystems 322
  of natural resources 294–8
  nonmarket 298
  surrogate market approach 297
vegetation, transitions 159, *160*
vegetative propagation, models 5
Verhulst–Pearl logistic equation 32–3, 34
virulence
  evolution of 118, 121–2
  use of term 119
von Neumann neighbourhood, four-cell 83, 84, 90

weaning, age of, over gestation period *236*
weather predictions, simulations 88
whales, harvesting, policies 302–3
wood storks, behavioural studies 14
World Wide Web, individual-based models on 5

yeast, growth studies 33

zooplankton
  ecosystem change effects 174–9
  food web studies *185*